D0820526

GeoDynamics

GeoDynamics

EDITED BY

Peter M. Atkinson, Giles M. Foody, Stephen E. Darby, and Fulong Wu

Boca Raton London New York Washington, D.C.

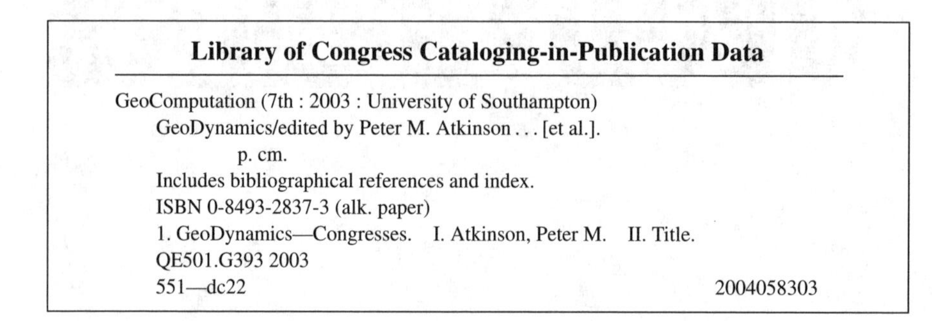
Library of Congress Cataloging-in-Publication Data

GeoComputation (7th : 2003 : University of Southampton)
GeoDynamics/edited by Peter M. Atkinson . . . [et al.].
p. cm.
Includes bibliographical references and index.
ISBN 0-8493-2837-3 (alk. paper)
1. GeoDynamics—Congresses. I. Atkinson, Peter M. II. Title.
QE501.G393 2003
551—dc22 2004058303

Direct all inquiries to CRC Press, 2000 N.W. Corporate Blvd., Boca Raton, Florida 33431.

Visit the CRC Press Web site at www.crcpress.com

International Standard Book Number 0-8493-2837-3
Printed in the United States of America 1 2 3 4 5 6 7 8 9 0
Printed on acid-free paper

Preface

GeoComputation 2003, held at the School of Geography, University of Southampton (SoG, UoS) and the Ordnance Survey on 8 to 10 September 2003, was the seventh in a series of successful international meetings concerned with solving geographical problems through the use of computers. The conference was organized by Peter M. Atkinson, Giles M. Foody, and David Martin. However, to take full advantage of the enthusiasm and interest of the large number of quantitative geographers at the SoG, UoS, the conference was advertised both as a whole and as a series of special interest sessions organized by individual staff in the institutions. Three of those special interest sessions were "Remote Sensing of Change" (Giles M. Foody), "Spatially Distributed Modelling of Land Surface Processes" (Peter M. Atkinson and Stephen E. Darby), and "Urban Dynamics" (Fulong Wu). These sessions were sculpted, for example, through clear guidance in the calls for papers, so as to develop academically strong, but focused collections of research papers. This was achieved both within and between the three sessions in order to assemble a critical mass of researchers interested in spatial dynamics. The result was a lively and interesting conference, for which this book provides a lasting record.

GeoComputation research arose from the application of computer power to solve geographical problems. The uppercase C in GeoComputation was retained to emphasize that each part (geography, computation) was equally important. Much was written at the end of the 1990s about the definition of GeoComputation and several books emerged on the topic. Importantly, whereas early proponents emphasized the efficacy of computational solutions, almost as an antidote to the strictures of model-based statistics, Longley (1998) was at pains to retain dynamics (and, thereby, process) as a component of GeoComputation. In many ways, this book focuses on that part of GeoComputation that deals with dynamics.

The conference itself was a great success both academically and socially and all those involved will no doubt have fond memories. Through very generous sponsorship, we were able to attract three very high profile academics as keynote speakers. First, John C. Davis (Kansas Geological Survey) opened the conference by giving the International Association for Mathematical Geology, IAMG, Distinguished Lecture. On the second day, during the afternoon session (organized by and held at the Ordnance Survey, Southampton) Mike Batty (CASA, UCL) delivered the Ordnance Survey keynote lecture and, on the final day, Peter Burrough (Utrecht, The Netherlands) presented the Taylor & Francis keynote lecture. All presentations were fascinating and maintained the "buzz" of the conference. The conference was also successful as a social event. The conference dinner was held at the Hampshire County Cricket ground at the Rose Bowl, Southampton. Equally memorable was the quiz night (organized by Gary Llewellyn and Matt Thornton) held the evening before. Delegates enjoyed answering a bizarre range of "pub quiz" style questions on the subject of (believe it or not) GeoComputation! One question involved actually deciphering some MATLAB code which drew the letters GC in raster image format. More than one team answered correctly!

All chapters in this book were presented within one of the three special sessions (above) at GeoComputation 2003, with the exception of the introduction (kindly written by Paul Longley — by invitation), the section introductions and the conclusion. The Foreword was kindly written by Peter Burough, also by invitation. All chapters have been subjected to peer review, each chapter being scrutinized by three referees (one selected from the set of authors within the book and two external to the book). While we do not name referees here for the sake of anonymity, we wish to record our deepest gratitude to those who gave their time to ensure the quality of the individual chapters and of the book as a whole. We are also grateful to the authors themselves for delivering their manuscripts and revisions to very strict deadlines. That the entire set of over 20 chapters was returned within the very tight schedule is remarkable. Further, the referees comments were often critical as befits scholarly work and the authors responded admirably to the challenge of dealing with such comment. As a result, a rigorous refereeing process was completed in a remarkably short time frame (6 months from conference to submission of the entire edited volume to the publisher).

Organizing and running a major international conference is a huge task and it would not have been possible without the help of many individuals. We would now like to take the opportunity to thank these individuals. First, we would like to thank our sponsors. These included Taylor & Francis, John Wiley & Sons, the Royal Geographical Society with the Institute of British Geographers Quantitative Methods Research Group (RGS-IBG QMRG), the Association for Geographic Information (AGI), and PCI Geomatics. We are very grateful to these bodies for their generous support. We also acknowledge the goodwill of our personal contacts within these organizations who continue to support important academic activities such as GeoComputation 2003. Second, many individuals helped with the actual day-to-day running of the conference. In particular, Karen Anderson, Richard Breakspear, Marie Cribb, Jana Fried, Pete Gething, Nick Hamm, Gary Llewellyn, Nguyen Quang Minh, Ajay Mathur, Aidy Muslim, Nick Odoni, Sally Priest, Isle Steyl, Matt Thornton, Debbie Wilson, and Matt Wilson provided invaluable assistance as the conference helpers. These individuals, all graduate students at the SoG, UoS, constantly overwhelm us with their generosity of time and effort. Their help is vital to running any sizeable conference and a great debt of gratitude is owed to them. We thank the Ordnance Survey for help with the smooth running of the conference at Ordnance Survey on the Tuesday afternoon. We thank, in particular, Vanessa Lawrence, Toby Wicks, Nick Groome, and David Holland. Also, we wish to thank the Geodata Institute and, in particular, Julia Branson who played a crucial role in handling the conference registration and the delegates packs, as well as liaising with the UoS Conference and Hospitality Office. Finally, we wish to thank all others who contributed to the production of the book. In particular, we thank Randi Cohen and Jessica Vakili (CRC Press) for their enthusiasm and positive attitude, Denise Pompa (SoG Office) for help with incorporating editorial revisions into manuscripts, and Adam Corres (SoG Office) for his help with finances, conference exhibitions and general running of the conference.

GeoComputation 2003 was a challenge to organize and run. We managed to achieve a successful and memorable conference without pain through the generous goodwill of all those associated with the conference. We hope that this book, which acts as a lasting record of the event, is found to be useful by academics and practitioners alike, and makes a valuable contribution to the growing literature on GeoDynamics.

REFERENCE

Longley, P.A., 1998, Foundations, In Longley, P.A., Brooks, S.M., McDonnell, R., and MacMillan, W. (Eds.), *GeoComputation: A Primer* (Chichester: Wiley), pp. 3–15.

Peter M. Atkinson
Giles M. Foody
Stephen E. Darby
Fulong Wu
Southampton March 2004

Foreword

This volume comprises a fine selection of the papers presented at the 7th International Conference on GeoComputation held at the School of Geography, University of Southampton on 8 to 10 September 2003. The term *GeoComputation* has been gathering strength in recent years: it embraces the links between numerical modeling (i.e., computing) and its application to explain and increase our understanding of a wide variety of the dynamic aspects of the physical (i.e., geological, geomorphological, hydrological, and ecological) and social (urban change and land use planning) processes operating on the earth's surface. Many new insights into these processes are becoming available as enthusiastic, computer-literate geographers continue to explore the opportunities for linking numerical models of landscape and urban dynamics to the increasingly large and detailed data sets that are being provided by both surface and satellite sensors. Indeed, many of the surface modeling applications developed for Earth are just as relevant for creating detailed views of distant Earth-like planets and moons (Pike, 2002). So, although the term *GeoComputation* may seem limited to our own planet, in fact it embraces a generic set of widely applicable tools for modeling surface processes. The main differences between planet Earth and the Earth-like moons and planets are that (a) in spite of the skills of rocket scientists and robot technology to persuade us otherwise, field work on the latter is severely limited, and (b) currently active hydrological, biogeographical, and urban processes seem to be of great importance only on the earth.

The editors have divided the book into three convenient sections: Remote Sensing, Physical Processes, and Human Processes. This division implies strong separation between these themes, but once one has read the contributions to this book, one gets the idea that there may be other generic structures and themes currently dominating the GeoComputational literature that better describe the shared aspects *within* rather than *between* the sections. These are (a) issues of accuracy and uncertainty, (b) dynamic modeling, and (c) applications of cellular automata and agents, and possibly combinations of all three.

In Section I, for example, the most common focus is on the spatial resolution and accuracy of the gridded satellite sensor image data, and this theme of data quality is repeated in the other sections, being applied to processes as physically diverse as flood modeling and urban change simulation. Several authors address the problems of making assessments of the accuracy of the classification of remotely sensed images. There is particular concern that few commercial image processing systems provide software to monitor classification quality, in spite of the large amount of knowledge of the problem. Why, they ask, are the vendors (and users) of commercial software systems so disinterested in error management and quality, which is such a recurring theme in the scientific side of the GeoComputation world? So, they take on the task of providing software tools for error analysis in a way that addresses both crisp and soft classification problems.

A related topic is the issue of how accuracy and error may vary with aggregation and spatial resolution. Although the problem is illustrated by a scanned aerial photograph of vegetation, the chapter addresses a general problem in the analysis of raster data that has wide application.

Change detection is also a commonly recurring theme. Holland and Tompkinson describe the approach being investigated by the U.K. Ordnance Survey to help a human photo interpreter pick out land use and cover change from aerial photographs using semi-automated user-guidance techniques. The more or less deterministic approach adopted here contrasts with and complements the chapter by Comber et al., who opt for a statistical approach to identifying land cover change. Brown addresses the problem of per-pixel uncertainty for change detection as revealed by airborne data. Which approach to change detection is better? What is in principle the difference between change detection (as a once off comparison of scenes) and a dynamic model of continuous change in a landscape or cityscape?

Further, what is the role of uncertainty in these studies? We shall have to wait and see, but my guess is that both the deterministic and the stochastic approaches are needed, and of course, accuracy and error analysis will be part of the game. For this, a study of patterns of noise could take spatial analysis a step further by examining the effects of visual texture (which is a surrogate for colored noise) in images and how the presence of different textural patterns in an image actually provides useful information.

Not everyone finds statistics necessary, and there is new material on the role that computer science and modern information theory may play in spatial analysis. The aim is to develop smart spatial systems around the CommonKADS approach, which recognizes new ways to deal with the complexity of the knowledge engineering process. This application of CommonKADS is new to spatial (remotely sensed) data and may provide interesting new insights.

One of the revelations to come out of GeoComputation in recent years concerns the ability to deal with geographical phenomena not just in space but over time as well. And then, time does not mean a single still point of the turning world but a vector that admits that the process includes the possibility of feedback, thereby implying continuous and often unpredictable change. The basic elements of space may be grid cells or vectors, but most GIS-based GeoComputational modeling implies dynamic change, whether the focus of study is a landslide, river flooding, or the intricacies of urban dynamics. In terms of GeoComputation, much is shared between models of physical and human processes, especially when the common link is provided by cellular automata (CA) or by agents.

As Fulong Wu discusses in his introduction to Section III, human processes provide the most dynamic forms of GeoComputation and active use of CA methods, at least in terms of visual material and model complexity. Is this because human processes are easier to deal with than geological or hydrological systems? This is almost certainly not so, but demographics dictate that more people are interested in urban dynamics than geomorphology or agriculture. Eighty percent of the world population live in cities and the changes they bring are so rapid, from the simple movements of millions of shoppers to geographical real-time, to urban landuse change management or problems of waste disposal or food supply requires effective networking and interaction on a real-time basis. Is it purely a question of numbers and concentrations of people and money that mean that urban problems attract more attention than geomorphology? As Keith Clarke points out, one important difference in modeling approach between the urban and the physical world is that urban models often aim explicitly to modify the future, whereas physical models usually only attempt either to explain past events or to give an indication of how a current situation has been reached. Only for short-term processes in the earth sciences (meteorology, hydrology, flooding) is the concept of "forward modeling" in geology and geomorphology really applicable, though the current geological use of the term covers periods from tens to hundreds of thousands of years.

Throughout the book there are many contrasting aspects concerning the practice of "good modeling." A recurrent theme is that both physical and urban modelers need powerful, complex, simple to operate, easy to understand, flexible, numerical models that can operate on a minimum of expensive, high spatial resolution, readily available data. An implicit aim is to keep the models simple so users and decision makers can understand what they are doing; data requirements must be modest so models are not too expensive to run. But the internal structures of the models must be as rich and diverse as possible so that the models will mimic the real-world complexity that people experience. In order to even begin to match these opposing requirements, the urban modelers have turned to cellular automata methods and agents in a big way, even though Michael Batty, Helen Couclelis, and colleagues have pointed out elsewhere (*Environment and Planning B* 1998) that CA and agents may not yet be up to the job. Perhaps we still have far to go. The difficulties with these approaches may not be so much that they are fundamentally unsound, but that initially at least, they have been supplied with deterministic rules. Given the stochastic nature of so much of the world, it seems that the next level of CA and agent-based models will need to include powerful methods for capturing uncertainties in the processes under study and appropriate methods of error propagation (or error recognition) in these models will need to be devised. Is it really likely that simplicity, honesty, and complexity can all be found together in a single model?

My own contribution to this symposium was to present the Taylor & Francis Lecture, which was a unique experience, because it was a keynote presentation that rounded off the meeting rather than introduce a new topic to one's colleagues. Having seen and heard the contributions before me, I decided to adopt an approach that dealt with the growth and decay over space and time of diverse subjects such as citation frequency of scientific publications and invading armies. Interestingly enough, citation frequency data published by Taylor & Francis in *Advances in Physics* (1950–2002) and the *International Journal of Geographical Science* display certain similarities. In fact, they adhere to what is well known as Zipf's law, which states that if the data are ranked from largest to smallest and are plotted on double-log axes together with the attribute of interest (e.g., number of citations), the plot is frequently a straight line. This means that the decay of citation index follows a power law, as do all other ranked data that yield a straight line on a double-log plot, such as the decay of Napoleon's Grand Armee enroute and back from his disastrous attempt to capture Moscow in 1812.

Many similar kinds of data seem to follow Zipf's law, and power-law hunting has become a regular sport (see Barabási, 2002). Large databases with hundreds of thousands of instances yield results that are similar to those derived from a couple of hundred. Batty (2003) has demonstrated that the model can also be applied without change to spatially distributed phenomena, such as local densities of researchers. The problem is, though the model (i.e., the "law") is simple, needs not too much expensive data and frequently produces robust results, no one really seems to know why. Some put it down to chance, but why should chance deliver similar forms? There must be a reason, but it has yet to be explained. A worthy challenge for a future GeoComputation!

Peter A. Burrough
Utrecht

REFERENCES

Barabási, A.-L. 2002. *LINKED: The New Science of Networks*. Perseus Publishing, Cambridge, MA, 280 pp.

Batty, M. 2003. Commentaries, *Environment and Planning A*.

Couclelis, H. 1998. GeoComputation and Space. *Environment and Planning B*. Planning and Design, 25(Anniversary Issue), 41–47.

Pike, R.J. 2002. *A Bibliography of Terrain Modeling (Geomorphometry), the Quantitative Representation of Topography*. USGS Open-File Report 02-465.

Contributors

Pragya Agarwal
School of Computing
University of Leeds
Leeds LS29JT,
UK

Manoj K. Arora
Department of Civil Engineering
Indian Institute of Technology
India

Peter M. Atkinson
Department of Geography
University of Southampton
UK

Joana Barros
Centre for Advanced Spatial Analysis
University College London
UK

Michael Batty
Centre for Advanced Spatial Analysis
University College London
UK

Michael Becht
Department of Physical Geography
University of Goettingen
Germany

Klaus-J. Beckmann
ISB
RWTH Aachen University
Germany

Itzhak Benenson
Department of Geography and Human Environment
Tel Aviv University
Israel

Kyle Brown
Environment Agency
National Centre for Environmental Data and Surveillance
UK

Yohay Carmel
Faculty of Civil and Environmental Engineering
Israel Institute of Technology
Israel

Keith C. Clarke
Department of Geography
University of California
USA

Alex J. Comber
Department of Geography
University of Leicester
UK

Paul Crowther
School of Computing and Management Sciences
Sheffield Hallam University
UK

Stephen E. Darby
Department of Geography
University of Southampton
UK

Charles K. Dietzel
Department of Geography
University of California
USA

John Elgy
School of Engineering and
Applied Sciences
Aston University
UK

Peter F. Fisher
Department of Geography
University of Leicester
UK

Giles M. Foody
Department of Geography
University of Southampton
UK

Noah C. Goldstein
Department of Geography
University of California
USA

Jürgen Gräfe
ZAIK
University of Cologne
Germany

Jacky Hartnett
School of Computing
University of Tasmania
Australia

David A. Holland
Research and Innovation
Ordnance Survey
UK

Xia Li
School of Geography and Planning
Sun Yat-Sen University
China

Paul Longley
Department of Geography
University College London
UK

Arko Lucieer
Department of Earth
Observation Science
ITC
Netherlands

George P. Malanson
Department of Geography
University of Iowa
USA

Rolf Moeckel
IRPUD
University of Dortmund
Germany

Heike Mühlhans
ISB
RWTH Aachen University
Germany

R. Nathan
Department of Evolution,
Systematics and Ecology
The Hebrew University
of Jerusalam
Israel

Guido Rindsfüser
ISB
RWTH Aachen University
Germany

Jochen Schmidt
Landcare Research
New Zealand

Mohamed A. Shalan
School of Engineering and
Applied Science
Aston University
UK

Alfred Stein
Department of Earth
Observation Science
ITC
Netherlands

Dirk Strauch
German Aerospace Center
Institute of Transport Research
Germany

Tal Svoray
Department of Geography
and Environmental Development
Ben-Gurion University
of the Negev
Israel

William Tompkinson
Research and Innovation
Ordnance Survey
UK

Paul M. Torrens
Department of Geography
University of Utah
USA

Richard A. Wadsworth
Department of Geography
University of Leicester
UK

Michael Wegener
S&W Urban and Regional
Research
Germany

Volker Wichmann
Department of Physical
Geography
University of Goettingen
Germany

Matthew D. Wilson
School of Geographical
Sciences
University of Bristol
UK

Fulong Wu
School of Geography
University of Southampton
UK

Yichun Xie
Institute for Geospatial
Research and Education
Eastern Michigan University
USA

Anthony Gar-On Yeh
Centre of Urban Planning and
Environmental Management
The University of Hong Kong
China

Yu Zeng
Department of Geography
University of Iowa
USA

Contents

GeoDynamics: An Introduction

Paul Longley

GeoDynamics is of fundamental importance to our ability to represent the world around us. Geographic information systems (GIS) and remote sensing provide us with the ability to abstract selectively from the seemingly infinite complexity of the real world: the accumulated experience of applications helps us to develop guiding principles, techniques, and conventions that will allow us to do still better next time. Remote sensing can provide us with increasingly detailed, and frequently updated, classifications of the surface (and shallow sub surface) of the Earth. But apparent success in depicting how the world *looks* is much less useful than understanding of how the world *works*, because only knowledge of how the world works enables us to *predict*. Most observers would concur that GIS and remote sensing are not only successful, but also strongly convergent, technologies: however, their use to represent the world often raises frustrating and profound questions — and this is often because of the limitations of cross-sectional analysis. GIS is important because "almost everything that happens, happens somewhere" (Longley et al., 2001, p. 1); understanding of GeoDynamics is important because almost everything that happens also entails a lifecourse of events and occurrences, and robust understanding necessarily requires represention as more than a single or limited sequence of cross-sectional images.

Almost any geographic representation of the world is necessarily incomplete, and the central art of cross-sectional GIScience entails informed judgment about what to retain and what to discard (Goodchild, 1992). Usually this decision is largely made for us — we have to "make do" with the raw data that we turn into usable information using GIS, simply because that is all that is available. Much activity in GIScience thus becomes concerned with managing the uncertainties that the process of "making do" generates. These tasks are compounded if, in building a representation, temporal as well as spatial granularity and completeness is compromised by circumstances beyond our control. In representing time as well as space we are faced with the totality of reality, and until recently, GIS and image processing software has simply not been up to this task. Yet, the contributions to the first part of this book illustrate the remarkable transformations that are now taking place in our abilities to represent, and hence understand, the dynamics of change. Remote sensing and image processing provide long-established approaches to representing surface characteristics that are geographically extensive, detailed and (most important of all in the present context) frequently updateable. This can reduce some of the uncertainty inherent in building representations of how the world works. The chapters in Section I of this book set out a number of current research perspectives upon management of the uncertainties that dog our representation of GeoDynamics. Here, fuzzy classification, management of scale and aggregation, increased ontological understanding and tactical user interventions are all mooted as ways of improving the depiction of change over space. Together, these various contributions engender the shift beyond image *processing* towards image *understanding* that is well underway in remote sensing, and illustrate how we are beginning to understand not just the state of geographically extensive areas but also their dynamics.

Richer depictions of the world are also necessary for improved understanding and simulating physical processes. Developments in the creation of geographic information are of undoubted importance here, and the last decade has seen the reinvention of large parts of geomorphology and geology as field disciplines, with the near ubiquitous deployment of portable measurement devices that are enabled with global positioning system (GPS) technology. If our representations are necessarily uncertain and incomplete in a static sense, so our representations of evolution are inherently more so, and together these provide significant challenges to the scientific community in the understanding of landscape change. But greater explanation and increased predictive ability also entails better models of the dynamic processes extant in three dimensions. The range of applications discussed in Section I of the book are essentially two-dimensional with the addition of time (2-D and 3-D): the emphasis upon geomorphologic process in Section II raises more explicitly the management of the third dimension, not least because outcomes such as flood inundation and vegetation spread are heavily focused upon such considerations. The range of applications described in this section are also particularly interesting because the temporal granularity of simulations is here much more coarse — the objective is still creation of predictions that work over short-time horizons but unlike the remote sensing contributions, the basis to measurement and the premises concerning evolutionary processes are necessarily examined over much longer periods.

If physical landscape modelling is inherently more uncertain than some remote sensing applications, so consideration of the human dimension to dynamic analysis introduces a host of additional problems. The work of Keith Clarke and colleagues at Santa Barbara (e.g., Clarke, 1998) has illustrated the ways in which the innovation of field computing has greatly increased the technical feasibility of measuring and monitoring human behaviour over time. The ways in which we all interact with machines, with increasing frequency, in our daily lives also leaves disparate yet telling digital traces of our activity patterns that can be subjected to analysis and generalisation. However, few individuals would wish all aspects of their behaviour to be monitored in these ways, and those that might acquiesce are likely to be unrepresentative of those who would not. The task of specifying and computing all possible interaction patterns between universes of human individuals also remains non-trivial. This poses significant problems of developing and managing data to specify and test models of systems whose dominant features are revealed through their dynamics. In analytical terms there are further problems manifest in finding appropriate, practicable, parsimonious ways of representing human agency. The chapters of Section III of this book describe state-of-the-art thinking about the representation of human agency in GIS, and establish a range of scientific perspectives on the processes through which agents use, occupy, and mould space. They illustrate tangible progress as well as state-of-the-art thinking, but also signal that we remain some way from consolidating a consistent basis to measurement and hence theory. In this quest, the accumulated experience of GIS actually provides rather few way markers, and age-old spatial analysis problems of defining measurable individuals and interfacing micro models of behaviour with macro scale system constraints recur with almost predictable frequency in the literature. In this context, the contributions to Section III of this book also provide a state-of-the-art overview of the new perspectives that are opening up on the problems of representing dynamics at fine levels of granularity where the agent rather than the aggregate is the prime concern.

Taken together, it is clear that considerable problems remain in the cumulative development of analysis that is grounded in fine levels of granularity and individual events rather than aggregate patterns. The data that are used to build representations of the world, the dynamic processes that define system behaviour, the interactions between system components and the networking of individual agents are all key to understanding the GeoDynamics of complex systems. The contributions to this book illustrate that these considerations take us well beyond the accumulated experience of GIS, and illustrate that GeoDynamics should be a central concern of GIScience. The development of GeoDynamics within GIScience will itself take time, and the contributions to the different sections of this book suggest cross-cutting strategies and priorities through which this might be achieved. One implication is that progress requires a science that complements contemporary GIS yet makes it possible also to assimilate our rapidly increasing understanding of GeoDynamics. Improvements in the quality, quantity, and delivery of disaggregate data are providing us with more detailed support for our existing theories. This initiates a virtuous circle, in which better data assists the development of better theory, which in turn stimulates demand for still better data. In this respect, the development of better theory is often indirectly driven by technologies that enable us to collect better data. High spatial resolution remote sensing, new kinds of geodemographics, and new methods of synthesising data from diverse sources are forcing the pace, and we should anticipate important advances in the various domains of our understanding of GeoDynamics in this decade.

Finally, it is important also to recognise that all of these developments are taking place in a scientific context that is itself changing. It is interesting to note that the scientific questions that are posed in this book are very different in type and detail from those that drove spatial theory development in geography's Quantitative Revolution (Berry and Marble, 1968), or even less than 10 years ago at the dawn of GeoComputation (Longley, 1998). Many of the chapters in this book illustrate the importance of interdisciplinary science, of clearly defined problems of practical concern, of sharing data and representations, of effective communication through visualisation, and of concern with the individual. In this context, the success of GIS through application provides a valuable template for further progress. But the essential contribution of this book is the recognition that it is no longer good enough to think of dynamics as an afterthought to the building of spatial representations. GeoDynamics is here to stay.

REFERENCES

Berry, B.J.L. and Marble, D.F. (Eds.), 1968. *Spatial Analysis: A Reader in Statistical Geography*. Englewood Cliffs NJ, Prentice Hall.

Clarke, K.C., 1998. Visualising different geofutures. In Longley, P.A., Brooks, S.M., McDonnell, R., and Macmillan, W. (Eds.), *GeoComputation: A Primer*. Chichester, Wiley, pp. 119–137.

Goodchild, M.F., 1992. Geographical information science. *International Journal of Geographical Information Systems*, 6, 31–45.

Longley, P.A., 1998. Foundations. In Longley, P.A., Brooks, S.M., McDonnell, R., and Macmillan, W. (Eds.), *GeoComputation: A Primer*. Chichester, Wiley, pp. 3–15.

Longley, P.A., Goodchild, M.F., Maguire, D.J., and Rhind, D.W., 2001. *Geographic Information Systems and Science*. Chichester, Wiley.

Part I

Remote Sensing

CHAPTER 1

Introduction — Remote Sensing in GeoDynamics

Giles M. Foody, Stephen E. Darby, Fulong Wu, and Peter M. Atkinson

The dynamic nature of geographical phenomena is widely recognised. Changes may be perceived as good or bad but are often of considerable importance. Environmental changes, for example, may have a marked impact on human health and well-being which, in addition to normal scientific curiosity, call for a greater understanding of phenomena undergoing change and of the effects of their dynamism. To support studies of dynamic geographical phenomena there is a need for information on environmental properties at a range of spatial and temporal scales. Various geospatial technologies have developed to provide the data needed to further the understanding of the environment (Konecny, 2003; Thurston et al., 2003). These geospatial technologies include the use of global position systems (GPS) for accurate information on location; geographical information systems (GIS) for data integration and analysis; and geostatistical tools for quantitative analyses, which recognise the spatial dependence that exists with most geographical data. In Section I of the book, however, the major focus is upon another major technology used in the study of GeoDynamics, remote sensing.

Remote sensing is a powerful means of acquiring data on environmental properties. It can provide data at a range of spatial and temporal scales, enabling dynamic phenomena with spatio-temporal characteristics ranging from local and rapid through to large and slow to be observed and characterised. Thus, it is unsurprising that remote sensing has had a major role to play in relation to numerous key application areas, including attempts to answer major science questions of the era. For example, in relation to studies of global warming, one of the most prominent environmental concerns for current generations upon which awareness amongst the general public

0-8493-2837-3/05/$0.00+$1.50

and politicians as well as scientists is high, remote sensing has made many important contributions. This is particularly important as climate change arising from global warming may represent a greater threat to the world than other politically high profile issues such as international terrorism (King, 2004). Global warming, driven partly by enhanced atmospheric carbon loadings, may lead to many negative environmental changes including increased frequency of extreme events such as flooding, drought, and crop harvest failures. Remote sensing may be used in many ways to help further the understanding of issues connected with global warming or in helping to mitigate or adapt to changes. It has, for example, been used in a diverse range of studies including those focused on important aspects of environmental change associated with the atmosphere and cryosphere (e.g., Laxon et al., 2003; van der A et al., 2003; Kelly, 2004). However, considerable attention has been directed at the use of remote sensing in studying terrestrial environmental properties that are associated with some of the major uncertainties in knowledge on environmental processes and change. For example, some of the greatest uncertainties in the global carbon cycle are associated with the terrestrial biosphere (Roughgarden et al., 1991; Houghton et al., 2000). It has been recognised for some time that remote sensing has the potential to provide key information on the terrestrial biosphere, notably the extent and biomass of forests, very basic environmental variables, at the spatial and temporal scales required (Dixon et al., 1994; Houghton et al., 2000; Royal Society, 2001; Achard et al., 2002). Thus, by providing information on apparently basic variables such as land cover and its properties remote sensing can help in answering major scientific questions.

Given that a remotely sensed image is simply a representation of the way that radiation interacts with the Earth's surface and most general land cover classes of interest differ in the way they interact with radiations commonly sensed, remote sensing should be an ideal source of land cover information. This is of immense value as land cover a key environmental variable. Moreover, land cover one of the most dynamic terrestrial features mapped (Belward et al., 1999) and its changes have an environmental effect that is at least as large as that associated with climate change (Skole, 1994). Remote sensing has considerable potential as a source of information on land cover properties and their dynamics. It has, for example, been used to map land cover classes ranging from individual tree species in forests (Turner et al., 2003; Carleer and Wolff, 2004) through broad classes over very large areas including globally (Hansen et al., 2000; Friedl et al., 2002; Latifovic et al., 2004) and in monitoring change in time (Wallace et al., 2003; Guild et al., 2004). Similarly, remote sensing has been used to estimate biophysical and biochemical properties over a range of spatial and temporal scales (Wessman et al., 1988; Curran et al., 1997; Blackburn, 2002; Roberts et al., 2003).

The immense potential of remote sensing as a source of information on land cover is, however, often not fully realised (Townshend, 1992; Wilkinson, 1996). A vast array of reasons may be suggested to explain this situation. These include the complex interactions of the atmosphere with the radiation sensed (e.g., attenuation by transient atmospheric components), the nature of the classes (e.g., class definition and degree of homogeneity), the properties of the sensor (e.g., spectral and spatial resolution), and the techniques used to extract the land cover information from the remotely sensed imagery (e.g., classification methods). There are also concerns that the maps derived from remote sensing may be presented at too fine a spatial resolution, with the minimum mapping unit typically defined implicitly as the spatial resolution of the sensor data from which the map was derived and many basic difficulties associated with issues such as class definition and scaling. The effect of these various issues, alone or combined in some manner, is to limit the accuracy of land cover data extraction, greatly limiting the ability to study geodynamic phenomena. For example, the IGBP map of global land cover, a major technical achievement, has an estimated accuracy of just 67% (Scepan, 1999). The meaning of this accuracy statement is, however, difficult to assess, as it varies with factors such as user requirements, the effect of differential error magnitudes and the confidence in the ground or reference data sets used in the accuracy evaluation (DeFries and Los, 1999; Estes et al., 1999; Foody, 2002). The low accuracy of this and other maps is, however, a cause for concern and has driven a research agenda focused on trying to reduce error.

One major source of error noted with land cover maps derived from remote sensing, especially in maps derived from coarse spatial resolution sensors such as those commonly used in the derivation of global maps, is associated with mixed pixels. Since the techniques traditionally used in analysing remotely sensed imagery assume pure pixels (i.e., the area represented by each pixel comprises a single land cover class), mixed pixels (representing an area covered by more than one land cover class) cannot be appropriately or accurately mapped. Since the use of conventional (hard) classification techniques in mapping land cover from remote sensor imagery must introduce error when mixed pixels are present, researchers have sought alternative approaches for mapping. One popular approach has been to use a soft or fuzzy classifier (Foody, 1996). These classifiers can give more accurate and realistic representations when mixing is common. Since mixing is a particular concern when using coarse spatial resolution imagery, that are typically dominated by mixed pixels, it is now common for soft approaches to be used. Although the representation derived from a soft classification approach may be more appropriate and accurate than a hard classification there are many concerns with this type of approach. One major concern with soft classification and related techniques is the assessment of classification accuracy, which is a more challenging task than for a conventional hard classification (Foody, 2002).

The problems encountered with mapping land cover may feed into and grow when attempting to monitor change. There are many methods by which land cover change may be studied with remotely sensed data (Mas, 1999). Fundamentally, however, the accuracy of change detection is often dependent on the accuracy with which land cover may be mapped. There are also other major non-thematic errors that creep into studies of change. Of these, the effect of image mis-registration is particularly important (Foody, 2002). A basic requirement in monitoring change is to register data sets spatially such that the same location in each time period is being studied. Precise spatial co-registration is difficult to achieve and hence it is common for a small, but important, mis-registration error to be introduced into analyses. The effect of mis-registration can be such that it may mask or greatly exaggerate the amount of change that has occurred (Roy, 2000). Additionally, as with mapping, most change detection studies are hard, commonly focusing on the binary variable of change or no-change. This may often be inappropriate as the change may be gradual or occur at a sub-pixel scale. The assessment of change on the basis of soft classifications, however, yields for instance a very different assessment of change to that derived from hard classifications, notably in the area of land changing class and the magnitude or severity of the change (Foody, 2001).

The great potential of remote sensing as a source of information on major environmental variables such as on land cover and its dynamics is, therefore, constrained by a suite of problems. This situation has driven the research agenda for many working in the remote sensing community. As a result, there has been a plethora of studies focused on different aspects of the problems from a variety of perspectives. Some have sought the use of hyperspectral sensor data sets in the hope that the fine spectral resolution will enhance class separability, while others have developed sub-pixel mapping approaches or have addressed basic philosophical and methodological issues relating to the nature of the land cover and the impacts this has in relation to the techniques used. The seven chapters that follow in Section I of this book all address important issues in the use of remote sensing as a source of data on dynamic geographical phenomena including some of the major issues noted above.

One of the key problems encountered in past studies has related to the apparently simple task of extracting the desired thematic information accurately from remotely sensed imagery. Issues of data extraction are the focus of several chapters that follow. Each of these chapters has addressed a different issue but makes an important contribution to the general aim of trying to exploit fully the potential of remote sensing. For example, Chapters 4 and 7 by Holland and Tompkinson, and Lucieer et al. respectively, include discussions on the use of image texture and context as a discriminatory variable. Although it is well established that textural and contextual information can be used together with spectral information to enhance class identification, numerous problems have been encountered and these two chapters focus on specific instances when such information may be useful. Lucieer et al., for example, present a supervised method for land cover mapping using textural information while Holland and Tompkinson use shadow as a variable to aid the identification of buildings and other objects in imagery. The chapters by Crowther and Hartnett and Shalan et al. (Chapters 5 and 2, respectively) address different issues in the extraction of information from imagery. Crowther and Hartnett discuss the use of spatial reasoning methods and illustrate the use of CommonKADS in the study of agricultural crops. Shalan et al. focus on the accuracy of thematic information extraction with particular regard to the methods that may be applied to assessment of hard and soft image classifications.

Uncertainty in geospatial data sets has been an issue of considerable interest over recent decades (Goodchild and Gopal, 1989; Atkinson, 1999; Foody and Atkinson, 2002). It is also an issue addressed in several chapters including those by Carmel, Lucieer et al., and Brown (Chapters 3, 7 and 8 respectively). In particular, issues associated with thematic and non-thematic uncertainty are addressed. These combined with additional problems encountered in change detection such as those raised by Comber et al. (Chapter 6) and Holland and Tompkinson are key foci of attention. To understand land cover change and use information on it to help mitigate against or adapt to negative environmental impacts the effects of mis-registration on change detection need to be addressed. Brown, Carmel, and Lucieer et al. all discuss issues associated with spatial uncertainty in studies of land cover change. Lucieer et al. show how thematic uncertainty may be largest in transition zones between classes. Similarly, Brown attempts to quantify thematic uncertainty in land cover classification but also presents early work on how this information may be combined with modelled positional uncertainties. From a different perspective, Carmel explores the effect of data aggregation on thematic and locational uncertainty, and critically illustrates a method by which an appropriate level of aggregation may be defined to reduce error. Finally, Comber et al. tackle an important but relatively rarely addressed set of issues related to class definition. With reference to two maps of British land cover produced approximately a decade apart Comber et al. highlight the effect of inconsistencies in the two mapping approaches, notably the meaning of the class labels used, on the evaluation of land cover change using the two maps.

The chapters that follow have, therefore, addressed a variety of major concerns in the study of dynamic variables such as land cover from remotely sensed data. This includes work focused on major concerns such as mixed pixels and mis-registration effects, often the largest sources of error in thematic mapping (Foody, 2002). Together the set of chapters provide an indication of the current position of the research frontier and point to some potential avenues for future research. Moreover, the chapters that follow discuss some important basic issues associated with the measurement of geographical phenomena upon which modelling and other activities discussed in later sections of the book may depend.

REFERENCES

Achard, F., Eva, H.D., Stibig, H.J., Mayaux, P., Gallego, J., Richards, T., and Malingreau, J.P., 2002, Determination of deforestation rates of the world's humid tropical forests, *Science*, 297, 999–1002.

Atkinson, P.M., 1999, Geographical information science: geostatistics and uncertainty, *Progress in Physical Geography*, 23, 134–142.

Belward, A.S., Estes, J.E., and Kilne, K. D., 1999, The IGBP-DIS global 1-km land-cover data set DISCover: a project overview, *Photogrammetric Engineering and Remote Sensing*, 65, 1013–1020.

Blackburn, G.A., 2002, Remote sensing of forest pigments using airborne imaging spectrometer and LIDAR imagery, *Remote Sensing of Environment*, 82, 311–321.

Brown, K., Chapter 8, this volume.

Carleer, A. and Wolff, E., 2004, Exploitation of very high resolution satellite data for tree species identification, *Photogrammetric Engineering and Remote Sensing*, 70, 135–140.

Carmel, Y., Chapter 3, this volume.

Comber, A.J., Fisher, P.F., and Wadsworth, R.A., Chapter 6, this volume.

Crowther, P. and Hartnett, J., Chapter 5, this volume.

Curran, P.J., Kupiec, J.A., and Smith, G.M., 1997, Remote sensing the biochemical composition of a slash pine canopy, *IEEE Transactions on Geoscience and Remote Sensing*, 35, 415–420.

DeFries, R.S. and Los, S.O., 1999, Implications of land-cover misclassification for parameter estimates in global land-surface models: an example from the simple biosphere model (SiB2), *Photogrammetric Engineering and Remote Sensing*, 65, 1083–1088.

Dixon, R.K., Brown, S., Houghton, R.A., Solomon, A.M., Trexler, M.C., and Wisniewski, J., 1994, Carbon pools and flux of global forest ecosystems, *Science*, 263, 185–190.

Estes, J., Belward, A., Loveland, T., Scepan, J., Strahler, A., Townshend, J., and Justice, C., 1999, The way forward, *Photogrammetric Engineering and Remote Sensing*, 65, 1089–1093.

Foody, G.M., 1996, Approaches for the production and evaluation of fuzzy land cover classification from remotely-sensed data, *International Journal of Remote Sensing*, 17, 1317–1340.

Foody, G. M., 2001, Monitoring the magnitude of land cover change around the southern limits of the Sahara, *Photogrammetric Engineering and Remote Sensing*, 67, 841–847.

Foody, G.M., 2002, Status of land cover classification accuracy assessment, *Remote Sensing of Environment*, 80, 185–201.

Foody, G.M. and Atkinson, P.M., 2002, Current status of uncertainty issues in remote sensing and GIS, In G.M. Foody and P.M. Atkinson (Eds.), *Uncertainty in Remote Sensing and GIS*, Wiley, Chichester, pp. 287–302.

Friedl, M.A., McIver, D.K., Hodges, J.C.F., Zhang, X.Y., Muchoney, D., Strahler, A.H., Woodcock, C.E., Gopal, S., Schneider, A., Cooper, A., Baccini, A., Gao, F., and Schaaf, C., 2002, Global land cover mapping from MODIS: algorithms and early results, *Remote Sensing of Environment*, 83, 287–302.

Goodchild, M.F. and Gopal, S., 1989, *Accuracy of Spatial Databases*, Taylor and Francis, London.

Guild, L.S., Cohen, W.B., and Kauffman, J.B., 2004, Detection of deforestation and land conversion in Rondonia, Brazil using change detection techniques, *International Journal of Remote Sensing*, 25, 731–750.

Hansen, M.C., DeFries, R.S., Townshend, J.R.G., and Sohlberg, R., 2000, Global land cover classification at 1 km spatial resolution using a classification tree approach, *International Journal of Remote Sensing*, 21, 1331–1364.

Holland, D.A. and Tompkinson, W., Chapter 4, this volume.

Houghton, R.A., Skole, D.L., Nobre, C.A., Hackler, J.L., Lawrence, K.T., and Chomentowski, W.H., 2000, Annual fluxes or carbon from deforestation and regrowth in the Brazilian Amazon, *Nature*, 403, 301–304.

Kelly, R. (Ed.), 2004, *Spatial Modelling of the Terrestrial Environment*, Wiley, Chichester.

King, D.A., 2004, Environment — climate change science: adapt, mitigate, or ignore?, *Science*, 303, 176–177.

Konecny, G., 2003, Geoinformation. *Remote Sensing, Photogrammetry and Geographic Information Systems*, Taylor and Francis, London.

Latifovic, R., Zhu, Z.L., Cihlar, J., Giri, C., and Olthof, I., 2004, Land cover mapping of north and central America — Global Land Cover 2000, *Remote Sensing of Environment*, 89, 116–127.

Laxon, S., Peacock, N., and Smith, D., 2003, High interannual variability of sea ice thickness in the Arctic region, *Nature*, 425, 947–950.

Loveland, T.R., Zhu, Z., Ohlen, D.O., Brown, J.F., Reed, B.C., and Yang, L., 1999, An analysis of the IGBP global land-cover characterisation process, *Photogrammetric Engineering and Remote Sensing*, 65, 1021–1032.

Lucieer, A., Fisher, P., and Stein, A., Chapter 7, this volume.

Mas, J-F., 1999, Monitoring land-cover changes: a comparison of change detection techniques, *International Journal of Remote Sensing*, 20, 139–152.

Roberts, D.A., Keller, M., and Soares, J.V., 2003, Studies of land-cover, land-use, and biophysical properties of vegetation in the large scale biosphere atmosphere experiment in Amazonia, *Remote Sensing of Environment*, 87, 377–388.

Roughgarden, J., Running, S.W., and Matson, P.A., 1991, What does remote sensing do for ecology?, *Ecology*, 76, 1918–1922.

Roy, D.P., 2000, The impact of misregistration upon composited wide field of view satellite data and implications for change detection, *IEEE Transactions on Geoscience and Remote Sensing*, 38, 2017–2032.

Royal Society, 2001, *The Role of Land Carbon Sinks in Mitigating Global Climate Change*, Policy document 10/01 (London: The Royal Society).

Scepan, J., 1999, Thematic validation of high-resolution global land-cover data sets, *Photogrammetric Engineering and Remote Sensing*, 65, 1051–1060.

Shalan, M.A., Arora, M.K., and Elgy, J., Chapter 2, this volume.

Skole, D.L., 1994, Data on global land-cover change: acquisition, assessment and analysis. In Meyer W.B. and Turner II B.L. (Eds.), *Changes in Land Use and Land Cover: A Global Perspective*, Cambridge University Press, Cambridge, pp. 437–471.

Thurston, J., Poiker, T.K., and Moore, J.P., 2003, *Integrated Geospatial Technologies: A Guide to GPS, GIS and Data Logging*, Wiley, Chichester.

Townshend, J.R.G., 1992, Land cover, *International Journal of Remote Sensing*, 13, 1319–1328.

Turner, W., Spector, S., Gardiner, N., Fladeland, M., Sterling, E., and Steininger, M., 2003, Remote sensing for biodiversity science and conservation, *Trends in Ecology and Evolution*, 18, 306–314.

van der A, R.J., Piters, A.J.M., Van Oss, R.F., and Zehner, C., 2003, Global stratospheric ozone profiles from GOME in near-real time, *International Journal of Remote Sensing*, 24, 4969–4974.

Wessman, C.A., Aber, J.D., Peterson, D.L., and Melillo, J.M., 1988, Remote-sensing of canopy chemistry and nitrogen cycling in temperate forest ecosystems, *Nature*, 335, 154–156.

Wallace, O.C., Qi, J.G., Heilma, P., and Marsett, R. C., 2003, Remote sensing for cover change assessment in southeast Arizona, *Journal of Range Management*, 56, 402–409.

Wilkinson, G.G., 1996, Classification algorithms — where next? In Binaghi, E. Brivio, P.A. and Rampini A. (Eds.), *Soft Computing in Remote Sensing Data Analysis*, World Scientific, Singapore, pp. 93–99.

CHAPTER 2

Crisp and Fuzzy Classification Accuracy Measures for Remotely Sensed Derived Land Cover Maps

Manoj K. Arora, Mohamed A. Shalan, and John Elgy

CONTENTS

2.1 INTRODUCTION

The value of any geodata set depends on its fitness for use. A critical measure of the fitness is the data quality, knowledge of which may significantly increase the confidence of the user in explaining and defending the results derived from analyses with the map (LMIC, 1999). Therefore, extensive information about the quality of geodata input to a GeoDynamics analysis and modeling process (refer chapters in Section II) is essential. Remotely sensed derived land use land cover information on temporal basis is a key data source on GeoDynamic phenomena and has been used in studies focused on modeling the effect of climatic variables on tree distributions (Svoray and Nathan, 2004) and land cover change detection (Comber et al., 2004).

0-8493-2837-3/05/$0.00+$1.50

Besides this, land use land cover is a significant variable for a variety of other earth science applications.

Remote sensing data are now being used on regular basis to produce land cover classifications. Both crisp and fuzzy classifications may be performed. Conventionally, in a crisp classification, each image pixel is assumed pure and is classified to one class (Lillesand and Kiefer, 2000). Often, particularly in coarse spatial resolution images, the pixels may be mixed containing two or more classes. Fuzzy classifications that assign multiple class memberships to a pixel may be appropriate for images dominated by mixed pixels (Foody and Cox, 1994).

Generally, supervised classification is adopted that involves three stages: training, allocation, and testing (Arora and Foody, 1997). The quality of the derived land cover data is expressed in terms of the accuracy of classification, which is determined in the testing stage and is the focus of this chapter. In essence, the assessment of accuracy involves comparison of the classified image with reference or ground data. The reference data may be gathered from field surveys, existing maps, aerial photographs, and datasets that are at finer spatial resolution than the image being classified. Typically, the accuracy of crisp classification is determined using traditional error matrix based measures such as the overall accuracy (OA), user's accuracy (UA) and producer's accuracy (PA), and kappa coefficient of agreement (Congalton, 1991). Smits et al. (1999) have recommended that the kappa coefficient may be adopted as the standard accuracy measure. Despite this, a number of other measures, for example, Tau coefficient (Ma and Redmond, 1995), and classification success index (Koukoulas and Blackburn, 2001) have also been proposed, though used sparingly. The development of a number of measures clearly indicates that there are many problems in the accuracy assessment of image classification (Foody, 2002). One of the major problems is the occurrence of mixed pixels in imagery. Fuzzy set theoretic approaches may be applied to classify the images dominated by mixed pixels to produce fuzzy classification outputs. To evaluate the accuracy of fuzzy classification, these outputs are often hardened so that the error matrix based measures may be used. Degrading a fuzzy classification to a crisp classification results in loss of information contained in the fuzzy outputs thereby hampering its proper evaluation. Moreover, the reference data are also not always error-free and may contain uncertainties, and therefore may be treated as fuzzy (Bastin et al., 2000). Hence, alternative accuracy measures that may appropriately include fuzziness in the classification outputs and/or reference data have been proposed. These include root mean square error, correlation coefficient (Foody and Cox, 1994), entropy and cross entropy (Maselli et al., 1994; Foody, 1995) distance-based measures (Foody, 1996; Foody and Arora 1996), fuzzy set-based operators (Gopal and Woodcock, 1994) and recently introduced fuzzy error matrix based measures (Binaghi et al., 1999).

The growth of so many accuracy measures both for crisp and fuzzy classification indicates clearly the current research potential of classification accuracy assessment procedures as no single measure may be adopted universally. Depending upon the nature of classification outputs, uncertainties in reference data and the quality of information desired by the end user, it may be necessary to adopt not one but a combination of accuracy measures.

Despite the considerable research undertaken on classification accuracy assessment and its importance, current image processing software are limited in providing sufficient accuracy information to the user. For example, the well known and the most widely used software namely ERDAS Imagine, ENVI, and IDRISI, contain accuracy assessment modules that can report only a few crisp accuracy measures notably OA, PA and UA, and kappa coefficient of agreement. No other competitive accuracy measures have been included, which may be of interest to the user in cases where the assumptions regarding the current accuracy measures are not met by the dataset. Also, there is no provision for accuracy assessment of fuzzy classification in these software. Only IDRISI has a measure called classification uncertainty that determines the quality of classification on a per-pixel basis, and thus may not be treated as a measure to indicate the accuracy of the whole classification. Thus, to evaluate the accuracy of fuzzy classification, users either have to depend on other statistical and mathematical software, where import/export of the data from one package to another may be a tedious task, or they may have to develop their own software. Further, to critically examine the usefulness of a particular accuracy measure *vis a vis* other measures, a dedicated software for classification accuracy assessment needs to be developed. In view of this, a stand alone software acronymed as CAFCAM (Crisp And Fuzzy Classification Accuracy Measurement) has been developed. Detailed information on this software may be found in Shalan et al. (2003a). Formulation of a number of accuracy measures discussed here has been included in this software.

The aim of this chapter is to apprise the readers with a set of accuracy measures, incorporated in this software, to evaluate the quality of both crisp and fuzzy classification of remote sensing data. The Section 2.1 has provided an introduction to the scope of this chapter. The Section 2.2 is a review on classification accuracy assessment of crisp and fuzzy classification. A working example is presented in the Section 2.3 that is followed by a summary of the chapter.

2.2 CLASSIFICATION ACCURACY ASSESSMENT

In its simplest form, classification accuracy refers to the correspondence between the class label assigned to a pixel and the true class as observed directly on the ground or indirectly from a map or aerial photograph (reference data). Since the beginning of satellite remote sensing, the problem of classification accuracy assessment has received an immense recognition within the remote sensing community (e.g., van Genderen et al., 1978; Congalton et al., 1983; Rosenfield and Fitzpatrick-Lins, 1986; Story and Congalton, 1986; Fung and LeDrew, 1988; Kenk et al., 1988; Congalton, 1991; Foody, 1992; Fitzgerald and Lee, 1994; Gopal and Woodcock, 1994; Janssen and van der Wel, 1994; Foody, 1995; Ma and Redmond, 1995; Verbyla and Hammond, 1995; Naesset, 1996a,b; Binaghi et al., 1999; Smits et al., 1999; Stehman, 2001; Foody, 2002). An inspection of these studies reveals that according to the nature of classification (i.e., crisp or fuzzy), proper accuracy measures may have to be adopted to derive meaningful land cover information from remote sensing data.

2.2.1 Crisp Classification Accuracy Assessment

A typical strategy to assess the accuracy of a crisp classification begins with the selection of a sample of pixels (known as testing samples) in the classified image based on a sampling design procedure (Stehman, 1999), and confirming their class allocation with the reference data. A number of statistically sound sampling schemes has been proposed in the literature and range from a simple random sampling, stratified random sampling to systematic and cluster sampling, each having their own merits and demerits (Congalton, 1988; Stehman, 1999). The formulation of many accuracy measures included in the software CAFCAM is based on simple random sampling scheme.

The pixels of agreement and disagreement are summarized in the form of a contingency table (known as error or confusion matrix), which can be used to estimate various accuracy measures (Congalton, 1991). Table 2.1 shows a typical $c \times c$ error matrix (c is the number of classes) with columns representing the reference data and rows the classified image, albeit both are interchangeable. For an ideal classification, it is expected that all the testing samples would lie along the diagonal of the matrix indicating the perfect agreement. The off-diagonal elements indicate the disagreements referred to as the errors of omission and commission (Story and Congalton, 1986).

The elements of the error matrix are used to derive a number of accuracy measures, which have been divided into three groups:

1. Percent correct measures
2. Kappa coefficients
3. Tau coefficients

The formulations of all the accuracy measures considered under these groupings are given in Table 2.2. Further details on these formulations can be found in the respective references cited in this table.

The first group consists of five accuracy measures — (OA), (UA), and (PA) (Story and Congalton, 1986), average and combined accuracy (Fung and LeDrew,

Table 2.1 Layout of Traditional Error Matrix

	Reference Data				
Classified Image	**Class 1**	**Class 2**	**…**	**Class c**	**Row Total**
Class 1	n_{11}	n_{12}	…	n_{1c}	N_1
Class 2	n_{21}	n_{22}	…	n_{2c}	N_2
⋮	⋮	⋮	…	⋮	⋮
Class c	n_{c1}	n_{c2}	…	n_{cc}	N_c
Column total	M_1	M_2	…	M_c	$N = \sum_{i=1}^{c} N_i$

Definition of terms: n_{ij} are the pixels of agreement and disagreement, N is the total number of testing pixels.

Table 2.2 Crisp Classification Accuracy Measures

Accuracy Metric	Formulation	Base Reference
Overall accuracy	$\frac{1}{N}\sum_{i=1}^{c} n_{ii}$	Story and Congalton (1986)
User's accuracy	n_{ii}/N_i	Story and Congalton (1986)
Producer's accuracy	n_{ii}/M_i	Story and Congalton (1986)
Average accuracy (user's)	$\frac{1}{c}\sum_{i=1}^{c}\frac{n_{ii}}{N_i}$	Fung and LeDrew (1988)
Average accuracy (producer's)	$\frac{1}{c}\sum_{i=1}^{c}\frac{n_{ii}}{M_i}$	Fung and LeDrew (1988)
Combined accuracy (user's)	$\frac{1}{2}[OA + AA_U]$	Fung and LeDrew (1988)
Combined accuracy (producer's)	$\frac{1}{2}[OA + AA_P]$	Fung and LeDrew (1988)
Kappa coefficient of agreement	$\frac{P_o - P_e}{1 - P_e}$	Congalton et al. (1983)
Weighted Kappa	$1 - \frac{\sum v_{ij} P_{oij}}{\sum v_{ij} P_{cij}}$	Rosenfield and Fitzpatrick-Lins (1986)
Conditional Kappa (user's)	$\frac{P_{o(i+)} - P_{e(i+)}}{1 - P_{e(i+)}}$	Rosenfield and Fitzpatrick-Lins (1986)
Conditional Kappa (producer's)	$\frac{P_{o(+i)} - P_{e(+i)}}{1 - P_{e(+i)}}$	Rosenfield and Fitzpatrick-Lins (1986)
Tau coefficient (equal probability)	$\frac{P_o - (1/c)}{1 - (1/c)}$	Ma and Redmond (1995)
Tau coefficient (unequal probability)	$\frac{P_o - P_r}{1 - P_r}$	Ma and Redmond (1995)
Conditional Tau (user's)	$\frac{P_{o(i+)} - P_i}{1 - P_i}$	Naesset (1996)
Conditional Tau (producer's)	$\frac{P_{o(+i)} - P_i}{1 - P_i}$	Naesset (1996)

Definition of terms: N is total number of testing pixels; n_{ii} is the number of samples correctly classified; N_i and M_i are the row and column totals for class i, respectively; $P_o = (1/N)\sum_{i=1}^{c} n_{ii}$ is the observed proportion of agreement $P_e = (1/N^2)\sum_{i=1}^{c} N_i M_i$ is the expected chance agreement; v_{ij} is the weight; $P_{o_{ij}}$ is the observed cell proportion; Pe_{ij} is the expected cell proportion; $P_{o(i+)}$ is the observed agreement according to user's approach computed from all columns in ith row of the error matrix; $P_{e(i+)}$ is the agreement expected by chance for ith row; $P_{o(+i)}$ is the observed agreement according to producer's approach computed from all rows in ith column of the error matrix; $P_{e(+i)}$ is the agreement expected by chance for ith column; $P_r = (1/N)\sum_{i=1}^{c} n_{i+}x_i$, where x_i is the unequal *a priori* probability of class membership; P_i is the *a priori* probability of class membership.

1988). While overall, average, and combined accuracy signify the quality of whole classification, UA, and PA indicate the quality of each individual class. PA is so called, because the producer of the classification is typically interested in knowing how well the samples from the reference data can be mapped using a remotely sensed image. In contrast, UA indicates the probability that a sample from the classification represents an actual class in reference data (Story and Congalton, 1986).

The OA is one of the most commonly adopted measures to assess the accuracy of a classification. OA weights each class in proportion to its area representation in the map. Its value may be affected by the relative abundance and separability of the classes. A way to resolve the problem of differences in sample size of an individual class may be to keep their sample sizes equal. Alternatively, the elements of the error matrix may be normalized to compute normalized accuracy (Congalton, 1991). However, Stehman and Czaplewski (1998) have not preferred normalization procedure since it may lead to bias and may have the effect of equalizing the UA and PA that may actually differ considerably.

Accuracy measures in the percent correct group do not take into account the agreement between the datasets (i.e., classified output and reference data) that arises due to chance alone. The kappa coefficient of agreement incorporates chance agreement into its formulation and thus provides a chance-corrected measure. It compensates for the chance agreement that results from misclassifications represented by the off-diagonal elements of the error matrix. The second group of accuracy measures consists of four measures from kappa family — kappa coefficient of agreement (Congalton et al., 1983), weighted kappa (Cohen, 1968; Rosenfield and Fitzpatrick-Lins, 1986; Naesset, 1996a), and conditional kappa — user's and producer's perspective (Rosenfield and Fitzpatrick-Lins, 1986; Naesset, 1995). Since kappa coefficient of agreement (K) is based on all the elements of the error matrix, and not just the diagonal elements (as is the case with OA), therefore, K may sometimes be an appropriate candidate for accuracy assessment of a classification. Conditional kappa coefficients may be derived to assess the accuracy of individual classes. Weighted kappa (K_w) may be thought as a generalization of K. When the misclassification of some classes is more serious than others, weighted kappa may be implemented since it does not treat all the misclassifications (disagreements) equally and tends to give more weight to the misclassifications that are more serious than others (Cohen, 1968; Hubert, 1978).

Nevertheless, as argued by Foody (1992) and later supported by Ma and Redmond (1995), the kappa family may overestimate the chance agreement that may result in an underestimation of accuracy. Therefore, alternatives to kappa family coefficients such as Tau coefficient have been proposed (see Brown (2004) this volume, for an example of its use in remote sensing). These form the third group of accuracy measures. The critical difference between the two coefficients is that Tau coefficient is based on *a priori* probabilities of class membership, whereas kappa uses the *a posteriori* probabilities. Unlike kappa coefficient, Tau coefficient compensates for the influence of unequal probabilities of classes on random agreement. Moreover, for classifications based on equal probabilities of class membership, Tau compensates for the influence of the number of groups (Ma and Redmond, 1995). A conditional

Tau coefficient may be used to indicate the accuracy of an individual class (Naesset, 1996b).

It may thus be seen that there are a number of measures that may be computed from an error matrix. Each measure may, however, be based on different assumptions about the data and thus may evaluate different components of accuracy (Lark, 1995; Stehman, 1997). Therefore, in general, it may be expedient to provide an error matrix with the classified image and report more than one measure of accuracy to fully describe the quality of that classification (Stehman, 1997).

2.2.2 Fuzzy Classification Accuracy Assessment

The traditional error matrix based measures inherently assume that each pixel is associated with only one class in the crisp classification and only one class in the reference data. Use of these measures to assess the accuracy of fuzzy classification may, therefore, under- or over-estimate the accuracy of a classification. This is due to the fact that the fuzzy classification outputs of pixels have to be degraded to produce a crisp classification to adhere to this assumption thereby resulting in the loss of information.

At the first instance, fuzzy classification outputs indicating the probability of class memberships in a given pixel may be used to compute entropy. Entropy is a measure of uncertainty in the classification (Table 2.3) and shows how the strength of class membership (i.e., fuzzy outputs) is partitioned between the classes for each pixel (Foody, 1995). The entropy for a pixel is maximized when the pixel has equal class memberships for all the classes. Conversely, its value is minimized, when the pixel is entirely allocated to one class. It, thus, shows the degree to which a classification output is fuzzy (i.e., uncertainty in class allocation) or crisp. Its value as an indicator of classification accuracy is based on the assumption that in an accurate classification each pixel will have a high probability of membership belonging to only one class. Another measure called classification uncertainty (Eastman, 2001) may also be used to indicate the uncertainty in class allocation of a pixel. The classification uncertainty is computed as the complement of the difference between the maximum class membership value and the total dispersion of the class memberships over all classes divided by the extreme case of the difference between a maximum proportion value of 1 (i.e., total commitment to a single class) and the total dispersion of that commitment over all classes (Table 2.3). This ratio expresses the degree of commitment to a specific class relative to the largest possible commitment that can be made. In reality, the classification uncertainty differs from entropy in the sense that it is not only concerned with the degree of dispersion of class membership values between classes, but also the total amount of commitment to a particular class present in a pixel. For example, if a pixel is assigned equal memberships to all classes then the classification uncertainty becomes 1, on the other hand if a pixel is assigned to only one class then the classification uncertainty becomes 0.

Both accuracy measures — classification uncertainty and entropy — however, are only appropriate for situations in which the output of the classification is fuzzy and the reference data are crisp (Binaghi et al., 1999). Since fuzzy reference data may

Table 2.3 Fuzzy Classification Accuracy Measures

Accuracy Metric	Formulation	Base Reference
Entropy	$-\sum_{i=1}^{c}(^2p_i)\log_2(^2p_i)$	Maselli et al. (1994)
Classification uncertainty	$1-\dfrac{\max(p)-((\sum p)/c)}{1-(1/c)}$	Eastman (2001)
Euclidean distance	$\dfrac{\sum_{i=1}^{c}(^1p_i-{}^2p_i)^2}{c}$	Kent and Mardia (1988), Foody (1996)
L_1 (City Block) distance	$\dfrac{\sum_{i=1}^{c}\lvert ^1p_i-{}^2p_i\rvert}{c}$	Foody and Arora (1996)
Cross-entropy or direct divergence	$D(^1p,{}^2p)=-\sum_{i=1}^{c}(^1p_i)\log_2(^2p_i)+\sum_{i=1}^{c}(^1p_i)\log_2(^1p_i)$	Foody (1995)
Measure of information closeness	$I(^1p,{}^2p)=D\left(^1p,\dfrac{^1p+{}^2p}{2}\right)+D\left(^2p,\dfrac{^1p+{}^2p}{2}\right)$	Foody (1996)
Correlation coefficients	$\dfrac{\mathrm{Cov}(^1p_i,{}^2p_i)}{\mathrm{Std}(^1p_i)\mathrm{Std}(^2p_i)}$	Foody and Cox (1994), Maselli et al. (1996)

Definition of terms: ^1p_i and ^2p_i is the proportion of ith class in a pixel from the fuzzy reference data (^{1}), and classification (^{2}), respectively; $\max(p)$ is the maximum proportion for a pixel in fuzzy classification; $\sum p$ is the sum of the proportions in a pixel; c is the number of classes in fuzzy classification; ^1p is the probability distribution of fuzzy reference data; ^2p is the probability distribution of fuzzy classification output; $\mathrm{Cov}(^1p_i,{}^2p_i)$ is the covariance between the two datasets; $\mathrm{Std}(^1p_i)$ and $\mathrm{Std}(^2p_i)$ are the standard deviations of the respective datasets.

also be generated on the basis of a linguistic scale (Gopal and Woodcock, 1994) or class proportions (Shalan et al., 2003b), a few fuzzy set based operators namely max, right, and difference may be applied to evaluate the accuracy of a crisp classification with respect to fuzzy reference data. For instance, in the working example in the Section 2.3, a land cover map produced from an IRS PAN image at 5 m spatial resolution has been used as reference data to assess the accuracy of fuzzy classification produced from an IRS LISS III image at 25 m spatial resolution. A pixel of the LISS III image corresponds to a number of 5 m pixels (in this case 25 pixels) from which the proportional coverage of a class in the 25 m spatial resolution image may be derived. These class proportions sum to one for each pixel are called as fuzzy reference data. The fuzzy reference data may be hardened to produce crisp reference data for accuracy assessment of crisp classification.

The accuracy assessment using the entropy and the fuzzy set based max, right, and difference operators account for fuzziness either in the classification or in the reference data, respectively. The ideal situation will be to compare fuzzy classification with fuzzy reference data. In this scenario, these accuracy measures may therefore not be appropriate. To accommodate fuzziness in both the classification output and the reference data, other measures are required. In this case, the accuracy of fuzzy classification outputs is determined by comparing these with fuzzy reference data. A number of accuracy measures may be used. For simplicity, these measures have been divided into three groups:

1. Measures of closeness
2. Correlation coefficient
3. Measures based on fuzzy error matrix

The formulations of all the accuracy measures considered under these groupings are given in Table 2.3. Further details on these formulations can be found in the respective references cited in this table.

The first group includes cross entropy, L_1 and Euclidean distances and a generalized measure of information closeness. These measures estimate the separation of two datasets based on the relative extent or proportion of each class in the pixel (Foody and Arora, 1996). The lower the value of these measures, the higher is the accuracy of the classification. The distance measures and cross entropy may be applicable when there is compatibility between the probability distributions of the classified outputs and reference data. On the other hand, the generalized measure of information closeness may be used even if the probability distributions of the two datasets are not compatible (Foody, 1996). The correlation coefficients for each class between fuzzy classification output and fuzzy reference data may also be used to indicate the accuracy of individual classes. The larger the correlation coefficient the more accurate the classification of the specific class considered.

All the above measures may be treated as indirect methods of assessing the accuracy of fuzzy classification, because the accuracy evaluation is interpretative rather than a representation of actual value as denoted by traditional error matrix based measures. Recently, Binaghi et al. (1999) proposed the concept of fuzzy error matrix, which can be generated on the lines of traditional error matrix used for the evaluation of crisp classification. The layout of a fuzzy error matrix is similar to the traditional error matrix with the exception that elements of a fuzzy error matrix can be any non-negative real numbers instead of non-negative integer numbers. The elements of the fuzzy error matrix represent class proportions corresponding to reference data (i.e., fuzzy reference data) and classified outputs (i.e., fuzzy classified image), respectively.

Let R_n and C_m be the sets of reference and classification data assigned to class n and m, respectively, where the values of n and m are bounded by one and the number of classes c. Note here that R_n and C_m are fuzzy sets and $\{R_n\}$ and $\{C_m\}$ form two fuzzy partitions of the testing sample dataset X, where x denotes a testing sample in X. The membership functions of R_n and C_m are given by,

$$\mu_{R_n} : X \to [0, 1]$$

and

$$\mu_{C_m} : X \to [0, 1]$$

where [0, 1] denotes the interval of real numbers from 0 to 1 inclusive. Here, $\mu_{R_n}(x)$ and $\mu_{C_m}(x)$ is the class membership (or class proportion) of the testing sample x in R_n and C_m, respectively. Since, in the context of fuzzy classification, these membership functions also represent the proportion of a class in the testing sample, the *orthogonality* or *sum-normalization* is often required, which for the fuzzy reference data may be represented as,

$$\sum_{l=l}^{c} \mu_{R_l}(x) = 1 \tag{2.1}$$

The procedure used to construct the fuzzy error matrix M employs fuzzy min operator to determine the element $M(m, n)$ in which the degree of membership in the fuzzy intersection $C_m \cap R_n$ is computed as,

$$M(m, n) = |C_m \cap R_n| = \sum_{x \in X} \min(\mu_{C_m}, \mu_{R_n}) \tag{2.2}$$

For clear understanding of the fuzzy error matrix, let us consider a simple example, where a pixel has been classified into three land cover classes. The corresponding class proportions from fuzzy reference data and fuzzy classification output for that pixel are given by,

$$\mu_{R_1}(x) = 0{\cdot}3, \quad \mu_{R_2}(x) = 0{\cdot}2, \quad \mu_{R_3}(x) = 0{\cdot}5$$

and

$$\mu_{C_1}(x) = 0{\cdot}5, \quad \mu_{C_2}(x) = 0{\cdot}2, \quad \mu_{C_3}(x) = 0{\cdot}3$$

The fuzzy error matrix for the above pixel is then created using min operator from fuzzy set theory, and is displayed in Table 2.4. The fuzzy error matrix for the whole classification is accumulated by summing up the elements of fuzzy error matrix of each testing pixel (i.e., testing sample). OA, PA, and UA may then be estimated in usual fashion from this matrix to indicate the accuracy of a fuzzy classification and of individual class.

Due to its correspondence to the traditional error matrix, the use of fuzzy error matrix to evaluate fuzzy classification may therefore be more appropriate than the distance based measures and correlation coefficients. Moreover, the formulation of fuzzy error matrix is such that it can also be used to assess the accuracy of crisp classification given the crisp reference data. Thus, from the point of view of standardizing

Table 2.4 Fuzzy Error Matrix for an Individual Testing Sample

	Fuzzy Reference Data			
Fuzzy Classification	**Class 1 (R_1)**	**Class 2 (R_2)**	**Class 3 (R_3)**	**Total Memberships**
Class 1 (C_1)	0·3	0·2	0·5	0·5
Class 2 (C_2)	0·2	0·2	0·2	0·2
Class 3 (C_3)	0·3	0·2	0·3	0·3
Total memberships	0·3	0·2	0·5	1·0

the accuracy assessment procedures for both crisp and fuzzy classification, fuzzy error matrix and the associated measures appear more suitable in assessing the quality of remotely sensed derived classifications.

2.3 A WORKING EXAMPLE

The software CAFCAM was written in MATLAB, which includes formulation of all the accuracy measures discussed in this chapter. To demonstrate the utility of some accuracy measures, a case study on accuracy assessment of fuzzy land cover classification from IRS 1C remote sensing data is briefly presented here. More details can be found in Shalan et al. (2003b). An IRS 1C LISS III image (Figure 2.1a) was used as the primary image to produce a fuzzy classification. Five dominant land cover classes in the region, namely agriculture, forest, grassland, urban, and sandy areas, were considered. Maximum likelihood derived crisp classification of the PAN image into five land cover classes was used as reference data (Figure 2.1b). The LISS image was registered to the PAN image derived land cover classification to an accuracy of one third of a pixel, using first order polynomial transformation and nearest neighborhood resampling. The registered LISS and PAN images were resampled to 25 and 5 m, respectively such that a LISS pixel corresponds to a specific number of PAN pixels (in this case 25 pixels) to facilitate generating fuzzy reference data in the form of class proportions.

Two markedly different classification algorithms, the conventional maximum slikelihood classifier (MLC) and the fuzzy *c*-means (FCM) clustering algorithm, were used to perform a fuzzy classification in this study. The MLC is the most widely used classifier in the remote sensing community. In a majority of studies, this classifier has generally been used to provide crisp classification output. However, the output

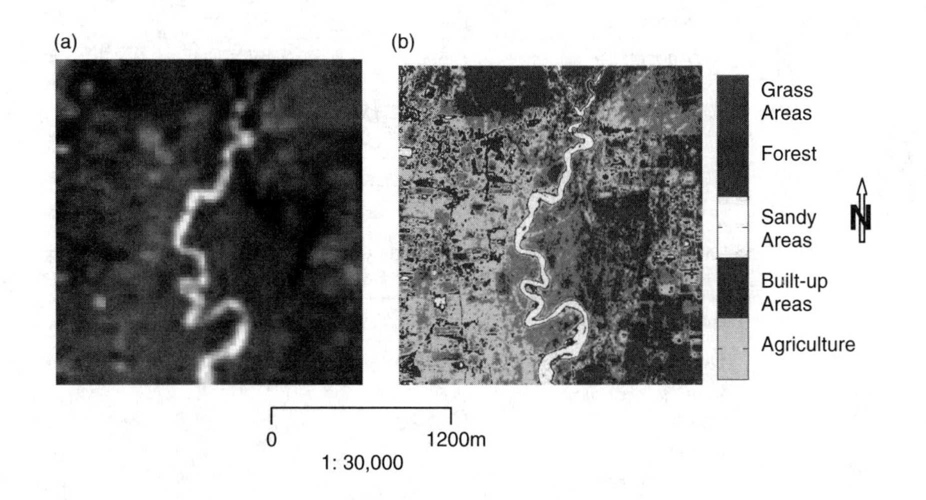

Figure 2.1 (see color insert following page 168) (a) IRS 1C LISS III FCC. (b) Classified PAN image used as reference data.

of MLC in the form of *a posteriori* class probabilities may be related to the actual class likelihoods for each pixel on the ground thereby providing fuzzy classification (Foody et al., 1992).

The FCM is based on an iterative clustering algorithm that may be employed to partition pixels of an image into class proportions (Bezdek et al., 1984). It is essentially an unsupervised clustering algorithm but can be suitably modified to run in supervised mode. In the supervised mode, cluster means are computed from the training areas created in the training stage.

For effectual comparison with MLC, the supervised version of FCM was applied here (Wang, 1990). In the formulation of FCM, a weighting factor m that describes the degree of fuzziness has to be provided. Similar to a study conducted by Foody (1996), where $m = 2{\cdot}0$ was found to produce the most accurate fuzzy classification, in this study also, m was set to 2·0 after several experiments on the dataset considered.

The training data consisted of 997, 286, 279, 1596, and 805 randomly selected pixels for agriculture, urban area, sandy area, forest, and grassland, respectively.

Table 2.5 Accuracy of Fuzzy Classifications Produced from MLC and FCM

Accuracy Metric (Average)	MLC	FCM
Entropy	0·526	0·565
Cross-entropy	0·262	0·287
Euclidean distance	0·057	0·060
Information closeness	0·145	0·160

Table 2.6 Correlation Coefficients of Classes from Fuzzy Classifications

Class	MLC	FCM
Agriculture	0·590	0·495
Urban area	0·507	0·626
Sandy area	0·854	0·860
Forest	0·708	0·583
Grassland	0·402	0·366

Table 2.7 Accuracy of Fuzzy Classification Derived from Fuzzy Error Matrix Based Measures

Accuracy Measure	MLC	FCM
Overall accuracy (%)	33·80	31·80
Average accuracy		
User's	31·20	29·20
Producer's	34·10	32·20
Combined accuracy		
User's	32·50	30·50
Producer's	33·90	32·00

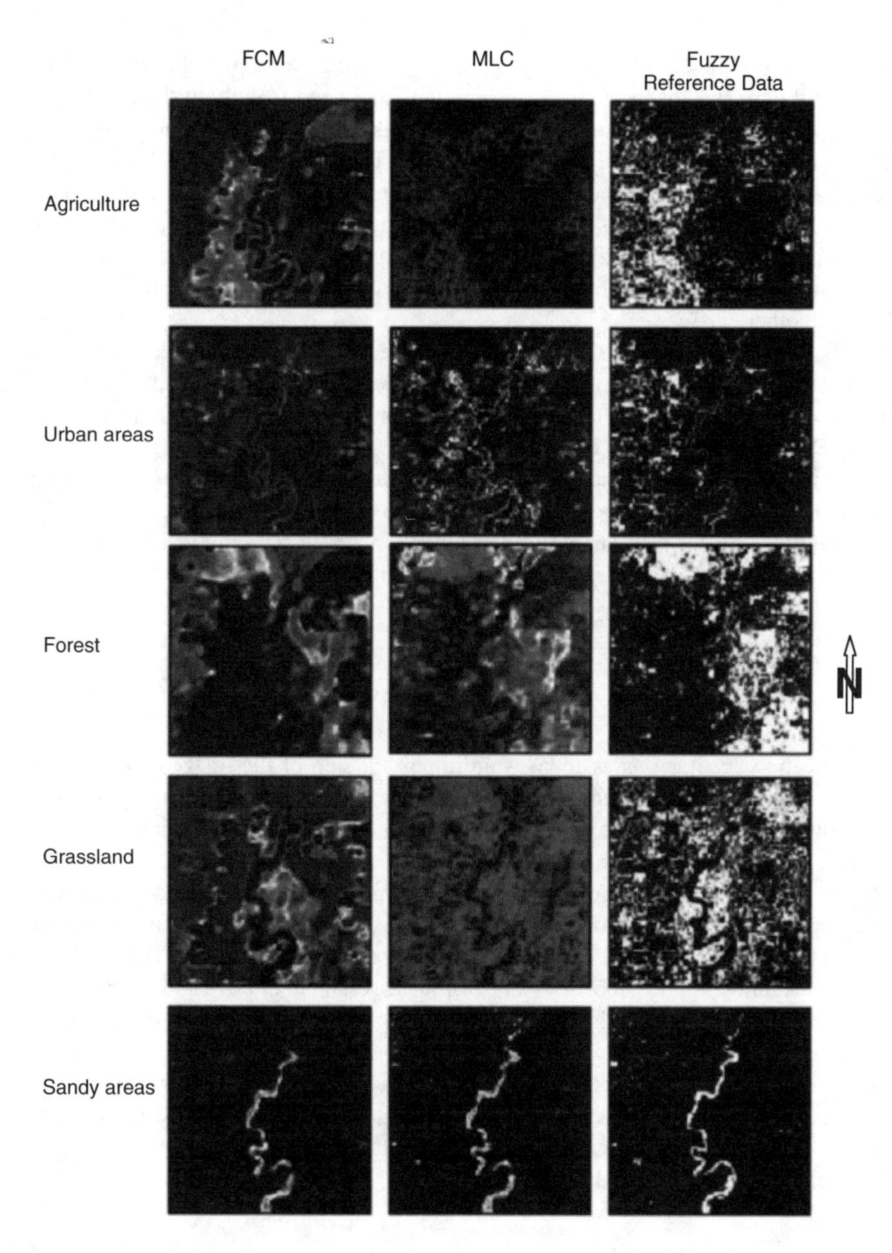

Figure 2.2 Fraction images portraying the spatial distribution of five land cover classes.

In the testing stage, a total of 650 testing pixels from the entire image were randomly selected for accuracy assessment using simple random sampling.

The accuracy of the fuzzy classifications was evaluated using a number of accuracy measures (Table 2.5). Entropy was used to examine the degree of uncertainty in the fuzzy classification outputs. For a five-class problem, the maximum value of entropy is 0·69. From Table 2.5, it can be seen that the average entropy values (computed over all the pixels) for the fuzzy classifications produced from both the classifiers, are very close to the maximum entropy value. This illustrates clearly the presence of class mixtures (or uncertainties) in the dataset. In Table 2.5, the lower values of cross entropy and Euclidean distance for MLC demonstrate that this classifier has produced more accurate classifications than FCM for the dataset considered. From correlation coefficients (Table 2.6), it can be seen that the class sandy area was the most accurately classified class in both classifications, as this class was very spectrally separable from the other classes.

The fuzzy error matrices were also generated for classifications produced from both classifiers. The resulting measures corresponding to this matrix are shown in Table 2.7. From this table, it can again be seen that MLC has produced more accurate fuzzy classification than the FCM.

To inspect fuzzy classifications visually, fraction images portraying the spatial distribution of five land cover classes were also produced (Figure 2.2). It can be observed that for all the classes, in particular for the class sandy areas, MLC has generally predicted class proportions more accurately than the FCM.

2.4 SUMMARY

Classification accuracy assessment is an important step of the image classification process. A number of accuracy measures for both crisp and fuzzy classifications have been proposed. No measure has been universally adopted. Often, a combination of accuracy measures may have to be used to describe the quality of classification completely. However, current image processing software lacks the provision of various accuracy measures. Therefore, a software, named CAFCAM was written exclusively for accuracy assessment of remotely sensed derived classifications. The software includes a number of accuracy measures for the assessment of crisp and fuzzy classifications. In this chapter, a discussion on the accuracy measures used in that software was presented. A working example demonstrating the usefulness of some accuracy measures to evaluate a fuzzy land cover classification from IRS 1C LISS remote sensing data was also provided. Amongst the two fuzzy classifiers namely MLC and FCM used, all the accuracy measures showed that the former produced the most accurate fuzzy classification for this dataset.

ACKNOWLEDGMENTS

This chapter was written when M.A. Shalan was at Aston University, U.K., under a scholarship from the Ministry of High Education, Syria. The authors would like to acknowledge the anonymous referees whose constructive comments have vastly contributed to the improvement of the quality of this chapter.

REFERENCES

Arora, M.K. and Foody, G.M., 1997. Log-linear modelling for the evaluation of variables affecting the accuracy of probabilistic, fuzzy and neural network classifications. *International Journal of Remote Sensing*, **18**, 785–798.

Bezdek, J.C., Ehrlich, R., and Full, W., 1984. FCM: the fuzzy c-means clustering algorithm. *Computers and Geosciences*, **10**, 191–203.

Bastin, L., Edwards, M., and Fisher, P., 2000. Tracking positional uncertainty in ground truth. In *Proceedings of the 4th International Symposium on Spatial Accuracy Assessment in Natural Resources and Environmental Sciences* (Delft: Delft University Press), pp. 39–42.

Binaghi, E., Brivio, P.A., Ghezzi, P., and Rampini, A., 1999. A fuzzy set-based accuracy assessment of soft classification. *Pattern Recognition Letters*, **20**, 935–948.

Brown, K., 2004. Chapter 8, this volume.

Cohen, J., 1968. Weighted kappa: nominal scale agreement with provision for scaled disagreement or partial credit. *Psychological Bulletin*, **70**, 213–220.

Comber, A.J., Fisher, P.F., and Wadsworth, R.A., 2004. Chapter 6, this volume.

Congalton, R.G., 1988. A comparison of sampling schemes used in generating error matrices for assessing the accuracy of maps generated from remotely sensed data. *Photogrammetric Engineering and Remote Sensing*, **54**, 593–600.

Congalton, R.G., Oderwald, R.G., and Mead R.A., 1983. Assessing Landsat classification accuracy using discrete multivariate analysis statistical techniques. *Photogrammetric Engineering and Remote Sensing*, **49**, 1671–1678.

Congalton, R.G., 1991. A review of assessing the accuracy of classifications of remotely sensed data. *Remote Sensing of Environment*, **37**, 35–46.

Eastman, J.R., 2001. *A Guide to GIS and Image Processing, Volume 2* (Clark University: Idrisi Production), pp. 66–67.

Fitzgerald, R.W. and Lee, W.G., 1994. Assessing the classification accuracy of multisource remote sensing data. *Remote Sensing of Environment*, **47**, 362–368.

Foody, G.M., 1992. On the compensation for chance agreement in image classification accuracy assessment. *Photogrammetric Engineering and Remote Sensing*, **58**, 1459–1460.

Foody, G.M., 1995. Cross-entropy for the evaluation of the accuracy of a fuzzy land cover classification with fuzzy ground data. *ISPRS Journal of Photogrammetry and Remote Sensing*, **50**, 2–12.

Foody, G.M., 1996. Approaches for the production and evaluation of fuzzy land cover classifications from remotely sensed data. *International Journal of Remote Sensing*, **17**, 1317–1340.

Foody, G.M., 2002. Status of land cover classification accuracy assessment. *Remote Sensing of Environment*, **80**, 185–201.

Foody, G.M. and Arora, M.K., 1996. Incorporating mixed pixel in the training, allocation and testing stages of supervised classification. *Pattern Recognition Letters*, **17**, 1389–1398.

Foody, G.M. and Cox, D.P., 1994. Sub-pixel land cover composition estimation using a linear mixture model and fuzzy membership functions. *International Journal of Remote Sensing*, **15**, 619–631.

Foody, G.M., Campbell, N.A., Trodd, N.M., and Wood, T.F., 1992. Derivation and applications of probabilistic measures of class membership from maximum likelihood classification. *Photogrammetric Engineering and Remote Sensing*, **58**, 1335–1341.

Fung, T. and Ledrew, E., 1988. The determination of optimal threshold levels for change detection using various accuracy indices. *Photogrammetric Engineering and Remote Sensing*, **54**, 1449–1454.

Gopal, S. and Woodcock, C., 1994. Theory and methods for accuracy assessment of thematic maps using fuzzy sets. *Photogrammetric Engineering and Remote Sensing*, **60**, 181–188.

Hubert, J.L., 1978. A general formula for the variance of Cohen's weighted kappa. *Psychology Bulletin*, **85**, 183–184.

Janssen, L.F. and Van Der Wel, F.J.M., 1994. Accuracy assessment of satellite derived land-cover data: a review. *Photogrammetric Engineering and Remote Sensing*, **60**, 419–426.

Kenk, E., Sondheim, M., and Yee, B., 1988. Methods for improving accuracy of thematic mapper ground cover classifications. *Canadian Journal of Remote Sensing*, **14**, 17–31.

Kent, J.T. and Mardia, K.V., 1988. Spatial classification using fuzzy membership models. *IEEE Transactions on Pattern Analysis and Machine Intelligence*, **13**, 917–924.

Koukoulas, S. and Blackburn, A., 2001. Introducing new indices for accuracy evaluation of classified images representing semi-natural woodland environments. *Photogrammetric Engineering and Remote Sensing*, **67**, 499–510.

Lark, R.M., 1995. Components of accuracy of maps with special reference to discriminant analysis on remote sensor data. *International Journal of Remote Sensing*, **16**, 1461–1480.

Lillesand, T.M. and Kiefer, R.W., 2000. *Remote Sensing and Image Interpretation* Fourth edition (New York: John Wiley and Sons).

LMIC, Minnesota Land Management Information Center 1999. Positional accuracy handbook: using the national standard for spatial data accuracy to measure and report geographic data quality.

Ma, Z. and Redmond, R.L., 1995. Tau coefficients for accuracy assessment of classification of remote sensing data. *Photogrammetric Engineering and Remote Sensing*, **61**, 435–439.

Maselli, F., Conese, C., and Petkov, L., 1994. Use of probability entropy for the estimation and graphical representation of the accuracy of maximum likelihood classifications. *ISPRS Journal of Photogrammetry and Remote Sensing*, **49**, 13–20.

Maselli, F., Rodolf, A., and Conese, C., 1996. Fuzzy classification of spatially degraded thematic mapper data for the estimation of sub-pixel components. *International Journal of Remote Sensing*, **17**, 537–551.

Naesset, E., 1995. Tests for conditional Kappa and marginal homogeneity to indicate differences between user's and producer's accuracy. *International Journal of Remote Sensing*, **16**, 3147–3159.

Naesset, E., 1996a. Use of the weighted Kappa coefficient in classification error assessment of thematic maps. *International Journal of Geographical Information Systems*, **10**, 591–603.

Naesset, E., 1996b. Conditional Tau coefficient for assessment of producer's accuracy of classified remotely sensed data. *ISPRS Journal of Photogrammetry and Remote Sensing*, **51**, 91–98.

Rosenfield, G.H. and Fitzpatrick-lins, K., 1986. A coefficient of agreement as a measure of thematic classification accuracy. *Photogrammetric Engineering and Remote Sensing*, **52**, 223–227.

Shalan, M.A., Arora, M.K., and Elgy, J., 2003a. CAFCAM: crisp and fuzzy classification accuracy measurement software. In *Proceedings of GeoComputation Conference 2003*, Southampton.

Shalan, M.A., Arora, M.K., and Ghosh, S.K., 2003b. Evaluation of fuzzy classifications from IRS 1C LISS III data, *International Journal of Remote Sensing*, **23**, 3179–3186.

Smits, P.C., Dellepiane, S.G., and Schowengerdt, R.A., 1999. Quality assessment of image classification algorithms for land-cover mapping: a review and a proposal for a cost-based approach. *International Journal of Remote Sensing*, **20**, 1461–1486.

Stehman, S.V., 1997. Selecting and interpreting measures of thematic classification accuracy. *Remote Sensing of Environment*, **62**, 77–89.

Stehman, S.V., 1999. Basic probability sampling designs for thematic map accuracy assessment. *International Journal of Remote Sensing*, **20**, 2423–2441.

Stehman, S.V., 2001. Statistical rigor and practical utility in thematic map accuracy assessment. *Photogrammetric Engineering and Remote Sensing*, **67**, 727–734.

Stehman, S.V. and Czaplewski, R.L., 1998. Design and analysis for thematic map accuracy assessment: fundamental principles. *Remote Sensing of Environment*, **64**, 727–734.

Story, M. and Congalton, R.G., 1986. Accuracy assessment: a user's perspective. *Photogrammetric Engineering and Remote Sensing*, **52**, 397–399.

Svoray, T. and Nathan, R., 2004. Chapter 10, this volume.

Van Genderen, J.L., Lock, B.F., and Vass, P.A., 1978. Remote sensing: statistical testing of thematic map accuracy. *Remote Sensing of Environment*, **7**, 3–14.

Verbyla, D.L. and Hammond, T.O., 1995. Conservative bias in classification accuracy assessment due to pixel by pixel comparison of classified images with reference grids. *International Journal of Remote Sensing*, **16**, 581–587.

Wang, F., 1990. Fuzzy supervised classification of remote sensing images. *IEEE Transactions on Geoscience and Remote Sensing*, **28**, 194–201.

CHAPTER 3

Aggregation as a Means of Increasing Thematic Map Accuracy

Yohay Carmel

CONTENTS

3.1 INTRODUCTION

Thematic raster maps are digital images where each pixel is assigned to one of a k discrete classes. This type of maps is ubiquitous, and numerous remote sensing applications yield products such as maps of land cover, vegetation, or soil, all in the format of thematic raster maps.

In remote sensing applications, aggregation is a process of laying a grid of cells on the image or raster map (cell size $>$ pixel size), and defining the larger cells as the basic units of the new map. This process is also referred to as image degradation. When pixels are aggregated into larger grid cells, the information on pixel-specific location is lost, regardless of the aggregation method used. However, aggregation retains some thematic information, and some aggregation methods can conserve the entire thematic information in the map. For example, a k-layer degraded map may be

0-8493-2837-3/05/$0.00+$1.50

constructed, where each layer corresponds to a specific class in the original map, and indicates its proportion in each cell (Carmel and Kadmon, 1999). In many cases, such maps are more useful than their precursors. In forestry applications, for example, users are sometimes interested in the proportion of trees in a given area. Note, however, that other aggregation methods do not retain full thematic information, such as the "majority algorithm," which is a common method of image degradation, where the grid-cell is assigned to its most frequent value. In this study, the specific aggregation process discussed conserves full thematic information, in the form of proportion cover of each class in each cell.

Several studies have suggested that a decrease in spatial resolution enhances thematic map accuracy significantly (Townshend et al., 1992; Dai and Khorram, 1998). On the other hand, the decrease in spatial resolution involves a loss of information that may be valuable for particular applications (Carmel et al., 2001). Thus, users could benefit from viewing the plot of thematic map accuracy as a function of spatial resolution, and choose the specific spatial resolution and its associated uncertainty level that best fits the needs of a specific application. The goal of this study is to explore the relationship between spatial resolution and thematic map accuracy, and to develop a model that quantifies this relationship for thematic ("classified") maps.

Accurate change detection is crucial for description of a variety of processes, such as vegetation dynamics (Svoray and Nathan, Chapter 10, this volume; Malanson and Zeng, Chapter 11, this volume), geomorphic processes (Wichmann and Becht, Chapter 12, this volume; Schmidt, Chapter 13, this volume). Two major approaches for change detection exist, image differencing (Stow, 1999) and post-classification analyses (Brown, Chapter 8, this volume). In the latter approach, thematic maps are used for change detection, and overall error is a product of classification errors in each time step, and may become large (Brown, Chapter 8, this volume). In addition, another source of error is introduced in change detection, namely location error, termed also misregistration (Brown, Chapter 8, this volume). The effect of misregistration on accuracy of change detection analyses is large (Stow, 1999), often surpassing the effect of classification error (Carmel et al., 2001).

In order to illustrate this point, consider a change detection analysis applied to the maps in Figure 3.1. The two maps portray a small area in an earlier (gray) and

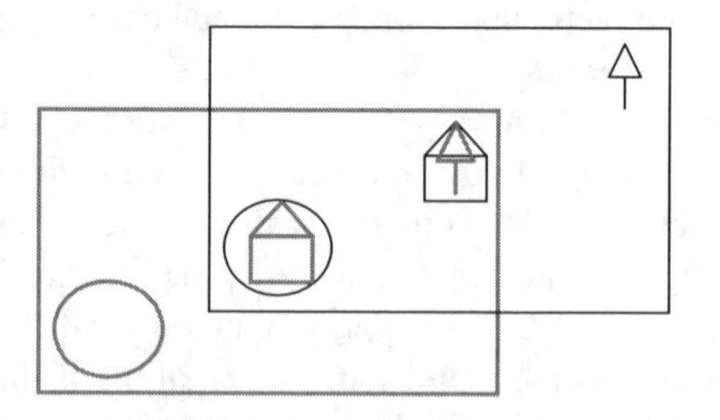

Figure 3.1 Two misregistered maps, representing a pond, a house, and a tree, in an earlier and a latter time steps (in gray and black, respectively). Although no change has occurred in the area, change detection analysis would indicate transitions.

latter (black) time steps, and are misregistered. One can tell that no change has occurred in this area. However, due to misregistration, change detection analysis would point out two transitions: a house in the earlier map became covered by pond in the latter, and a tree in the earlier map gave way to a house in the latter. In the aggregation procedure, the whole area portrayed in Figure 3.1 may be considered as a single grid cell, and proportion cover of each land-use may be estimated. A change detection analysis of the aggregated map would avoid those false transitions and is likely to be more accurate.

3.2 THE MODEL

3.2.1 Location Accuracy

For a single pixel, misregistration is translated into thematic error if its "true" location is occupied by a pixel belonging to a different class. Let us define $p(\text{loc})$, the probability that a pixel is assigned an incorrect class due to misregistration:

$$p(\text{loc})_{rc} = p(i_{rc} \neq i_{r+e(x),c+e(y)}) \tag{3.1}$$

where r and c are pixel coordinates, i_{rc} is the class assigned to the pixel, $e(x)$ and $e(y)$ are the x and y components of location error, respectively (measured in pixels). Thus, $p(\text{loc})_{rc}$ depends on the magnitude of location error, and on landscape fragmentation (map heterogeneity). $p(\text{loc})$ can be estimated empirically for a given map, based on map pattern and the magnitude of location error. Location error may not be uniform across an image (Brown, Chapter 8, this volume; Carmel, 2004a). Thus, the general model scheme presented here does not assume a constant location error, but allows error to vary across the image. Location error is assumed, however, to be constant in each grid cell (the unit that the pixels are aggregated into).

Considering a larger cell size A, let us define a similar probability, $p^A(\text{loc})$, which is the probability that a pixel within the framework of a larger cell was misclassified due to misregistration. For cell sizes larger than location error, this probability would be lower than the original probability $p(\text{loc})$, since misregistration would shift a certain proportion of the pixels only within the grid cell, and for those pixels, thematic error is cancelled at the grid cell level (Figure 3.2). Here, a conceptual model is presented, in which this probability is denoted by:

$$p^A(\text{loc}) = \alpha \cdot p(\text{loc}) \tag{3.2}$$

where α is the proportion of a cell of size A in which pixels are misplaced into neighboring cells, and may thus result in thematic error (Figure 3.2).

This proportion, α, is termed here the effective location error. It is a function of cell size A and of location error magnitude. The effective location error α is the proportional area of the dark gray region in the cell (Figure 3.2) and is denoted by:

$$\alpha = \frac{(A \cdot e(x) + A \cdot e(y) - e(x) \cdot e(y))}{A^2} \tag{3.3}$$

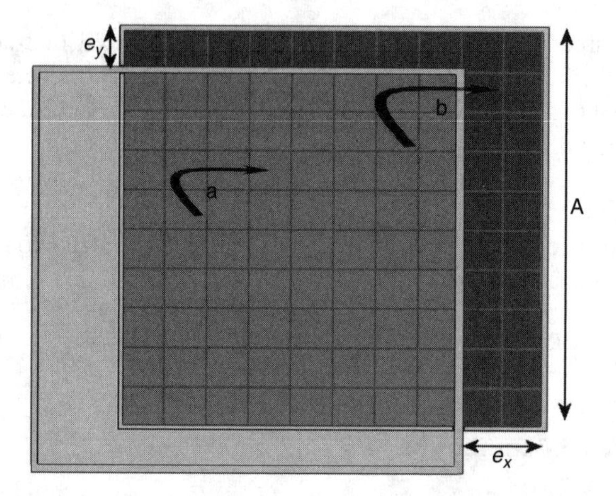

Figure 3.2 Quantification of the effect of location error on thematic accuracy. An image is "shifted" against itself. The results are shown here for a single grid cell within the image. Here, cell size $A = 10$ pixels. The original image is shown in light gray and the shifted image is shown in dark gray. $e(x)$ and $e(y)$ denote the x and y components of location error, respectively (in this figure $e(x) = 2$ pixels and $e(y) = 1$ pixel). Pixels in the area of overlap between the two images would remain within the cell and thematic accuracy at the cell level would not be affected (see arrow a). Pixels in the dark gray region are shifted into neighboring cells, and may result in thematic error (see arrow b). The effective location error α is the proportional area of the dark gray region within the cell (Equation 3.3).

where location error components $e(x)$ and $e(y)$ are assumed constant for all pixels within a single cell. Using Equation 3.3, the reduction in effective location error when cell size increases may be illustrated easily (here, for the special case where $e(x) = e(y)$, Figure 3.3). Effective location error α declines rapidly from 1 for cell sizes $\leq$ the magnitude of location error to 0·36 for cell sizes five times the magnitude of location error (Figure 3.3).

Location error may vary largely across an image (since its major sources, topography and quality of ground-control points imply inherent spatial pattern). Thus, α, $p(\text{loc})$, and $p^A(\text{loc})$, should ideally be estimated for each grid cell in the map. This requires that location error components $e(x)$ and $e(y)$ are available at the grid-cell level. Typically, location error information is available for several locations (test points) only. Interpolation methods such as kriging may be used to construct location error surfaces for both $e(x)$ and $e(y)$ (Fisher, 1998). Following Equation 3.1, a specific cell in coordinates r, c, is shifted by $e(x)_{rc}$ and $e(y)_{rc}$ on the x and y axes, respectively. $p(\text{loc})$ is estimated as the proportion of pixels that are 'misclassified' due to the shift. Equations 3.2 and 3.3 may then be used to derive cell-specific α and $p^A(\text{loc})$. This process is repeated for each cell in the map, and the global mean of these parameters can then be calculated.

An alternative to this process, that may be somewhat less accurate but much simpler to apply, is to assume a constant location error across the image. The average location error is typically defined as RMSE, Root Mean Square Error (see Brown, Chapter 8, this volume), decomposed here into its x and y components. The x and y

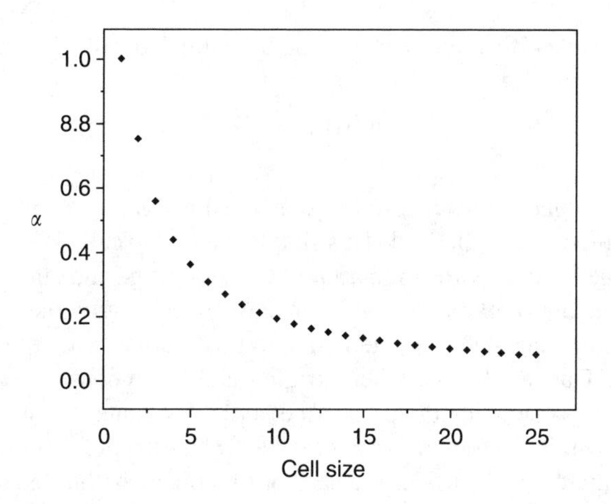

Figure 3.3 Effective location error α as a function of cell size. Cell size is shown here in multiples of pixels, where average location error is set to 1 pixel in both *x* and *y* directions. Cell size varies between 1 and 25 times the magnitude of location error. (After Carmel, 2004b; *IEEE Geoscience and Remote Sensing Letters*, reproduced with permission from the IEEE.)

components of RMSE are denoted by:

$$\text{RMSE}(x) = \sqrt{\frac{\sum_{g=1}^{n} e(x)_g^2}{n}} \quad \text{RMSE}(y) = \sqrt{\frac{\sum_{g=1}^{n} e(y)_g^2}{n}} \tag{3.4}$$

where $e(x)_g$ and $e(y)_g$ are the x and y components of the deviation between the true location of a test point g and its location on the map, respectively, and n is the number of test points.

In order to determine $p(\text{loc})$, α and $p^A(\text{loc})$, the practice of "shifting" a map against itself can be employed (Townshend et al., 1992; Dai and Khorram, 1998). This is a simulation approach that enables one to estimate the thematic consequences of misregistration between two maps. The process results in two identical maps, where one is spatially "shifted" off the other by a lag equal to the estimated location error (here, the map is shifted against itself by RMSE(x) and RMSE(y) on the x and y axes, respectively). A typical thematic error analysis can then be applied, to form an "error matrix." This matrix contains essentially two types of pixels: (1) those that were not affected by the shift between maps, and (2) those for which the shift resulted in "thematic error." $p(\text{loc})$ is derived as the proportion of type (2) pixels in the map. Finally, α and $p^A(\text{loc})$ are calculated for the entire image, using Equations 3.2 and 3.3.

3.2.2 Classification Accuracy

The probability that a pixel is misclassified, $p(\text{cls})$, may be estimated as the proportion of misclassified pixels in the map. The probability that a pixel of class i is assigned to class j due to misclassification, $p(\text{cls})_{ij}$, can be estimated from the error matrix as:

$$p(\text{cls})_{ij} = \frac{n_{ij}}{n_i} \tag{3.5}$$

where n_i is the number of pixels of class i in a cell and n_{ij} is the number of class i pixels misclassified as j in that cell. This simple method to calculate $p(\text{cls})_{ij}$ follows the common practice in classification accuracy assessments, and ignores the spatially heterogeneous nature of classification error (e.g., error is more likely to occur near boundaries between classes). An alternative to this method was recently suggested (Kyriakidis and Dungan, 2001), where kriging is used to construct a classification error surface. The same method may be used here to estimate cell-specific $p(\text{cls})_{ij}$.

Considered within the framework of grid cells, classification error may be reduced when cell size increases. Consider the case of two pixels within the same grid cell, class i pixel misclassified as j and class j pixel misclassified as i. At the cell level, where pixel information is reduced to proportion cover of each class in the cell, both errors cancel each other. The probability for a pixel to be misclassified within the frame of a larger cell A, $p^A(\text{cls})$, can be calculated as:

$$p^A(\text{cls}) = \beta \cdot p(\text{cls}) \tag{3.6}$$

where β is the proportion of misclassified pixels in the cell that were not cancelled out at the grid-cell level. β is dependent on cell size and on the spatial pattern of land covers shown in the map (since it is a function of the number of pixels of each class in each grid cell). Thus, β should be estimated for each classification error pair ij separately. β_{ij} is dependent on the number of both ij and ji misclassification types. The abundance of these misclassifications is denoted by n_{ij} and n_{ji}, respectively. From this point on, n_i, n_{ij}, n_{ji}, and β, are cell-specific. Consider a cell that contains many class i pixels and many class j pixels. It is expected that some misclassified ij pixels (n_{ij}), as well as several ji misclassified pixels (n_{ji}) will be present in that cell. n_{ij} is calculated locally as the product of the number of class i pixels in the cell n_i and $p(\text{cls})_{ij}$:

$$n_{ij} = n_i \cdot p(\text{cls})_{ij} \tag{3.7}$$

The calculation of β_{ij} requires information on the spatial relationship between ij and ji misclassified pixels in each grid cell. If $n_{ij} < n_{ji}$ then all ij misclassifications are cancelled out, and an equal quantity of ji misclassified pixels is cancelled out as well. In that case, the effective ij misclassification rate, β_{ij}, is 0, and the effective ji misclassification rate, β_{ji} is $(n_{ji} - n_{ij})/n_{ji}$. Thus, β_{ij} is denoted by a conditioned term as follows:

$$\begin{aligned} \beta_{ij} &= 0 && \text{if } n_{ij} \leq n_{ji} \\ \beta_{ij} &= \frac{n_{ij} - n_{ji}}{n_{ij}} && \text{if } n_{ij} > n_{ji} \end{aligned} \tag{3.8}$$

Using Equations 3.5, 3.7, and 3.8, β_{ij} can be calculated for each aggregated cell in the map. Next, β can be determined as the weighted average of all β_{ij}:

$$\beta = \sum_{i=1}^{k} \sum_{j=1}^{k} \left(\beta_{ij} \cdot \frac{n_{ij}^{A}}{\sum_{i=1}^{k} \sum_{j=1}^{k} n_{ij}^{A}} \right) \tag{3.9}$$

Average β can be calculated for the whole map, for a range of cell sizes, and the reduction in effective classification error that accompanies the aggregation process can be illustrated.

3.2.3 Model Application

In order to exemplify the utility of this model, it was applied to a vegetation map derived from a 1995 aerial photograph of Carmel Valley, California (Figure 3.4). This map was produced as a part of a study of Mediterranean vegetation dynamics (Carmel et al., 2001; Carmel and Flather, 2004). In the original image, pixel size is 0·6 m. Location error was estimated using 40 test points that were compared with a known reference (independent of the control points used in the rectification process). Location error components RMSE(x) and RMSE(y) were 1·86 and 1·68 m, respectively. Classification error was assessed using 325 test-points, most were collected in the field and some were determined with a stereoscope-aided manual photo-interpretation.

These data served to construct a classification error matrix. The map was shifted against itself by 1·86 and 1·68 m, in the x and y axes, respectively. The original and shifted maps served to construct a location error matrix, similar to the classification

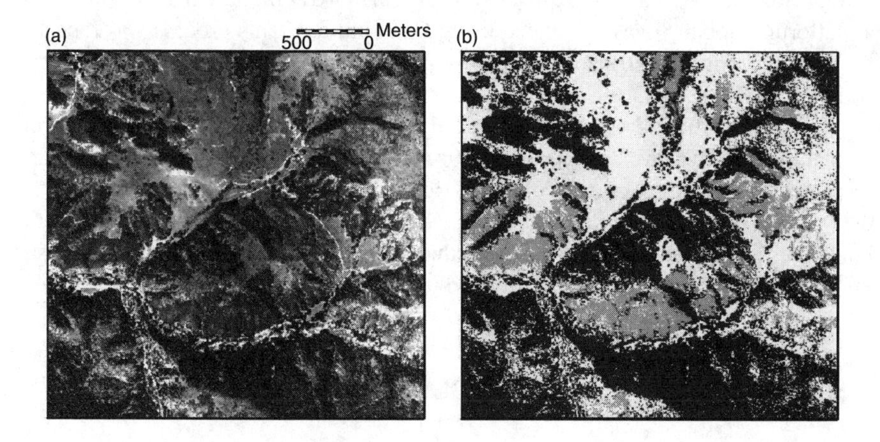

Figure 3.4 An aerial photograph (a) and a vegetation map (b) of Hastings Nature Reserve, in Carmel Valley, California. The white, light gray, and dark gray shades in the map represent grass, chaparral, and trees, respectively. (After Carmel, 2004b, *IEEE Geoscience and Remote Sensing Letters* reproduced with permission IEEE.)

Table 3.1 Location and Classification Error Matrices for the 1995 Vegetation Map of Hastings Nature Reserve, California

	Location			Classification		
1995	**Forest**	**Chaparral**	**Grassland**	**Forest**	**Chaparral**	**Grassland**
Forest	7,260,180	173,030	1,492,576	169	3	2
Chaparral	177,796	1,328,014	328,503	5	29	1
Grassland	1,457,213	333,164	4,903,500	14	5	107
PCC		0·77			0·91	

Source: Modified after Carmel et al. (2001). Reproduced with permission, The American Society for Photogrammetry and Remote Sensing. Carmel et al. combining location and classification error sources for estimating multi-temporal database accuracy, *Photogrammetric Engineering and Remote Sensing*, 67, 865–872.

error matrix. The proportion of pixels classified correctly (PCC) is the sum of the diagonal cells divided by the sum of all cells. Table 3.1 shows both matrices.

Model parameters $p(\text{loc})$ and $p(\text{cls})$ were estimated globally, assuming error homogeneity for both error types. The former was estimated from the location error matrix as the proportion of "mismatched" pixels due to the shift, while the latter was estimated from the classification error matrix as the proportion of misclassified pixels (in both cases those are the sum of non-diagonal cells divided by the sum of all cells, in the respective matrices). The probabilities of error at the pixel level $p(\text{loc})$ and $p(\text{cls})$ were 0·23 and 0·09, respectively. Using the same classification error matrix, $p(\text{cls})_{ij}$ was estimated for each ij cell in the classification error matrix.

Using Equation 3.3, α was estimated for a range of cell sizes, between 3 and 60 m. α decreased notably from 0·83 to 0·07, when cell size changed from 3 to 60 m. Model parameters n_{ij} and β_{ij} were then estimated for each cell in the image, using Equations 3.7 and 3.8, respectively, and the GRID module of ARC-Info™ as a platform. Global β was calculated using Equation 3.9. This process was repeated for the same range of cell sizes (3 to 60 m). β decreased moderately across this range, from 0·95 to 0·8.

Finally, Equations 3.2 and 3.6 were used to derive the probabilities of error for this map at the grid cell level, $p^A(\text{loc})$ and $p^A(\text{cls})$, for location and classification errors, respectively, for a range of grid cell sizes. Results show that $p^A(\text{loc})$ diminished from 0·19 to ~0·01 in the range of 3 to 60 m, while the decrease in $p^A(\text{cls})$ for the same range was negligible (Figure 3.5). Note that in this figure, unlike Figure 3.3, a real image is analyzed, and cell size is defined in meters, not in pixels.

3.3 DISCUSSION

The results of this study support previous indications that the impact of misregistration on map accuracy is large (Townshend et al., 1992; Dai and Khorram, 1998; Stow, 1999), and reveal that aggregation is a very effective means of reducing this impact. This study develops a conceptual model to quantify the effect of aggregation on map accuracy. Given the tradeoff between spatial resolution and accuracy, potential utility

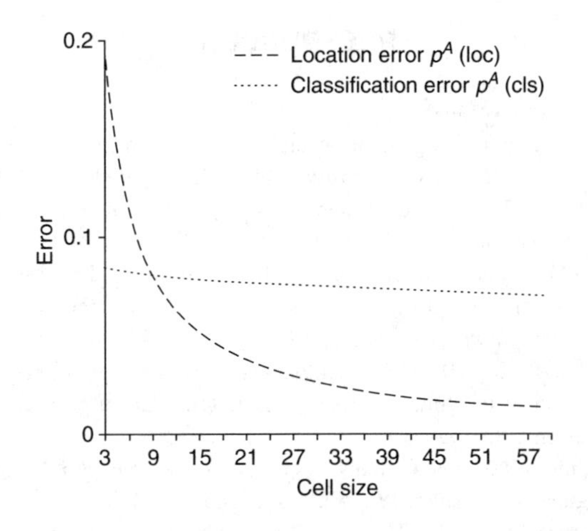

Figure 3.5 Cell-level error probabilities, p^A(loc) for location error and p^A(cls) for classification error, as a function of cell size. Cell size is in meters. (After Carmel, 2004b, *IEEE Geoscience and Remote Sensing Letters*.)

of this model would be to help the user choose an appropriate cell size for a specific application. A first approximation of the gain in accuracy with increased aggregation level can be visualized using a simple procedure: solve Equation 3.3 for a range of relevant cell values, and portray α as a function of cell size (Figure 3.3). This procedure is particularly easy to apply if RMSE is taken to represent e.

In addition, the impact of aggregation on classification accuracy can be viewed by drawing β, the effective classification error, as a function of cell size. This study found that the impact of aggregation on classification error was negligible. In maps of highly fragmented landscapes, β may be more prominent. Further information can be gained by estimating the actual probabilities of error, p^A(loc) and p^A(cls), for various aggregation levels. This stage requires spatially explicit simulations that manipulate the actual map.

The model was applied to an actual case study, assuming a uniform error across the map. An on-going research in our laboratory (http://envgis.technion.ac.il/) studies the spatially explicit modifications for estimating model parameters, and their impact on the accuracy of these estimates.

In conclusion, this methodology provides an effective tool for assessing the impact of aggregation on thematic map accuracy, and evaluating it against information loss, in order to decide on a proper level of map aggregation. The most effective reduction in error was achieved when cell size was in the range of 3 to 10 times the size of average location error. Map-specific error rates may somewhat alter this conclusion.

ACKNOWLEDGMENTS

This study is supported by The International Arid Lands Consortium (IALC) project 02R-17. Avi Bar Massada helped with model verification. The author is grateful for improvements suggested by Giles M. Foody and two anonymous reviewers.

REFERENCES

Brown, K., Chapter 8, this volume.

Carmel, Y., 2004a, Characterizing location and classification error patterns in time-series thematic maps. *IEEE Geoscience and Remote Sensing Letters*, 1, 11–14.

Carmel, Y., 2004b, Controlling data uncertainty via aggregation in remotely sensed data. *IEEE Geoscience and Remote Sensing Letters*, 1, 39–41.

Carmel, Y., Dean, D.J., and Flather, C.H., 2001, Combining location and classification error sources for estimating multi-temporal database accuracy. *Photogrammetric Engineering and Remote Sensing*, 67, 865–872.

Carmel, Y. and Flather, C.H., 2004, Comparing landscape scale vegetation dynamics following recent disturbance in climatically similar sites in California and the Mediterranean basin. *Landscape Ecology*, 19, 573–590.

Carmel, Y. and Kadmon, R., 1999, Grazing, topography, and long-term vegetation changes in a Mediterranean ecosystem. *Plant Ecology*, 145, 239–250.

Dai, X.L. and Khorram, S., 1998, The effects of image misregistration on the accuracy of remotely sensed change detection. *IEEE Transactions on Geoscience and Remote Sensing*, 36, 1566–1577.

Fisher, P., 1998, Improved modeling of elevation error with geostatistics. *GeoInformatica*, 2, 215–233.

Kyriakidis, P. and Dungan, J., 2001, A geostatistical approach for mapping thematic classification accuracy and evaluating the impact of inaccurate spatial data on ecological model predictions. *Environmental and Ecological Statistics*, 8, 311–330.

Stow, D.A., 1999, Reducing the effects of misregistration on pixel-level change detection. *International Journal of Remote Sensing*, 20, 2477–2483.

Townshend, J.R.G., Justice, C.O., Gurney, C., and McManus, J., 1992, The impact of misregistration on change detection. *IEEE Transactions on Geoscience and Remote Sensing*, 30, 1054–1060.

CHAPTER 4

Improving the Update of Geospatial Information Databases from Imagery Using Semi-Automated Change Detection and User-Guidance Techniques

David A. Holland and William Tompkinson

CONTENTS

0-8493-2837-3/05/$0.00+$1.50

4.1 INTRODUCTION

In the United Kingdom, as in many parts of the world, the demand for up-to-date geospatial information continues to grow rapidly. One of the foremost user demands is that the geospatial information reflects the real world situation as it stands at the time they purchase the product and that any changes in the environment (such as the building of new houses, or the clear-felling of woodlands) are incorporated into the product as soon as possible after they occur. In order to meet this demand, national mapping agencies require efficient methods of capturing change; effective spatial data management techniques; and rapid, user-targeted, methods of data dissemination. This chapter concentrates on the first of these — the rapid and efficient detection and capture of change, to update an existing geospatial database.

There are several possible methods, which may be used to determine when change occurs on the ground. For a mapping agency, change is often identified by local observation by field surveyors or notification of planning applications by local authorities. These methods are perhaps least effective in rural areas, where a small number of surveyors must cover a large area of land, and where many types of change are not subject to planning legislation. Many mapping agencies, including the Ordnance Survey in Great Britain, have used remotely sensed imagery as a source of information on change detection for several decades. However, the processes of detecting the change and subsequently updating the spatial information from imagery have remained very manually intensive. One long-term goal of a mapping agency is to develop or procure a system that can detect change and capture feature information from imagery, with little or no human intervention. While this goal of full automation remains unrealised, there are ways in which automatic techniques can be used to streamline at least some elements of the data capture process.

Almost all the large-scale spatial information products supplied by mapping agencies are based on vector datasets. These datasets are derived from a variety of sources, to meet specifications defined at a regional or national level. Most of these products are stored in geospatial databases composed of points, lines and polygons, which represent both natural and artificial features in the landscape. Aerial photography is used extensively by mapping agencies as a source of raw data from which such national geospatial datasets are derived and maintained. In most cases, the process of capturing any changes in the underlying spatial dataset relies solely on the ability of a photogrammetrist to manually identify and accurately capture the features that have changed.

One way to make the data capture process more efficient would be to automate the *detection* of change. Ideally, this would be followed by an automatic extraction of all the features for which a change has been detected. Although this subject has been an active area of research for many years (e.g., see Brenner et al., 2001; Niederöst, 2001 and other papers in Baltsavias et al., 2001), the transfer of results from a research to a production environment is still quite limited. In this chapter, a prototype change detection system is described, for use in a data collection production flowline. Also presented are examples of methods which may be used to aid the automated change detection processes embedded within such a system.

4.2 A PROTOTYPE CHANGE DETECTION SYSTEM

4.2.1 Rationale

Change detection, as performed by a mapping agency, often involves the use of an image pair — one image showing the current state of the region to be mapped; the second image showing the state at a previous point in time — often the time of the last update. In addition, many mapping agencies will already have a geospatial database, in which topographic features are stored. One frequently encountered problem is that the real requirement is to determine the differences between the features in the geospatial database and the new image; rather than the simpler requirement of determining the differences between two images. The problem lies in the fact that one dataset is a vector model, determined by a specification designed to show cultural information; while the other is a raster model, dependent on the physical characteristics of the sensor and the conditions extant at the time of capture. One way to overcome this problem is to map one of the data sources onto the characteristics of the other — that is, the vector data must be rasterised; or the imagery must be segmented into polygons. Either approach leads to some loss of integrity in the data, and it is generally better to use a like-for-like comparison. The method first described in this chapter is similar to an image-to-image comparison; but differs in that the system first performs a segmentation of each image, then compares the two sets of segments. This automatic stage is followed by a machine-assisted manual capture process.

Many researchers in the field of change detection wish to develop systems that can not only identify change, but can also capture the exact outlines of all the features that have changed. This has proved to be more difficult than one might expect. For example, despite over 20 years of research in digital photogrammetry and computer vision, no universal edge detector has been developed that can both identify and track edges with sufficient success (Agouris et al., 2000). The capacity of the human eye and brain to infer the presence of a feature from rather imprecise visual cues has proved very difficult to reproduce in an automated system. Despite the lack of a complete solution, it is possible to develop a system that provides a marked improvement on a purely manual data collection system. Such a system takes into account the fact that much time is potentially wasted by human image-interpreters, as they scan two or more images and identify those areas that require a closer look. Once this process is complete, the image-interpreter must revisit each area of potential change, and employ different criteria to determine exactly what has changed, and which features must be edited, deleted or created in the geospatial database.

4.2.2 Choice of Source Data

Many different types of remotely sensed imagery are currently available, at a range of different resolutions and from an increasing number of different technologies, including laser and radar as well as the more familiar optical sensors. Mapping agencies have traditionally used aerial photography as the data source; either in analogue form (diapositives or prints) or in digital form. While analogue aerial photography still forms the basic source material in current data capture operations, it is now often scanned to convert it to digital form before being interpreted.

There has been a remarkable growth in recent years in the use of digital cameras in the home consumer market, but the transfer of this technology into the professional photogrammetric market has been slower. Traditional photogrammetric cameras are expensive to purchase, and most of those currently in use have many years of service left in them. To discard such a system in favour of an equally expensive but relatively un-tested digital system is perhaps too great a risk for many mapping agencies at the present time. However, as the technology matures and the costs reduce, it can only be a matter of time before most film-based aerial camera systems are replaced by their digital equivalents.

In the field of satellite remote sensing, the last 5 years have seen a radical change in the type of imagery available to the general consumer. Technologies developed for the defence and intelligence market have been developed for commercial use, enabling anyone with sufficient funds to purchase satellite sensor imagery at spatial resolutions of 1 m or less (Petrie, 2003a). There are now three main U.S. companies providing such imagery — Space Imaging (IKONOS satellite, operational since 1999), Digital Globe (QuickBird, operational since 2001), and OrbImage (OrbView 3, launched in 2003). Several other countries have launched military satellites that may serve a dual purpose by providing imagery to the commercial market (Petrie, 2003b).

The current photogrammetric processes within Ordnance Survey use scanned colour aerial photographs. However, the use of digital imagery from both airborne and spaceborne sensors has been a major topic of research in recent years (Holland and Marshall, 2003). In this chapter, two types of imagery are used. Since the existing update systems use scanned aerial photography, it was natural to use such imagery in the prototype system described in this section. Also, digital imagery has not been readily available for very long, so obtaining two sets of such images, separated by a sufficient period of time, can be difficult. The work described later in this chapter, on shadow analysis, was partly inspired by the observation of prominent shadows in QuickBird sensor imagery, hence this imagery was used in the research. In each case, the choice of imagery is, therefore, somewhat arbitrary and the authors do not mean to imply that this is the most appropriate type of imagery for the application under discussion.

4.2.3 System Design

The prototype system described in the following sections is the focus of ongoing research investigating the extent to which the software interface of a data capture production system can help to reduce the time taken to update a geospatial database. The underlying premise of this research is that there needs to be a strong appreciation of the fact that the system needs to be considered in terms of *both* "software" and "user" subsystems. As a result, the research concentrates upon the information flow between the software and user. In this chapter, the focus is directed on the visual cues presented to the user by the system. This system represents a software interface that can have components inserted, removed, or changed, as and when new methods for enhancing such information are developed.

The first incarnation of the system (described in more detail in Tompkinson et al., 2003) relies on the availability of imagery captured at two instances in time (usually separated by several years). Once the sensor distortions, chromatic differences and the different viewing geometries have been accounted for, the two images are compared and any differences are highlighted. In a traditional change detection system, the process stops at this point, and it is left to the operator to search for all the areas which the system has identified as areas of potential change, to study these areas, and to capture the data manually. Figure 4.1 illustrates how such a change detection system might be organised according to a generic tiered structure commonly used in system design (e.g., as described in Bennett et al., 1999).

In the proposed system, the user is given more information than before and is guided automatically to the places where potential changes have been identified. In this system, the following functions are executed:

- An automatic image-to-image change detection process is performed:
  - First the two images are segmented into polygons. This gives a set of polygons for each of the two instances in time.
  - Polygons from the two time instances are then compared and overlapping polygons are detected.

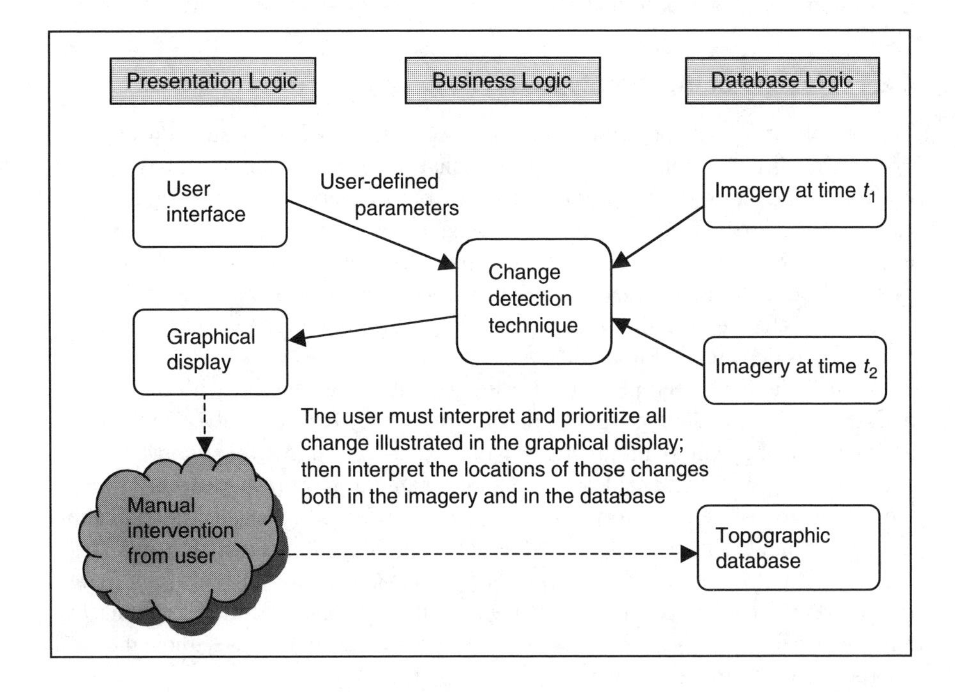

Figure 4.1 A tiered system-design diagram illustrating a typical change detection system within a data capture software environment.

- The overlapping polygons are then compared using their compactness ratios (Equation 4.1)

$$\frac{p^2}{4\pi A} \tag{4.1}$$

where p is the perimeter of the polygon, A is the area of the polygon.
- Those with very different compactness ratios are flagged as potential areas of change.

- The process described above is similar to that described by Ware and Jones (1999). (Note: Although image segmentation was used in the prototype, any change detection technique could be performed at this stage.)
- Via a graphical user interface, the operator is guided to each potential area of change in turn. The user is presented with a view of the changed areas in both the images, together with a representation of the geospatial data currently held in the topographic database.
- The user is also presented with a visual cue of the polygons that are most likely to have changed, as determined by the system.

By placing change detection within the context of the system in which it is applied, this research places an emphasis upon the application of computational methods to simulate the thought processes of a manual interpreter. In the prototype system, emphasis is placed upon utilising cues of change detection to assess where change has taken place and presenting this information to the user (Figure 4.2). This information is then used to iteratively "drive" the user to the location of the change. The user and the system work together to make the data collection process more efficient.

4.2.4 Results of the Prototype

The prototype system presented here has been developed in Visual Basic using ESRI ArcObject^{TM} technology, with a change detection technique implemented in C++. Since it produces vector segments from imagery in a consistent and efficient manner, the eCognition software package (from Definiens Imaging) is used to segment the images and generate the input vector data. The process flow in this system, from the input of the image pair, via the segmentation and the implementation of the change detection technique, through to driving the user to the location of change, is depicted in Figure 4.3.

Initial tests have been performed using a small image subset, with dimensions of 250 m × 210 m. Demonstrations of this prototype, to those involved in the management of a large photogrammetric flowline, have been successful in conveying the concept of a system that enables efficient interpretation of change locations through the automation of pan and zoom functions. It is envisaged that such an application has the potential to increase the speed of data capture from image blocks that cover even larger areas of land. This becomes of greater importance if a mapping organisation were to capture features from fine spatial resolution satellite sensor imagery rather than aerial photography, since such imagery covers large areas without the need to produce aerial image mosaics.

Experimentation using larger (4 km × 1 km) areas of aerial orthophotographs is currently underway. It is intended that such trials will enable a more complete

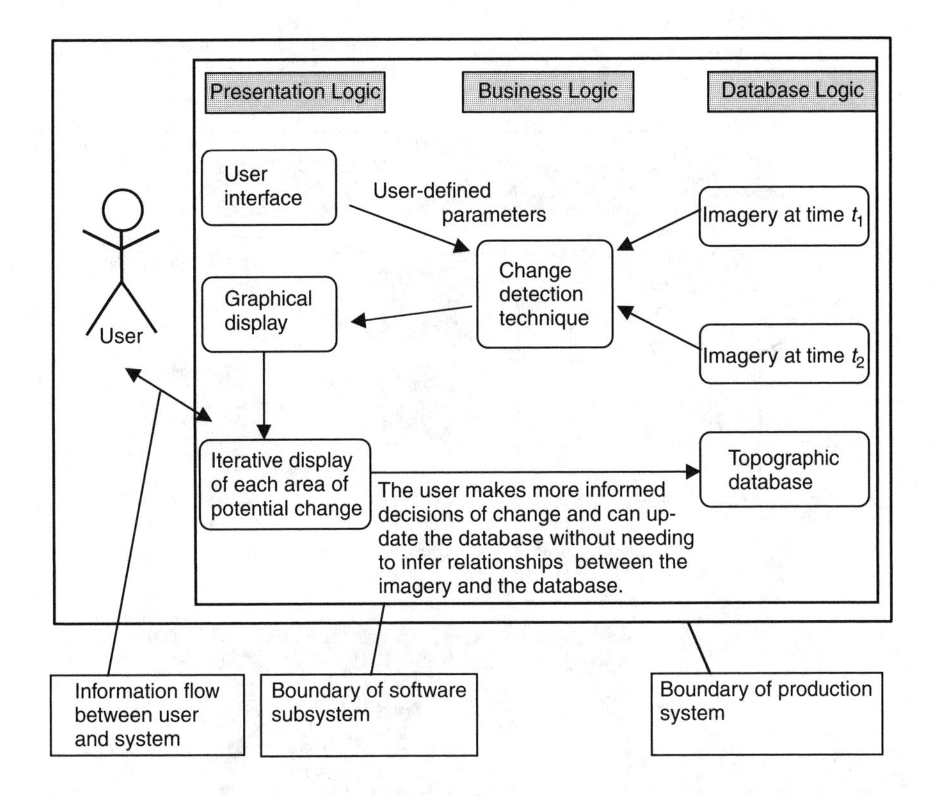

Figure 4.2 A tiered system design diagram illustrating the proposed change detection system within a data capture software environment.

assessment of whether simply implementing the methods described in this system really do increase efficiency. These trials have indicated that limitations in the software platform and data structures used mean that the prototype can be slow to function.

Research is also underway that addresses the problems of edge effects, which currently present difficulty when processing sets of images taken from different orientations. These problems notably occur in the region between the convex hull and the envelope of a line of photographs, or in a region where photography is only available for one set of images. In such areas, polygons are often misclassified as having changed.

4.3 OTHER CUES OF CHANGE DETECTION

4.3.1 Types of Change Detection Techniques

The change detection technique in the prototype system described above uses a shape comparison between image segments. While this has produced encouraging results in the area tested, the system is not dependent on the technique. The system could be improved by allowing a variety of different change detection techniques to be

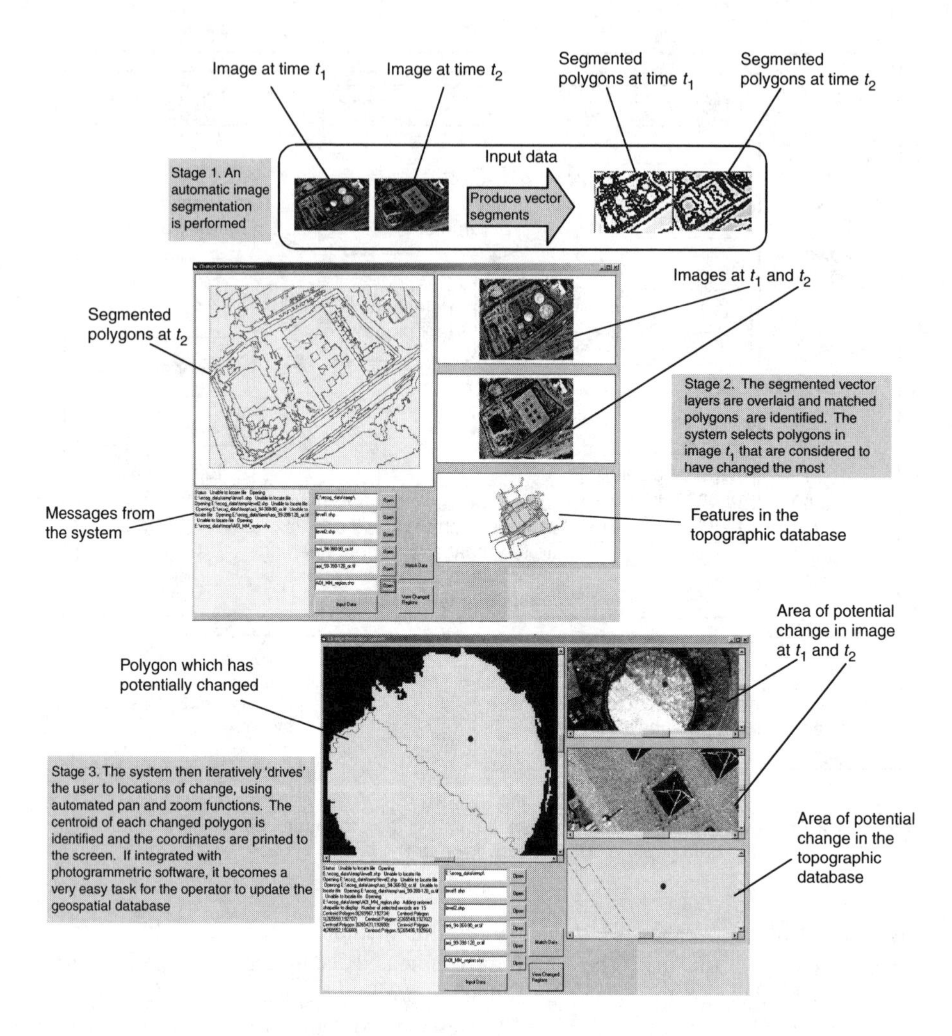

Figure 4.3 (see color insert following page 168) Process flowline through the prototype change detection system.

incorporated into the flowline, each technique being adapted to a particular type of change. In a manual change detection process, the photogrammetrist will be skilled in identifying changes in the landscape from remotely sensed imagery. During manual comparisons between temporally separated image pairs, or between an image and map data, changes may be identified in several ways. In one method, an operator may use indicators such as alterations in the shape of landscape features; in another the operator may concentrate on the detection of artefacts within large homogeneous topographic objects. A new building development may be recognised by a change in texture over a large area of land; whilst a new wall may be identified through indirect evidence such as its shadow. If existing topographic data are available, these too can be used to supply extra information, directly either to the user or to the system, to help identify significant areas of change. In Section 4.3.2, some other techniques which may be used to indicate the incidence of change are discussed. These techniques could be incorporated into future versions of the change detection system.

4.3.2 Using Shadows to Detect Change

4.3.2.1 Shadows as Indicators of Topographic Features

Shadows within an image can provide an indication of the presence of buildings and other objects, and they are usually easier to detect than the often-complex multi-textured objects which cast them (Irvin and McKeown, 1989). Many authors have used shadows observed in aerial photographs to estimate the height of buildings (Hinz et al., 2001). Until recently, the analysis of shadows in satellite sensor imagery was of very little interest to researchers, since the length of a shadow cast by a typical object (e.g., a suburban house) would often be smaller than the size of an image pixel. The latest satellite sensors from Space Imaging (IKONOS) and Digital Globe (QuickBird), however, provide imagery in which shadows are quite easy to detect. As previously mentioned, the main catalyst for this research was the observation that shadows appeared prominently within a QuickBird sensor image of Christchurch in Dorset, supplied to Ordnance Survey by Eurimage. The image from the QuickBird sensor used in this research has pixels with a ground sample distance of 64 cm. Even in the late morning, when the image of the study area was acquired, the shadows cast by a small house comprise 40 to 50 pixels (Figure 4.4). In this study, no attempt was made to extract 3D information from these shadows, they were simply used as an indicator of the presence of a solid object (such as a building) or a "semi-solid" object (such as a tree in leaf).

4.3.2.2 Detecting Shadows

One way of finding shadows is to simply threshold the image at a certain gray level, so that any pixels brighter than a given value are removed from the image, leaving only the areas of potential shadows. The most suitable threshold to use for this purpose depends on the characteristics of the image (e.g., what proportion of the image is in shadow and how much contrast is present in the image). After experimenting

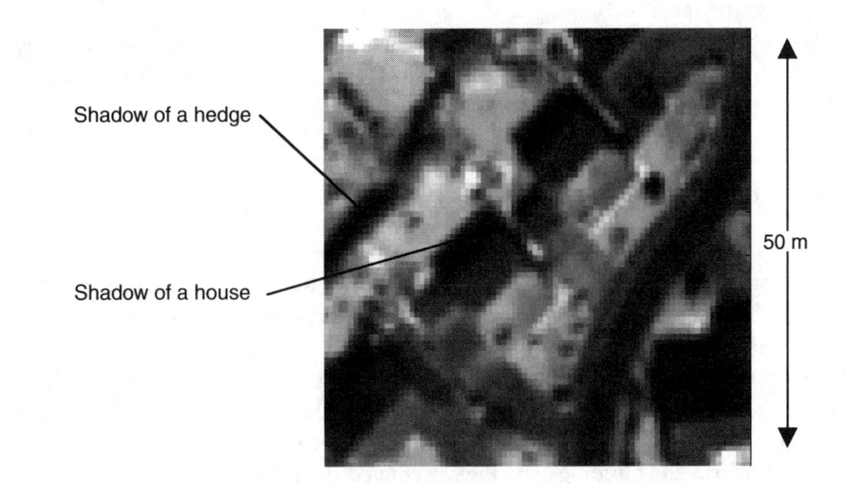

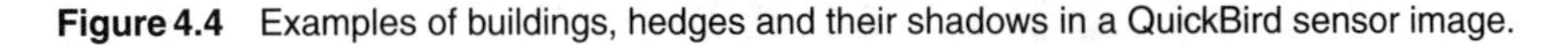

Figure 4.4 Examples of buildings, hedges and their shadows in a QuickBird sensor image.

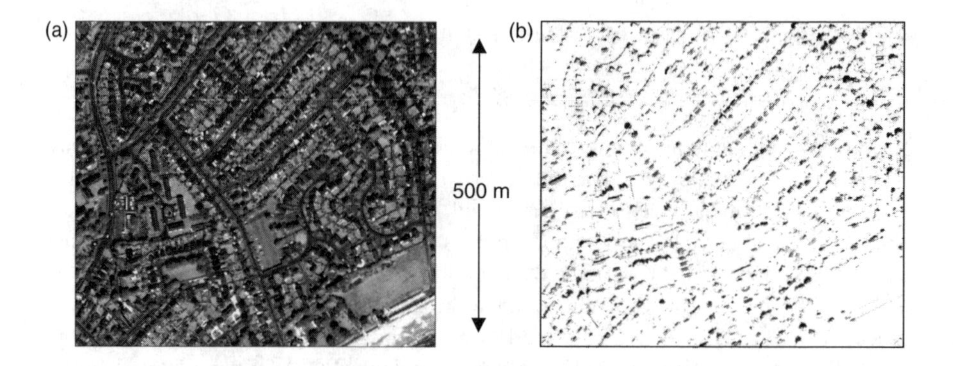

Figure 4.5 Image of the study area, comprising a mainly residential area of Christchurch, centred on National Grid coordinates SZ191929: (a) the original image, (b) the image after thresholding.

with various values, a suitable threshold was chosen, calculated as the mean minus one standard deviation (these being the statistics of the pixel values in the image). For the test area of Christchurch in Dorset (Figure 4.5A), this results in the thresholded image shown in Figure 4.5B.

After thresholding, the potential shadows were enhanced by passing a shadow detection filter over the image. The filter used was an ellipse, with its minor axis aligned with the direction of the Sun at the time of image capture (23° east of south, as recorded in the image metadata) and its values weighted as the inverse distance from the central pixel, modified by the angular distance from the Sun angle. The filter values were positive on the sunward side of the ellipse and negative on the leeward side. A diagram of the filter is shown in Figure 4.6, with the positive values in the filter represented by white-to-mid-gray and the negative values represented by black-to-mid-gray. The characteristics of this filter were determined empirically — further work is required to determine whether this filter is appropriate for other images, and to investigate other types of filter (e.g., a simple linear filter, oriented in the direction of the Sun).

The filtering process highlighted the parts of the original image where the boundaries between light areas and dark areas are aligned with the predicted shadow angle (Figure 4.7A). These correspond to features such as buildings, walls, bushes, and trees.

4.3.2.3 Detecting Shadows Cast by "New" Features

To detect the features that are not present in the topographic database, a set of "expected shadows" was generated from the building footprints already held in the database. The first step in the process was to assign a nominal height of 6 m to each building (the typical eave-height of a standard house in the study area, determined by local observation). Then the shadows cast by these buildings, at the time and date on which the image was taken, were predicted using the Sun azimuth and elevation angles recorded in the image metadata (azimuth 23° east of south, elevation 60°). Figure 4.7B shows the buildings and their predicted shadows.

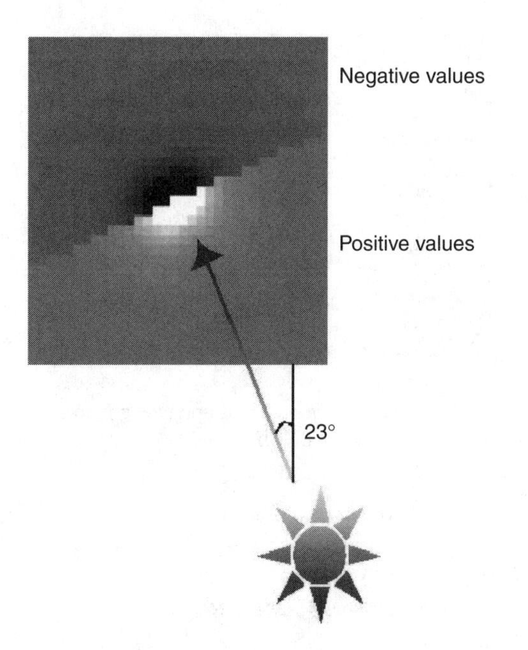

Figure 4.6 The shadow filter, with a Sun azimuth of 23°.

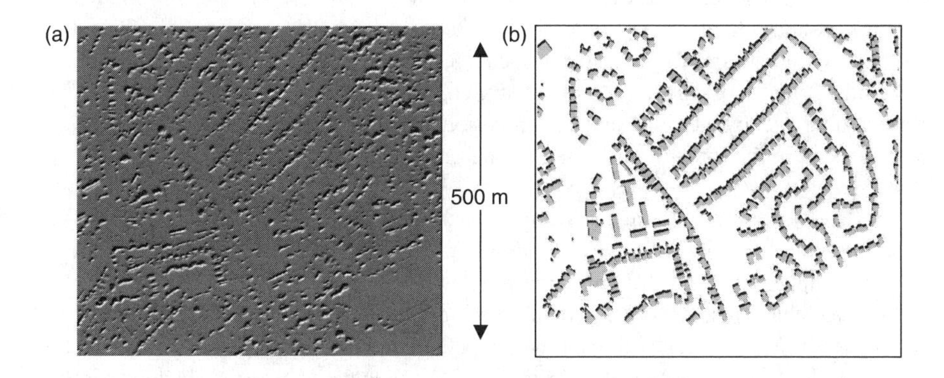

Figure 4.7 Buildings, trees, and their shadows: (a) the image after thresholding and filtering, (b) the buildings in the database, together with their predicted shadows.

The areas covered by the buildings and their predicted shadows were removed from the filtered image. This effectively took out of the image all the features that were already known. Many very small areas of shadow were present in the resulting image. This occurs for a variety of reasons, not least of which is the imprecision inherent in the prediction of shadows from the building features in the database. If the true heights of buildings were recorded in the topographic database, a more realistic prediction of shadows would be possible. During this research, such height information was not available, so only a nominal typical building height could be applied. To remove the small shadow areas from the filtered image, a morphological "opening" (dilation

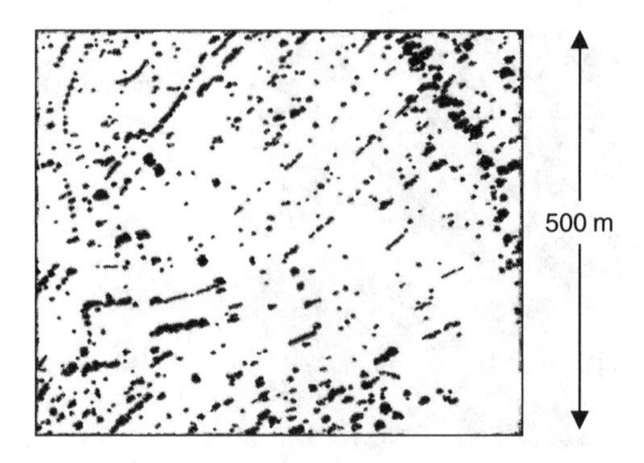

Figure 4.8 The filtered image after known buildings and their shadows have been removed and the resulting image morphologically opened.

followed by erosion) was performed on the image. The final filtered image is shown in Figure 4.8.

4.3.2.4 Results of the Shadow Detection

The processes described above leave an image of "unidentified shadows" (Figure 4.8) that could not be predicted from the existing spatial data. In an ideal world, these would indicate the presence of new features, which could then be manually or automatically extracted and added to the topographic database. In reality, many of the detected shadows were found to be cast by bushes and trees, even in the urban area under consideration. Most of these "new" features, although real enough, are not part of a mapping agency's typical data specification. Since individual trees and bushes are not present in the topographic database, neither these features nor their shadows can be predicted.

At the southern edge of the study area the topographic data were not used, so the final image in this area shows shadows cast by both buildings and by vegetation. A close-up view of part of this area is shown in Figure 4.9. From this example, it is clear that those shadows which were cast by buildings were very difficult to separate from those cast by trees. It was expected that the building shadows would be characterised by straight lines and right angles, while the vegetation shadows would be less regular. In fact, the vegetation in this urban area is often arranged in regular patterns, and a shadow of a large hedge looks very similar to the shadow of a row of houses. Conversely, a house with a gable roof can cast a shadow that looks remarkably tree-like.

To quantify the results, an analysis of the shadows in the filtered image (Figure 4.7A) was undertaken. Recall that this image contained all the shadows identified by the thresholding and filtering processes, before any predicted building shadows are removed. From this image, all the shadow features greater than 100 m^2 in area were manually classified as either buildings or trees/hedges (without reference to the original image, or to the topographic vector data). Each of these classified shadows was then identified in the original image and compared with the topographic data. The results for every classified shadow are shown in Table 4.1. Note that there were

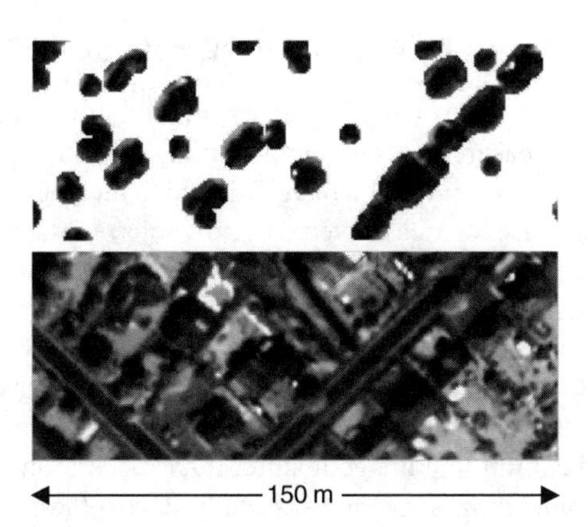

Figure 4.9 A comparison of the predicted shadows and the original image, showing shadows cast by both trees and buildings.

Table 4.1 Results of the Classification of Shadows into Buildings or Trees

	Predicted				
Actual	**Building**	**Tree**	**Not Classified**	**Classified Features Correctly Identified in Percent**	**Total Features Correctly Identified in Percent**
Building	181	69	60	72·4	58·4
Tree	32	93		74·4	
Predicted features correctly classified in percents	85·0	57·4			

Note: The table includes buildings in the topographic database that could not be classified (as either building or tree) from the shadow data alone. A similar statistic for trees is not included, since trees are not present in the topographic database.

60 buildings in the topographic data that were not classified at all (i.e., they did not appear in the shadow data, or the shadow was too small to be classified). The most encouraging aspect of these results is that 85% of the shadows predicted to be buildings were found to be present as buildings in the topographic data. Conversely, only 58.4% of the total number of buildings present in the topographic data were correctly predicted from the shadow data.

The above analysis indicates that, without extra information, an automatic technique would have limited success in identifying new buildings from the shadow data alone. Suitable extra information may, for example, be obtained from a multispectral image captured at the same time as the panchromatic image used to identify the shadows. The combination of multispectral image classification and shadow detection algorithms will be further investigated. Also, further analysis of the shadows of potential features could provide enough evidence to guide a semi-automatic process, such as the one described earlier in this chapter.

4.3.3 Detecting Change by Examining "Empty Space"

A second indicator of change is currently under investigation. This research focuses upon utilising texture measurements to compare imagery with topographic data. The analysis is concentrated on areas in the image for which there are no existing features in the topographic database (or areas covered by polygons recorded as "general surface land" or similar). In these areas, since there are no noteworthy topographic features, the presence of image-texture differences will provide a further cue to the presence of changes to the features on the ground. However, as with the shadow detection process, initial results of this research show that trees in these areas severely restrict the effectiveness of the procedure. While one might expect that areas of "general surface land" would exhibit low image texture, it has been found that image texture can often be very complex in these areas, usually due to the presence of trees and bushes. Paradoxically, it is often the case that features such as buildings in the image will exhibit less texture than the "general surface land" features. Buildings often show little texture difference within their large, flat areas of roofs (and also within vertical walls if the sensor is pointing off-nadir), while there will be large texture differences at the boundary between these features and their surroundings.

Further research will examine these areas more thoroughly, and will look at other cues, which could be used to indicate change, such as the detection of straight lines and corners within areas recorded in the topographic database as featureless.

4.4 SUMMARY

This chapter presented a prototype change detection system that could be used to increase the speed of spatial data capture from imagery. The prototype has been tested on sample images of a small area and has received favourable comments from personnel involved in the photogrammetric data capture area of a national mapping agency. In the initial prototype, image segmentation and polygon-matching techniques were used to determine the locations of change. Related research work has investigated the use of other cues of change, including shadow detection and the analysis of regions in the image which are "empty" in the geospatial database. These techniques could be used in later incarnations of the change detection system.

Current research concentrates on enhancing the prototype to take larger images; to overcome edge effects in the data processing; and to optimise the system to make it more useable in a production environment.

4.5 CONCLUSIONS

This chapter puts forward the view that a technique that includes the user as an integral part of the change detection process can make the geospatial data capture process more efficient.

The use of shadows to detect change between features in a geospatial database and features in an image has met with limited success. While new buildings *can* be detected from their shadows, the presence of shadows from trees and other vegetations make it difficult to determine whether or not the identified shadows indicate changes to the features required in the database. Further work is needed to determine whether other cues, such as shadows with straight edges, could be used to discriminate between building features and vegetation.

Initial research into the use of image texture to indicate the presence of features in areas recorded in a database as "empty space" has shown that many such areas are in fact rich in image texture, again mainly due to the presence of vegetation. Further research is needed in this area to determine whether the technique can be modified to produce more reliable indicators of change.

ACKNOWLEDGMENT

The authors wish to thank Eurimage for providing the QuickBird imagery used in this research.

REFERENCES

Agouris, P., Mountrakis, G., and Stefanidis, A., 2000, Automated spatiotemporal change detection in digital aerial imagery, *SPIE Proceedings Vol. 4054, Aerosense 2000, Orlando, FL*, pp. 2–12.

Baltsavias, E.P., Gruen, A., and van Gool, L. (Eds.), 2001, *Automatic Extraction of Man-Made Objects from Aerial and Space Images (III)* (Lisse, The Netherlands: AA Balkema).

Bennett, S., McRobb, S., and Farmer, R., 1999, *Object-Oriented Systems Analysis and Design Using UML* (Maidenhead: McGraw-Hill).

Brenner, C., Haala, N., and Fritsch, D., 2001, Towards fully automated 3D city model generation, in Baltsavias, E.P. et al. (Eds.), *Automatic Extraction of Man-Made Objects from Aerial and Space Images (III)* (Lisse, The Netherlands: AA Balkema), pp. 47–57.

Hinz, S., Kurz, F., Baumgartner, A., and Wasmeier, P., 2001, The role of shadow for 3D object reconstruction from monocular images, in *5th Conference on Optical 3-D Measurement Techniques, Vienna*, pp. 354–363.

Holland, D.A. and Marshall, P., 2003, Using high-resolution satellite imagery in a well-mapped country, in *Proceedings of the ISPRS/EARSeL Joint Workshop on High Resolution Mapping from Space, Hannover*.

Irvin, R.B. and McKeown, D.M., 1989, Methods of exploiting the relationship between buildings and their shadows in aerial imagery, *IEEE Transactions on Systematic and Management Cybermatics*, 19(6), 1564–1575.

Niederöst, M., 2001, Automated update of building information in maps using medium-scale imagery, in Baltsavias, E.P. et al. (Eds.), *Automatic Extraction of Man-Made Objects from Aerial and Space Images (III)* (Lisse, The Netherlands: AA Balkema), pp. 161–170.

Petrie, G., 2003a, Eyes in the sky — Imagery from space platforms, in *GI News*, 3(1), 42–49.

Petrie, G., 2003b, Current developments & future trends in imaging & mapping from space, in *Proceedings of the ISPRS/EARSel Joint Workshop on High Resolution Mapping from Space, Hannover*.

Tompkinson, W., Seaman, E., Holland, D., and Greenwood J., 2003, An automated change detection system: making change detection applicable to a production environment, *Proceedings of Multitemp 2003, Ispra.*

Ware, M.J. and Jones, C.B., 1999, Error-constrained change detection, *Proceedings of GeoComputation 1999, Fredericksburg*, available at http://www.geovista.psu.edu/sites/geocomp99/Gc99/064/abs99-064.htm.

CHAPTER 5

Developing Smart Spatial Systems Using CommonKADS

Paul Crowther and Jacky Hartnett

CONTENTS

5.1 INTRODUCTION

In 1997, a project was initiated to investigate the feasibility of using satellite-based sensor data to produce classified crop images in North West Tasmania. The domain of

0-8493-2837-3/05/$0.00+$1.50

crop identification is dynamic, with the spectral characteristics of a paddock changing during the growing season. Data acquisition continues throughout the season, but at irregular intervals due to cloud. Despite this, the study resulted in the development of techniques that identified the major economic crops at an accuracy of >85%.

Finding algorithms that produced a classified image was the first objective of this initial study, which was met (Barrett et al., 2000). There were, however, two problems. First, it became apparent that the structure of the system that was being developed was ad hoc and the original notion of the task comprised both the task specification and the knowledge base. Effectively, the problem solving method was fixed and formed an integral part of the implementation. Second, the organizations providing the funding support had different ideas of how the resulting classified image could be used. The initial results were classified images, which then required further processing to be useful to the organizations.

To solve these problems, a CommonKADS model was developed for the next iteration of the system. One application, that of planning pyrethrum crop plantings based on existing crops and crop rotation knowledge, will be presented here. The components of that system were then made available to develop further systems, which met other objectives. Reuse in this context has been successfully achieved (Crowther et al., 2002), but not in a geographic domain.

The Knowledge Acquisition and Design System (KADS) was initiated as a "Structured methodology for the development of knowledge based systems" (Schreiber et al., 1993, p. xi). The limitations of production rules, combined with their inherent non-reusability contributed significantly to the impetus to develop methodologies like KADS. KADS modeled knowledge, but it became apparent that a model of knowledge was only one component of an overall knowledge-based systems (KBS). This lead to CommonKADS. The two central principles that underlie the CommonKADS approach are the introduction of multiple models as a means of coping with the complexity of the knowledge engineering process, and the use of knowledge-level descriptions as an intermediate model between expertise data and system design.

Motta (1997) coins the term "knowledge modelling revolution," which refers to the paradigm switch from symbol level (rule based) approaches to knowledge level task centered analysis. This heralded the necessary decoupling of the task specification and the problem solving method.

5.2 COMMONKADS

Figure 5.1 gives an overview of the interconnecting CommonKADS models. Each of the models has a series of associated templates that, when filled in, provide detailed documentation of the system and its requirements (Schreiber et al., 2000). In this chapter only the organizational, task, agent, and knowledge model will be discussed in detail as the communication and design model are very platform specific.

The *Organizational Model* is a model that documents the objectives of the system and identifies opportunities of value to the organization. One of the advantages of CommonKADS is that the organizational model provides an analysis of the socio-organizational environment that the knowledge-based system (KBS) will have to

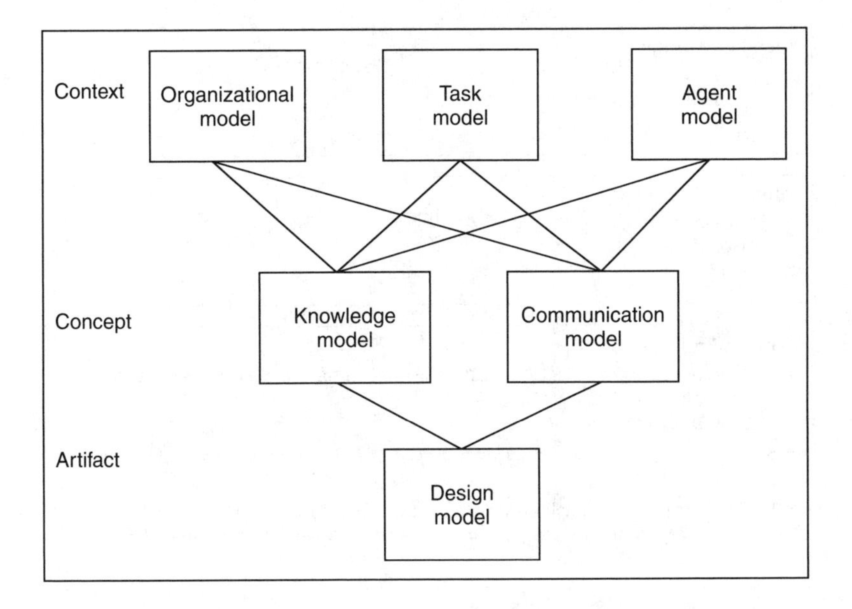

Figure 5.1 The CommonKADS knowledge classification scheme (after Schreiber et al., 2000; *Knowledge Engineering and Management*, reproduced with permission from the MIT Press.)

function in. This includes descriptions of functions within the organization (Wielinga and Schreiber, 1993). The organization model is also used to identify risks of fielding a KBS. These risks tend not to be technical as KBS's are now widely used and there is less risk of KBS failure through technical issues (De Hoog et al., 1993). This is as true of spatial KBS's as non-spatial ones.

The *Agent Model* provides an understanding of the systems users and identifies how these users or agents will perform their tasks (Gustafsson and Menezes, 1994). An agent in this context is a person, hardware or software that interfaces with the KBS. In a spatial system, for example, it would be likely that an agent could be a geographic information system (GIS).

The CommonKADS *Task Model* specifies how the functionality of the system is achieved (Gustafsson and Menezes, 1994). The task model links to the agent model to identify the people (or the roles that individuals have), hardware or systems that perform tasks. These tasks operate in the domain defined in the organizational model.

The *Communication Model* models the interaction of the system with the user and other system components. It models how software and users of the system work together and specifies the environment in which KBS must work.

The *Knowledge Model* (Figure 5.2) defines the knowledge necessary to meet the objectives specified in the organizational model and the tasks in the task model. It is split into three layers.

The CommonKADS *Domain Layer* is knowledge describing a declarative theory of the domain. Knowledge at this level should be represented in a way that is independent of the way in which it is to be used. Generally, this is described using the Unified Modeling Language (UML) notation (Rumbaugh et al., 1999). It defines

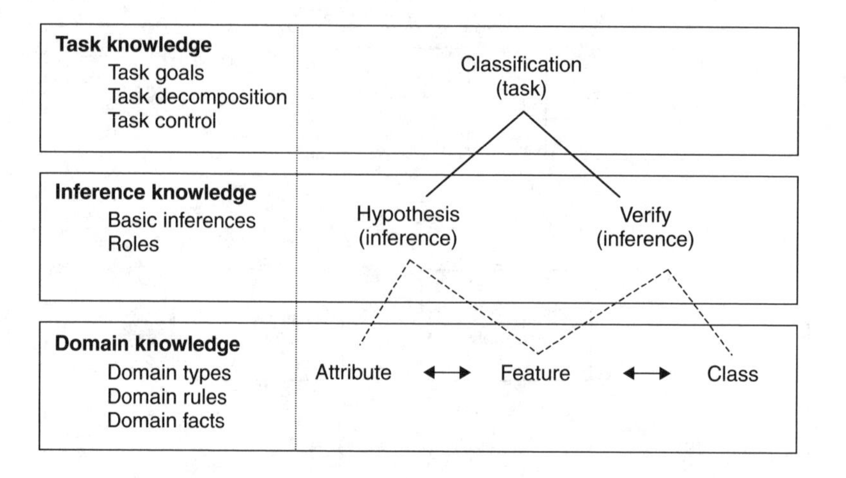

Figure 5.2 The three-layer knowledge model (after Schreiber et al., 2000; *Knowledge Engineering and Management*, reproduced with permission from the MIT Press.)

the conceptualization and declarative theory of the problem domain and provides the knowledge to carry out given tasks. In other words it contains facts, rules, and domain types. The other layers contain knowledge to control the use of knowledge from the domain layer (Fensel and Van Harmelen, 1994).

The *Inference Layer* specifies how to use the knowledge from the domain layer. It restricts the use of the domain layer and abstracts from it. A formal specification language has been developed to record knowledge in each of the layers (Schreiber et al., 1994).

The *Task Layer* represents a fixed strategy for achieving problem solving goals. It includes a specification of the goals related to a specific task and indicates how these goals can be decomposed into subgoals.

The task layer in the geographical classification scheme shows how to apply the problem solving strategy to a whole image set. Depending on the objective of the system, different strategies can be used, including the masking out of areas not relevant in a particular domain. For example, in a geographical knowledge classification scheme, knowledge about specific objects, like an instance of the *crop* class *potato*, would be held at the domain level. Inference knowledge would be used to apply the domain knowledge. At the task level there would be a task to identify all *paddocks* containing *potato crops*. This could be defined in more abstract terms (apply crop labels to paddocks) and reused to identify other crops.

The original KADS methodology proposed a four-layer knowledge model (Schreiber et al., 1993). The CommonKADS methodology (Schreiber et al., 2000) amended this to the three-layer model, which is shown in Figure 5.2.

In the earlier four-layer model, the top level of knowledge was the *Strategy Layer*, which involves knowledge of how to choose between various tasks that when completed successfully achieve the same goal. This layer has been removed in the CommonKADS model, but may have application in spatial systems. In geographic knowledge applications the strategy layer could contain alternate ways of classifying images. This could be important when choices need to be made between using rules developed from machine learning, neural networks or rules elicited from domain experts to classify images. It should be noted once again that the strategy layer is not part of the CommonKADS model.

5.3 A SPATIAL APPLICATION

The rest of this chapter will present an example based on the analysis of satellite sensor imagery in an agricultural domain where expert knowledge was captured and applied. A series of CommonKADS models will be presented and evaluated.

In 1997, a project was initiated to create crop identification and monitoring system using remotely sensed images from Landsat 5 TM, SPOT HRV 2, and SPOT HRV 4 satellite systems over the major agricultural region of North West Tasmania (Figure 5.3). The aim was to establish crop distributions, monitor crop health, and estimate yield. Crops of interest were restricted to poppies, potatoes, peas, pyrethrum, and onions.

The Landsat TM provided seven spectral bands of information and the SPOT HRV, three (the 1580 to 1750 nm sensor of SPOT 4 was not used). All images were co-registered and resampled to cover the same area at 20 m resolution (Barrett et al., 2000).

Figure 5.4 shows a UML class model (Rumbaugh et al., 1999) defining the overall problem domain. This model forms the basis for the design of the database, which is part of the system. The automated part of the system deals with assigning crop labels to paddocks. Some research has been carried out into automated boundary (fence) extraction (Anderson et al., 1999), but at the moment this is not accurate enough to use. Paddock boundaries were therefore manually digitized and added to the system as an overlay. A soil distribution map and a digital elevation model were also added as map overlays and linked to paddock once the boundary digitization has been completed. Classes such as *Farmer*, *AgRegion*, *Fertilizer*, and *Spray* were

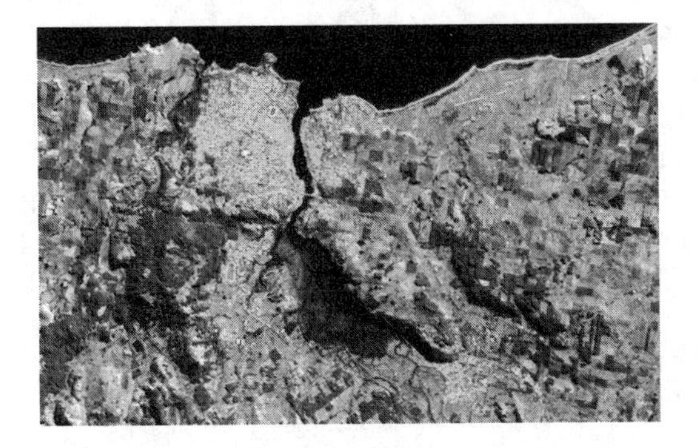

Figure 5.3 Composite image of part of the study area around Devonport in North West Tasmania.

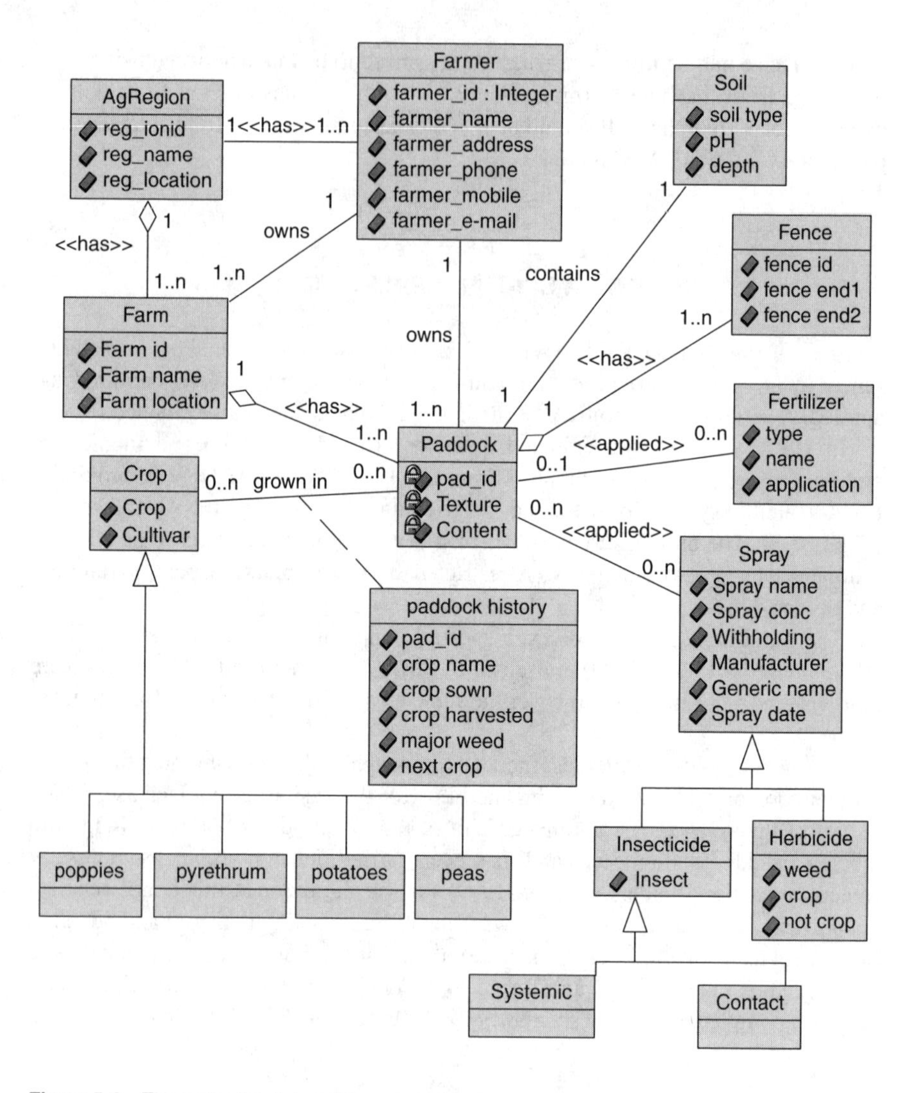

Figure 5.4 The agricultural domain.

all entered manually. Once entered, these were displayed as overlays based on their association with the class *paddock*.

In this system, a *paddock* is a polygonal object that is central to the study. A *paddock* contains a *crop*, although at any one time there is only one *crop* in a *paddock*. The different types of crop of interest to the case study have been shown as specializations of *crop* and inherit all of *crop*'s attributes and operations. Specific attributes can be added to these specializations as required. In this system, it is the contents of the *paddock* (*crop*) that is identified and then assigned to an instance of *paddock*.

In the initial stages of the study, the crop identification techniques were paramount, with only a broad definition of how the final system would be used (Barrett et al., 2001).

This study addresses that problem by building models for individual applications using CommonKADS models. Components of the crop planning system can be reused. Reuse is one of the main advantages of using the CommonKADS approach (Valente et al., 1998).

5.3.1 Organizational Model

The aim of producing an organization model is to establish the objectives of the system in terms of the organization (De Hoog, 1994). This model also establishes the organizational setting, problems, opportunities, and available resources for meeting these objectives.

Figure 5.5 is a skeleton of the organization model for a pyrethrum processing company that wants to use a satellite sensor image classified into different crop categories. This information is then used as a basis for planning future crop plantings.

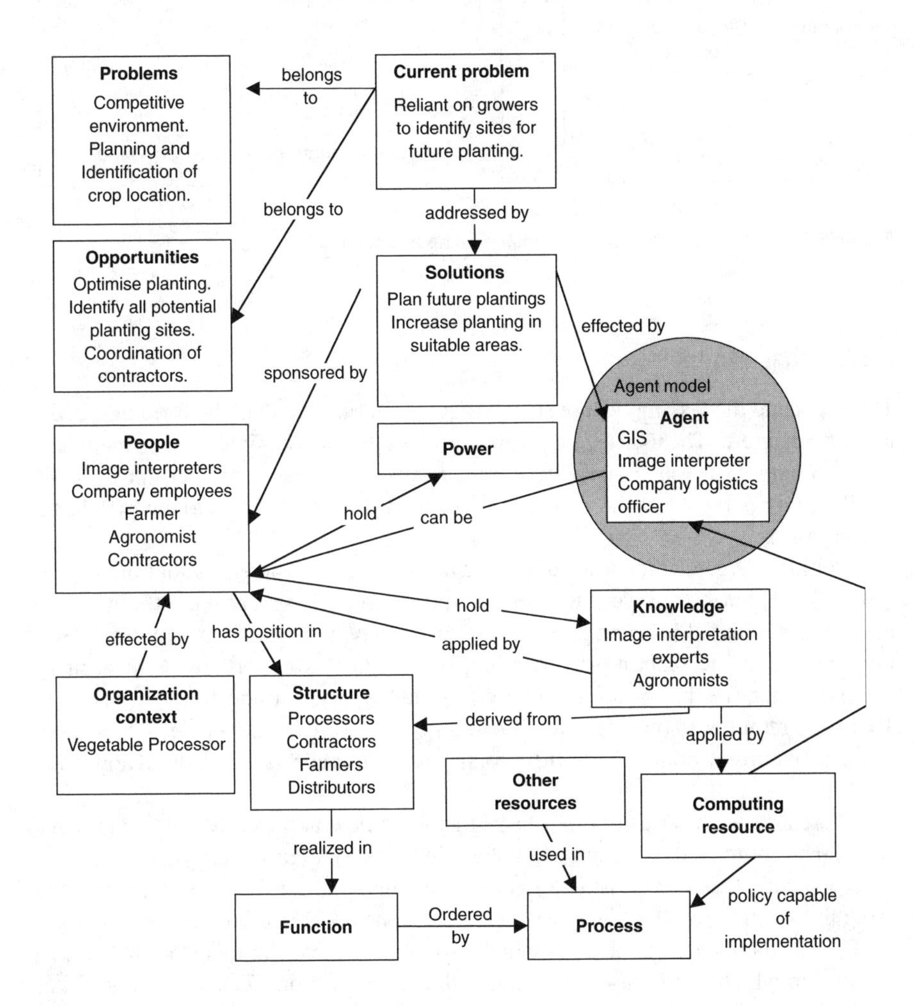

Figure 5.5 Organizational model.

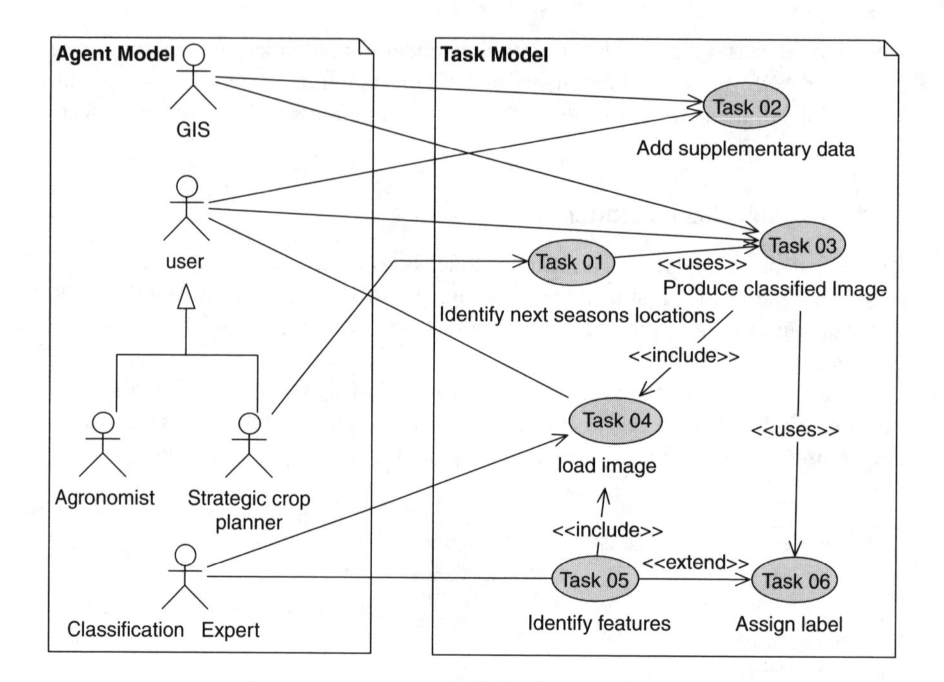

Figure 5.6 Task model showing relationship to the agent model.

5.3.2 Task Model

The agent and the task model are shown together in Figure 5.6 in the form of a UML use-case diagram. The use cases describe the tasks that are required. The primary task of the system, as it was initially developed, was to produce a classified image. This is still required, but a classified image on its own is of limited value to the pyrethrum processor.

From the Organizational Model view, one identified problem is that the organization is reliant on growers to identify future crop sites. The opportunity is to identify all potential planting sites. From that it may be possible to optimize planting and coordinate planting contractors. To realize this opportunity, the primary task is to identify the location where crops can be sown in the following season. This task requires knowledge of crop rotation as well as knowledge about how to identify current crops and will be found in the domain layer of the Knowledge Model.

There are also a series of other tasks that must be completed before the primary task can be completed. For example, to be able to identify current crops, an image set needs to be loaded, then classified. In the task model, this dependency is documented using the <<uses>>, <<include>>, and <<extend>> stereotypes from UML.

Finally, there needs to be tasks which provide a mechanism to add knowledge to the system which can be used to produce that classification. Each task in the Task Model is described by a template (Schreiber et al., 2000). As an example Task 01 from the Figure 5.6 would be described as follows:

TASK: 01 Identify next seasons locations
ORGANIZATION agronomist, contractor coordinator, strategic crop planner
GOAL AND VALUE identify paddocks where the crop can be grown next season
Find optimal locations which will result in a good crop yield
DEPENDENCY AND FLOW
Input tasks: specific crop type distribution,
Output tasks: paddock location for sowing
OBJECTS HANDLED
Input objects: classified image (showing crop which precedes the target planting crop)
Output objects: classified image with target paddocks for sowing
Internal objects: this is shown by Figure 5.4
TIMING AND CONTROL
Frequency: yearly.
Timing: after collection of growing season images and before planting time
(i) *preconditions* crop ready for harvesting;
(ii) *postconditions* paddocks ready for sowing.
AGENTS agronomist, GIS
KNOWLEDGE AND COMPETENCE Competent with GIS
RESOURCES As required
QUALITY AND PERFORMANCE Increase the total area to be harvested

The task template describes what is required. The knowledge required to carry out the task is *Task Knowledge* and is defined in the knowledge model, described below in Section 5.3.4.

5.3.3 Agent Model

The agent model describes the actors that interact with the overall system. In the case of the agricultural system these are the end users of the system, generally those who want a classified image, an image classification expert, and finally a GIS package which may be useful to manipulate the classified images. Individual end users can have multiple roles and each role is shown as a separate agent. For example, the *strategic crop planner* and the *agronomist* may be the same person, and both roles fall under the more generalized role of *user*. There is interaction between these agents as well. For example, the strategic crop planner could use a GIS to view the classified image and next season's potential crop sites (not all of which may be used).

The following is an example of an agent template from the model defining the strategic crop planner agent

NAME strategic crop planner
INVOLVED IN Load image, identify next seasons locations
COMMUNICATES WITH agronomist, GIS
KNOWLEDGE how to select between suggested crop locations
OTHER COMPETENCES use of a GIS or image processing software
ORGANIZATION consultant
RESPONSIBILITIES AND CONSTRAINTS select paddocks where the next planting of the crop is to be carried out. Do not select too many locations to avoid a shortage of sites for the year following the planned year

5.3.4 Knowledge Model

The original crop identification system had no structuring of the knowledge, which it contained. The system was examined and knowledge at all levels was extracted and placed into the CommonKADS knowledge model structure. The feasibility of this approach has been shown in industrial process systems (Crowther et al., 2001). As part of this approach the validity of existing knowledge was checked and new knowledge required to meet the objectives defined in the task model added.

5.3.4.1 Task Knowledge

Task knowledge is not the same as the task model, although there is a relationship. In the CommonKADS methodology high level abstract classes would be associated with the task level of the knowledge model. For example, a key task in the task model is *identify next seasons planting locations*. The task knowledge necessary to achieve this would be held at the abstract class *image primitive* level rather than the concrete level such as *poppy*. An example of task knowledge specification is as follows with a naming convention from UML and stored as an operation in the *image_primitive* class:

```
TASK: identify next seasons planting locations
      ROLES:
      INPUT: object: paddock with current crop label
      OUTPUT: decision (will plant, could plant, will not
      plant)
END TASK

TASK-METHOD assign-decision-label
REALIZES: identifying next season planting locations
```

```
DECOMPOSITION
    INFERENCES: select, assign
    TRANSFER-FUNCTIONS: obtain;
ROLES
    INTERMEDIATE
    Goal: establish the suitability of a paddock for
    planting pyrethrum
    establish suitability;
    establish level of suitability;
    Result: a label of suitability

CONTROL-STRUCTURE
FOR each paddock;
DO
    establish current crop;
        retrieve previous crop;
        compare(observed-rotation + expected
        rotation -> result);
        IF result = = equal
               THEN label = suitable;
               ELSE label = unsuitable
        END IF
        IF label = = suitable
                THEN establish suitability level
        END IF
        END FOR
END TASK-METHOD assign-decision-label
```

Task knowledge is therefore the knowledge required to fulfil the tasks identified in the task model. Hence, each task in the task model would have at least one equivalent template in the task layer of the knowledge model.

5.3.4.2 Inference Knowledge

Inference Knowledge is how to apply domain knowledge to meet a particular task. In the agricultural domain there are three main types of feature that require inference. These are:

- image primitives, such as polygons, points, and lines,
- features which are determined through their relationships to other features,
- more complex features which are a combination of image primitives.

In Figure 5.7, these have been identified as *Primitive Knowledge*, *Relationship Knowledge*, and *Assembly Knowledge* respectively (Crowther, 1998).

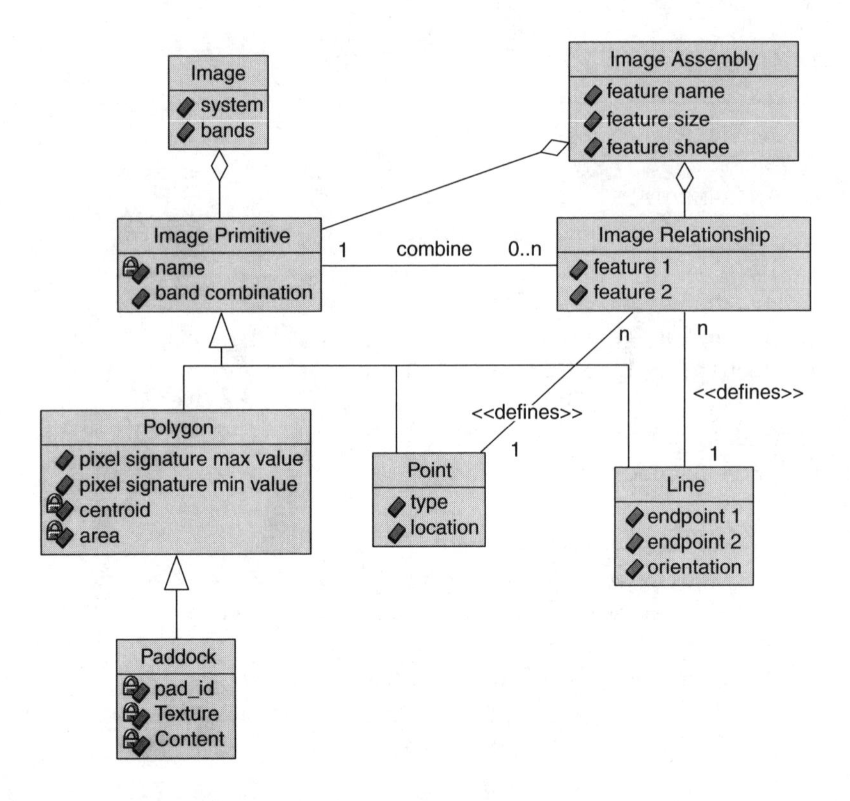

Figure 5.7 CommonKADS Inference Knowledge.

For example, the inference *select* was identified in "TASK: identify next seasons planting locations" which is one element of the task knowledge discussed in Section 3.4.1. The domain class involved is *paddock* which is a subclass of *polygon* which itself is a specialization of the *image primitive* class. The inference select would be defined:

```
INFERENCE select
ROLES
    INPUT paddock, paddock history
    OUTPUT availability
    STATIC rotation-model
SPECIFICATION:
       Each time the inference is invoked it generates a
       Boolean for the availability of the paddock for
       planting
END INFERENCE
```

This knowledge could be reused in other domains which require classification of scenes. The detail of how the result was derived would be held in the static

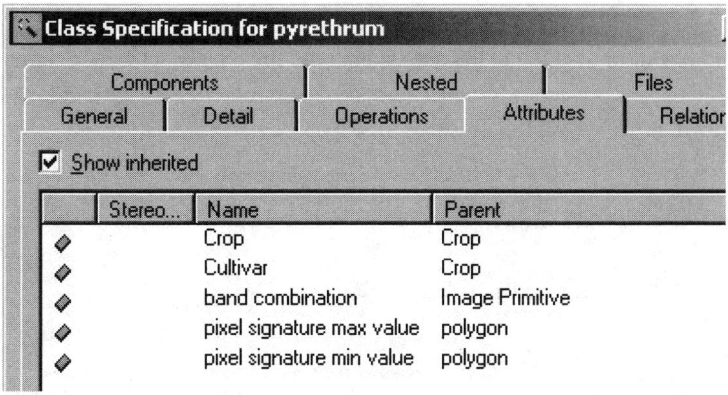

Figure 5.8 UML attribute definition from concept *pyrethrum* showing inherited attributes.

classification model defined at the domain knowledge level. The inference knowledge is, therefore, used to achieve the task goal of classifying the image.

5.3.4.3 Domain Knowledge

Domain Knowledge is about individual instances of objects in the problem domain. This is shown at a high level in Figure 5.4 where the relationships between the various domain classes are described. From a CommonKADS domain knowledge stance, these classes are termed *concepts*. Each of these concepts has *attributes* and *value types* associated with it. Hence the concept *pyrethrum* in Figure 5.8 has five attributes associated with it. Each attribute has a value type defined in the form:

```
VALUE-TYPE pixel signature maximum value
VALUE: 79
TYPE: NUMBER
END VALUE-TYPE pixel signature maximum value
```

This information is still very similar to that which would be found in a class diagram, however, there are also *rule types* associated with each concept. For example, an important piece of domain knowledge is about crop rotation, which would be defined as:

```
RULE-TYPE crop-rotation
ANTECEDENT: current crop
    CARDINALITY: 1;
CONSEQUENT: suggested crop
    CARDINALITY: 1;
CONNECTION-SYMBOL
    precedes
END RULE-TYPE crop-rotation
```

For each of the rule-type definitions individual rules are defined using the connection symbol to connect the antecedent and the consequent, for example,

```
PADDOCK-HISTORY.crop_name = peas
precedes
PADDOCK-HISTORY.next crop = pyrethrum;

PADDOCK-HISTORY.crop_name = potato
precedes
PADDOCK-HISTORY.next crop = peas;
```

The final knowledge base contains instances of the rule types defined in the domain model. The CommonKADS domain knowledge in the knowledge base is then used by the inference knowledge which in turn is required by the task knowledge.

5.4 DISCUSSION

CommonKADS provides a series of models in which any KBS can be developed. In a general sense it provides the organizational objectives of the system, the organizational setting, the users, both human and system, and the specific goals. The knowledge model provides the knowledge required to meet these goals and objectives.

The crop identification component of the system which applied crop labels to each of the paddocks in the domain exceeded 85% in the 1998 to 1999 growing season. The results for the crops of commercial interest being: poppies (producer's accuracy 92.2%), onions (88.6%), peas (94.2%), and pyrethrum (87.1%). The results were then used to plan the following seasons planting of pyrethrum. Since the system was specifically requested by the pyrethrum processor, there was potential conflict for paddocks with other processors. For example, paddocks suitable for pyrethrum are also suitable for poppies. As a result a paddock was still only potential until negotiations with individual farmers were completed. A further complication in this was that pyrethrum is a perennial crop committing a grower to having it in their paddock for at least 4 years.

5.4.1 Spatial Extensions to CommonKADS

There are no direct extensions needed to the CommonKADS model to develop smart or expert spatial systems. However, at the knowledge model level it is useful if a template for different types of spatial knowledge is used. The one suggested here is of primitive, relationship, and assembly knowledge (Crowther, 1998). This provides an overall framework which can be used for the inference layer in the knowledge model.

5.4.2 A Case for Strategy Knowledge in a Spatial Domain

There are many ways of producing a classified satellite image including the use of neural networks (Openshaw, 1993), knowledge-based approach (Crowther et al., 1997), and clustering algorithms (Richards, 1993). The classification method used in this system was a knowledge-based approach using a pixel level classification. Although this was the method that was used in the planning of pyrethrum crop locations, other inference methods could have been used and may be more suitable in other applications. Strategy knowledge from the original KADS system would allow modeling of this high-level knowledge.

5.4.3 CommonKADS Library

An advantage of building a CommonKADS series of models is that an ontology of the domain is developed. An ontology defines the constraints of possible objects expressed in the model in addition to the constraints imposed by the syntax. Wielinga and Schreiber (1993) use the following basis for a description of ontologies. A knowledge base can be viewed as a model of some part of the world, that allows for reasoning to take place in that world model, given some inference mechanism. The model is described in a particular language and has a vocabulary and syntax.

The ontology developed as part of this project could be reused on similar types of projects, but to do that, it would have to be made available as part of a library. The CommonKADS Expertise Modeling Library (Valente et al., 1998) is an example of such a library, as is Ontoligua server (Farquhar et al., 1996). One possible future development is the setting up of a library of spatial ontologies. In this example that is unlikely to happen as the pyrethrum processor regards this knowledge as proprietary.

5.5 CONCLUSIONS

Many spatial KBS have been developed, generally with a lot of attention to reasoning strategies, but few seem to have been developed using an overall set of interrelated models. CommonKADS provides these models without specific spatial extensions, which can be easily represented using the now widely accepted UML notation. These document the goals and objectives of a system and place it within the organizational context where it will be used.

The advantage of using a series of integrated models to develop knowledge-based geographic systems is that not only is knowledge modeled, but also the organizational requirements and the environment it will operate in. This is an aspect that appears to have been lacking in the development process of most spatial KBS. By using this type of modeling, an information system that meets a specific users needs can be developed in a form that encourages structure and reuse of components, be they knowledge components or interface components.

Given that CommonKADS is designed to exploit ontologies in the form of the CommonKADS Library (Valente et al., 1998), collaborative systems could be built reusing knowledge from the domain layer and inference layer of the knowledge model.

In other words, there is no need to build specific spatial extensions to CommonKADS, rather a library of model components of a spatial nature could be developed. These components could then be used for rapid development of other systems with a spatial base.

ACKNOWLEDGMENTS

The work reported in this chapter has been supported by a Horticultural Research and Development Grant (Project No. HRDC VG97011).

REFERENCES

Anderson, S.J., Williams, R.N., and Crowther, P., 1999, Automatic Detection of Agricultural Field Boundaries Using High-Resolution Visible Light Satellite Images, *Proceedings of the 27th Annual International Conference and Technical Exposition of the Australasian Urban and Regional Information Systems Association*, Fairmont Resort, Blue Mountains NSW.

Barrett, R., Crowther, P., Laurence, R., and Lincolne, R., 2000, Agricultural Crop Identification Using Spot and Landsat Images in Tasmania, *International Archives of Photogrammetry and Remote Sensing (IAPRS), XXXIII*, pp. 133–139.

Crowther, P., Hartnett, J., and Williams, R.N., 1997, A Visual Tool for Capturing the Knowledge of Expert Image Interpreters: A Picture is Worth More than a Thousand Words, *Proceedings of the International Geoscience and Remote Sensing Symposium (IGARSS) 1997,* Singapore.

Crowther, P., 1998, Knowledge Acquisition from Satellite Images for Geographic Expert Systems — Show or Tell?, *Proceedings of the International Conference on Modeling Geographic and Environmental Systems with Geographic Information Systems*, Hong Kong, June.

Crowther, P., Berner, G., and Williams, R.N., 2002, Re-usable Knowledge: Development of an Object Oriented Industrial KBS and a Collaborative Domain Ontology, *Proceedings of the International Workshop on Intelligent Knowledge Management Techniques (I-KOMAT) 02*, Crema, Italy, 16–18 September 2002, pp. 1539–1543.

De Hoog, R., Benus, B., Metselaar, C., Vogler, M., and Menezes, W., 1994, *Organisation Model: Model Definition Document.* Deliverable DM6.2c, ESPRIT Project P5248 KADS-II/M6/M/UvA/041/3.0, University of Amsterdam and Cap Programmator.

Farquhar, A., Fikes, R., and Rice, J., 1996, *The Ontolingua Server: A Tool for Collaborative Ontology Construction.* Stanford Knowledge Systems Laboratory. [Online] Available at http://www.ksl.stanford.edu/software/ontolingua/project-papers.html (last accessed 8th July 2003).

Fensel, D. and Van Harmelen, F., 1994, A Comparison of Languages which Operationalize and Formalise KADS Models of Expertise, *Knowledge Engineering Review*, 9, 105–146.

Gustafsson M. and Menezes, W., 1994, *CommonKADS Reference Manual*, Esprit Project P5248 KADS-II, KADS-II/P2/WP/CP/001/1.0, 05/1994.

Motta, E., 1997, Trends in Knowledge Modeling: Report on the 7th KEML Workshop. *The Knowledge Engineering Review*, 12, 209–221.

Openshaw, S., 1993, Modelling Spatial Interaction Using a Neural Net, *GIS Spatial Modelling and Policy Evaluation* (Springer-Verlag), pp. 147–164.

Richards, J.A., 1993, *Remote Sensing Digital Image Analysis*, Springer-Verlag, New York, Chs. 4, 6 and 8.

Rumbaugh, J., Jacobson, I., and Booch, G., 1999, *The Unified Modelling Language Reference Manual*, Addison Wesley.

Schreiber, G., Wielinga, B., and Breuker, J., 1993, *KADS—A Principled Approach to Knowledge-Based System Development*, Academic Press, San Diego.

Schreiber, G., Wielinga, B., Akkermans, H., Van de Velde, W., and Anjewierden, A., 1994, CML: The CommonKADS Conceptual Modelling Language, *Proceedings of the 8th European Knowledge Acquisition Workshop*, Hoegaarden, Belgium.

Schreiber, G., Akkermanns, H., Anjewierden, A., De Hoog, R., Shadbolt, N., Van de Velde, W., and Wielinga, B., 2000, *Knowledge Engineering and Management: The CommonKADS Methodology*, MIT Press.

Valente, A., Breuker, J., and Van de Velde, W., 1998, The CommonKADS library in perspective, *International Journal of Human-Computer Studies*, 49, 391–416.

Wielinga, B.J., Schreiber, A.T., and Breuker, J.A., 1992, KADS: A Modelling Approach to Knowledge Engineering, *Knowledge Acquisition*, 4(1), 5–53.

Wielinga, B.J. and Schreiber, A., 1993. Reusable and Sharable Knowledge Bases: A European Perspective, *Proceedings of the International Conference on Building and Sharing of Very Large-Scaled Knowledge Bases '93.*

CHAPTER 6

Identifying Land Cover Change Using a Semantic Statistical Approach

Alex J. Comber, Peter F. Fisher, and Richard A. Wadsworth

CONTENTS

0-8493-2837-3/05/$0.00+$1.50

6.1 INTRODUCTION

Land resource inventories are often subject to changes in methodology due to technological innovation and developments in policy (Comber et al., 2002, 2003a). The net result is that different data nominally purporting to record the same processes may do so in different ways, identifying different objects, using different labels, and different definitions. Subjectivity derives from the use of different pre-processing techniques, classification algorithms, and class definitions. There is an increasing recognition of this phenomenon within remote sensing, acknowledging the subjective nature of much information classified from remotely sensed data. This is exemplified by the increased acceptance of concepts such as fuzzy classifications, soft accuracy measures, as described by Shalan et al. (Chapter 2, this volume) and fuzzy land cover objects Lucieer et al. (Chapter 6, this volume).

The central feature of change detection is that change is a subset of inconsistency between two signals. This inconsistency may be as a result of artefactual differences between the data (thematic or raw) or due to actual differences on the ground. In order to identify any signal of actual change it must be separated from the noise of data difference caused by differential processing and classification. The approach presented by Brown (Chapter 8, this volume) requires thematic and registration errors to be quantified so that they can be used as an input to any change model. In this chapter, an approach for integrating different data based on the way that the land cover objects contained within each data are conceptualised is described. The way that a dataset conceives, specifies, measures, and records objects defines the data ontology (Guarino, 1995). For some landscape processes (e.g., geological) the interval between surveys is small relative to the timescale at which changes occur. For others, such as land cover, differences in ontology may obfuscate any signal of change. This situation hinders the development time series models, environmental monitoring and the identification of locales of change.

This generic problem is exemplified by the land cover mapping of Great Britain in 1990 (LCM1990) and in 2000 (LCM2000), descriptions of which can be found in Fuller et al. (1994) and Fuller et al. (2002), respectively. LCM1990 records 25 target land cover classes defined mainly on pixel reflectance values, identified using a *per-pixel* maximum likelihood classification algorithm. Some knowledge-based corrections were included but no metadata was included to record their application. In contrast, LCM2000 records land cover information on a *per-parcel* basis at three levels of detail, including 26 Broad Habitat classes. Parcel reflectance values were classified, assisted by ancillary data (soils) and Knowledge-based Corrections. Extensive metadata are attached to each parcel as described in Smith and Fuller (2002). The net result is a drastic change in the *meaning* behind the class labels between LCM1990 and LCM2000. Changes in ontology make it difficult to relate the information contained in the previous dataset to that of the current. To overcome this problem, Comber et al. (2003b) have proposed a semantic statistical approach for reconciling the ontological difference between datasets.

The generic problem of spatial data integration has been addressed by various workers under a series of different headings (Bishr, 1998; Devogele et al., 1998; Ahlqvist et al., 2000; Visser et al., 2002; Frank, 2001; Kavouras and Kokla, 2002).

A common theme to emerge from this work is the identification of semantics as the bottleneck to translation between different data ontologies. The problem of semantic translation between data concepts can be tackled using formal ontologies. An ontology in this context is an explicit specification of a conceptualisation (Guarino, 1995). The integration "problem" is that an object described in one conceptualisation may not correspond or translate to another conceptualisation. For instance, Bennett (2001) describes the problem of vague definitions of "forest" that can result in uncertainty due to a lack of clearly defined criteria. Various definitions of forest are illustrated at http://home.comcast.net /~gyde/DEFpaper.htm. For instance, a forest may be defined on the basis of the physical properties of tree height and crown canopy cover, and many countries include definitions of land that could be under trees, or where there is a probable intention to replant in the foreseeable future; also that many countries include bamboo and palms into their definition of what constitutes a "tree" and hence a forest. In the "ontologies approach," the capacity for data integration and sharing depends on understanding the way that these data are conceptualised — what the data objects mean. To this end Pundt and Bishr (2002) describe how members of different information communities (e.g., remote sensing, ecology, landscape ecology) are able to access to the meaning (in the wider sense) of other data if they can unravel the ontologies that have been developed by those who collected the data.

This chapter describes the results of using the semantic statistical approach to identify locales of land cover change from LCM1990 and LCM2000. The approach combines expert opinion of how the semantics of the two datasets relate with metadata descriptions of LCM2000 parcel spectral heterogeneity. This approach explores semantic and ontological links between datasets, exposes weakness in expert understanding of the data concepts, and provides a methodology for integrating different data.

6.2 SEMANTIC STATISTICAL APPROACH FOR DETECTING CHANGE

The hypothesis behind the semantic statistical approach is that by characterising each parcel in 1990 and 2000 it is possible to identify change between LCM1990 and LCM2000. The 1990 characterisation was based on the distribution of LCM1990 classes that intersected with the parcel. The 2000 characterisation was based on the distribution of spectral subclasses contained in the parcel spectral heterogeneity metadata attribute "PerPixList." Parcels whose characterisations were different were possible areas of change. Note that due to methodological similarities the parcel spectral heterogeneity information is equivalent to LCM1990 classes.

An expert data user, familiar with both datasets, was asked to describe the pair-wise relations between LCM1990 and LCM2000 classes in terms of their being *unexpected* (U), *expected* (E), or *uncertain* (Q). This was done under a scenario of idealised semantic relations — that is, without considering any noise in the pair-wise relations from such things as spectral confusions or likely transitions. These relations were tabulated into the semantic Look Up Table (LUT). The parcel characterisation for 1990 was calculated by interpreting each intersecting LCM1990 pixel via the semantic LUT. For each parcel a $(U, E, Q)_{1990}$ triple was determined by summing the number of pixels in each type of relation.

A spectral LUT was used to characterise the parcel in 2000. The LUT was constructed from information about the expected spectral overlap between the spectral subclass in the spectral heterogeneity attribute ("PerPixList"; PPL) and the broad habitat classes that accompanied the release of the LCM2000 data. The pairwise relations in the spectral LUT between the broad habitat classes and spectral subvariant classes were again described in terms of their being *expected*, *unexpected*, or *uncertain*. Each element in the spectral heterogeneity attribute was interpreted through the spectral LUT to produce a $(U, E, Q)_{2000}$ triple for the parcel.

A vector in a feature space of U and E was generated by comparing the two characterisations (after normalising one to the other: 2000 was based on percentage of the parcel area, 1990 on the number of pixels). Note that the Uncertain scores (Q) were ignored because the objective was to identify parcels whose characterisation had changed most, that is, whose changes in U and E represented definite shifts rather than uncertain ones. Parcels whose descriptions varied the most between the two dates (i.e., had the largest vectors) formed the set of hypothesised land cover changes. The 100 parcels with the largest vectors were identified for three LCM2000 broad habitat classes, representing three cases of ontological difference between LCM1990 and LCM2000:

1. "Broadleaved Woodland" was similarly defined in 1990 and in 2000 and represented minor ontological change.
2. "Suburban and Rural Development" was likely to have been affected by the use of the parcel structure in 2000 to delineate homogenous areas. Therefore, this class represented a moderate level of ontological change between 1990 and 2000.
3. "Acid Grassland" did not exist as a land cover mapping concept in 1990. It is a class designed to address specific post-Rio habitat monitoring obligations. It was differentiated from other grassland land covers by applying a soil acidity mask. Thus this class represented major ontological change between 1990 and 2000.

Because it was possible to calculate the *gross* vector for each parcel based on the *number of pixels* or the *percentage* vector based on the *percentage of pixels*, two sets of 100 parcels were identified for each broad habitat class. Of the 600 parcels, 100 were randomly selected and of those 93 were visited. The numbers of parcels visited and identified in the different directions for each LCM2000 broad habitat are shown in Table 6.1.

6.3 RESULTS

The semantic statistical method has at its core the notion that change can be identified by comparing two characterisations of the LCM2000 parcel, one from the intersecting LCM1990 pixels, and the other from the spectral subclass ("PerPixList") attribution. Parcels that were hypothesised to have changed were those whose characterisations

Table 6.1 The Number of Parcels of Each Broad Habitat Land Cover Type Visited Against the Number Identified using the Semantic Statistical Method

	LCM2000			
Direction	**Broadleaved Woodland**	**Acid Grassland**	**Suburban and Rural Development**	**Total**
NE	—	3/20	—	3/20
SE	2/7	19/157	11/56	32/220
SW	5/7	5/13	3/3	13/23
NW	21/186	2/10	22/141	45/337
Total visited	28	29	36	93

Table 6.2 The Number of Parcels Identified by the Magnitude of their Vector Found to have Changed in the Field for Each Broad Habitat

	Change Observed			
LCM2000 Class	**No**	**Yes**	**Total**	**Proportion Correct**
Broadleaved Woodland	23	5	28	0·22
Acid Grassland	26	3	29	0·12
Suburban and Rural Development	25	11	36	0·44
Total	74	19	93	0·26

were most different. This process generated a vector between 1990 and 2000 and it was possible to analyse the results with reference to the direction and magnitude of the 1990 to 2000 vector. Placing unexpected scores on the x-axis and expected ones on the y, movements were grouped into four basic directions corresponding to quadrants of a compass: NE, SE, SW, and NW.

6.3.1 Analysis of Change Detection Results

The reliability of change detection was evaluated, by broad habitat, by vector direction, and by magnitude. Some examples of actual changes are also included by way of illustration.

6.3.1.1 Broad Habitat

A simple binary analysis of whether a field survey confirmed (or denied) that the parcels suspected of change had actually been subject to some change on the ground is shown in Table 6.2. From Table 6.2, 26% of the hypothesised changes were found to be actual change on the ground and predictions about the "Suburban and Rural Development" were the most reliable.

Table 6.3 The Number of Parcels Identified by the Magnitude of their Vector Found to have Changed in the Field for Each Direction

Direction	Change Observed No	Change Observed Yes	Total	Proportion Correct
NE	3	—	3	0·00
SE	28	4	32	0·14
SW	11	2	13	0·18
NW	32	13	45	0·41
Total	74	19	93	0·26

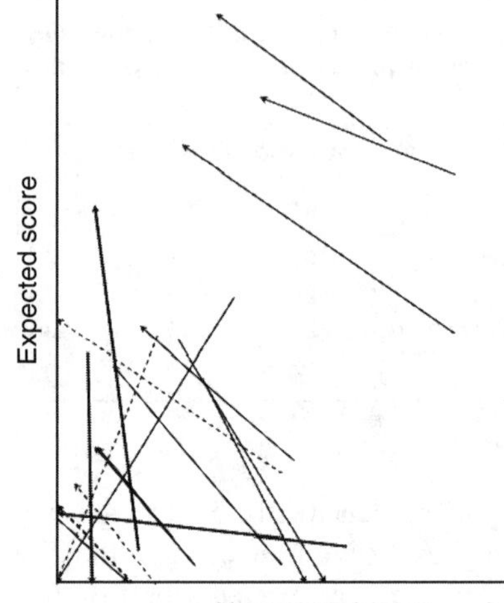

Figure 6.1 The distance and direction of the parcel vectors for which actual change was found on the ground. Vectors for Suburban and Rural Development are in black lines, Broadleaved Woodland are dashed and Acid Grassland in bold.

6.3.1.2 Vector Direction

Table 6.3 presents results by vector direction. It is apparent that most of the successfully identified changes had a vector direction NW, that is moving from (high U, low E) to (low U, high E). These are shown in Figure 6.1.

6.3.1.3 Comparing Change and "No-Change" Vectors

The magnitude and direction of the vector of those parcels found to have changed can be compared with the remainder of the 100 parcels, to see whether they are systematically different. These are shown in Figure 6.2 for Suburban/Rural Development. Changes to this class were the most reliably identified and parcels identified on the *number* of pixels as opposed to the *percentage* of pixels are shown for clarity.

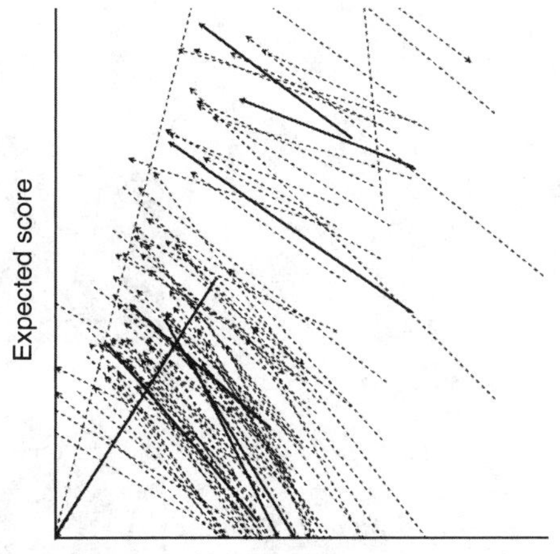

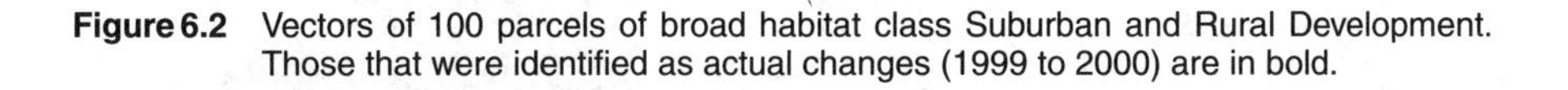

Figure 6.2 Vectors of 100 parcels of broad habitat class Suburban and Rural Development. Those that were identified as actual changes (1999 to 2000) are in bold.

6.3.1.4 Example Changes

Figure 6.3 shows six example parcels, with vectors that moved NW and that were found to have changed by way of illustration.

6.3.2 Analysis of False Positive Results

The second set of results is a breakdown of the reasons for the false positive results (i.e., those parcels identified as a potential change area which when visited were found not to have done so). Table 6.4 shows the origins of the large vector between the 1990 and 2000 characterisations for the visited "no change" parcels and the reasons why the false positives were identified as change parcels using this methodology:

1. 46% of the parcels that were found not to have changed either had a large number of LCM1990 pixels that were inconsistent with "no change" or the LCM2000 parcels were incorrectly classified.
2. 7% of the parcels had empty PerPixList fields.
3. 35% of the parcels had a PerPixList attribution that was inconsistent with the parcel classification.
4. 12% of the parcels were erroneously identified because of expert misunderstanding of the changed ontologies between 1990 and 2000 combined with inconsistent or empty PerPixList fields.

The impact of all of these artefacts was to create a large vector between 1990 and 2000 which resulted in the polygon being identified as a potential change area. From Table 6.4 it is evident that false positives for Acid Grassland, the class with the greatest change in ontology between 1990 and 2000, were all related to PerPixList (spectral attribution) anomalies. Artefacts for the other two classes are evenly spread.

LCM2000 class, 1990 to 2000 change process, *1990 classes*	**Geo-Context**	**Picture**
Broadleaved Woodland, Maturation, from *Grass Heath*	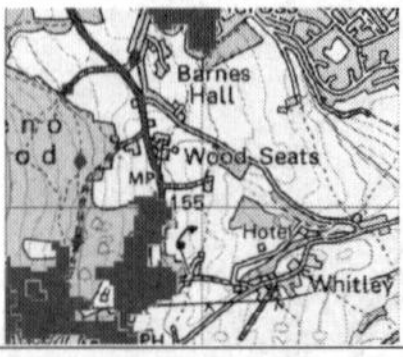	
Broadleaved Woodland, Maturation, from *Bracken* and *Moorland Grass*	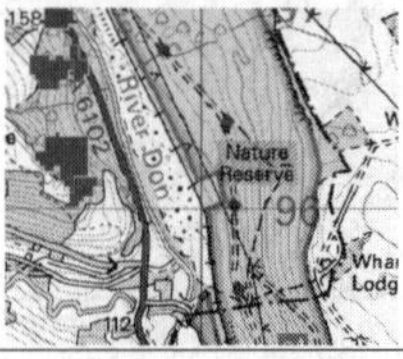	
Acid Grassland, degraded land quality by overstocking, from *Mown* and *Grazed Turf* and *Pasture*	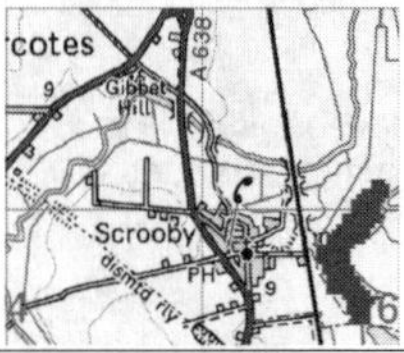	
Acid Grassland, changes in land quality, let go, degraded from *Tilled Land*		
Suburban and Rural Development, new housing development, from *Tilled Land*	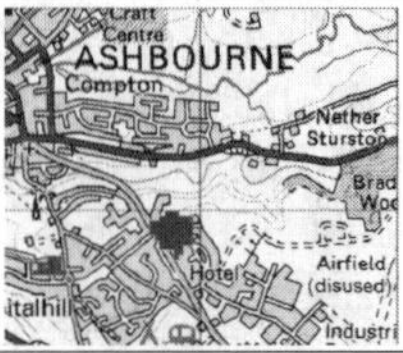	
Suburban and Rural Development, new housing development, from *Mown* and *Grazed Turf* and *Pasture*	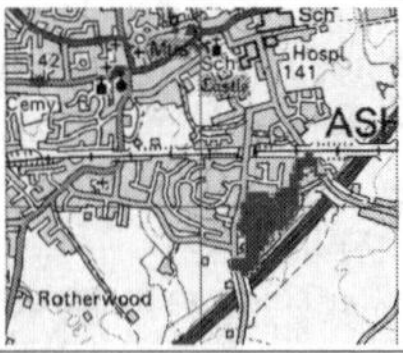	

Figure 6.3 **(see color insert following page 168)** Six example change parcels, a brief description of the nature of the change, the original (1990) class and some context from OS 1:50,000 Raster scanned maps (© Crown Copyright Ordnance Survey. An EDINA Digimap/JISC supplied service).

6.3.3 Explanation of Negative Results Using Vector Directions

The false positive results were then analysed by vector direction as shown in Table 6.5, to identify any further rules or filters.

Table 6.4 The Broad Habitat Classes and Origins of Large Vector Size for Parcels Suspected of Change but When Visited Found Not to Have Done So (i.e., The False Positive Results)

	LCM1990	LCM2000	Empty PPL	Incorrect PPL	Expert and PPL	Total
Broadleaved Woodland	5	12	4	2	—	23
Acid Grassland	—	2	—	15	9	26
Suburban and Rural Development	6	9	1	9	—	35
Total	11	23	5	26	9	74

Note: LCM1990: the LCM1990 pixels were not consistent with the actual land cover.
LCM2000: the LCM2000 parcel classification was not consistent with the actual land cover.
Empty PerPixList (PPL): empty spectral heterogeneity attribute field.
Incorrect PPL: spectral heterogeneity attributes not related to parcel class.
Expert and PPL: expert error where the expert failed to fully understand the change ontology *and* either an empty or incorrect spectral heterogeneity attribute.

Table 6.5 The Vector Directions and Origins of the Large Vector for the Parcels Suspected of Change but When Visited Found not to Have Done So (i.e., The False Positive results)

Direction	LCM1990	LCM2000	Empty PPL	Incorrect PPL	Expert and PPL	Total
NE	—	—	—	—	3	3 (**0**)
SE	—	2	—	23	3	28 (**4**)
SW	—	—	5	3	3	11 (**2**)
NW	11	21	—	—	—	32 (**13**)
Total	11	23	5	26	9	74

Note: True positive results in bold.

Analysis of the false positive results by direction presents a much clearer picture than analysis by land cover class (Table 6.3). In the sections below, the anticipated explanations for movement in each direction between 1990 and 2000 are described in the subsections below, followed by analysis of the field validation.

6.3.3.1 Movement NE: Increased E, Increased U

Parcels with a NE vector were anticipated to contain conflicting spectral attribute information. This may have been due to a changed ontologies or parcels with high spectral heterogeneity. All of the parcels identified in this category were Acid Grassland. Movement was due to a failure by the expert to understand the links with the LCM1990 class of Moorland Grass and due to major ontological change. The effect of these was a low starting point in 1990 (due to high uncertainty scores) and conflicting attribute information in 2000.

6.3.3.2 Movement SE: Decreased E, Increased U

Parcels with a SE vector were expected to contain spectral attribution inconsistent to the parcel class. Despite some change being found (4/32 parcels) these were believed to be by chance rather than design. All of the parcels with vectors moving in this direction had anomalous spectral attribution. Most of this was due to the use of single date imagery, rather than composite two date imagery, resulting in an attribute of "100% Water, variant c" for 20 out of the 32 parcels. Of the other 12:

i) seven were indications of genuine heterogeneity (one Acid Grassland parcel and six Suburban with "Urban" variants classes with no *expected* relation to Suburban);
ii) three were parcels classified incorrectly and therefore their attribution did not match the class (two Acid Grassland, one Suburban);
iii) two were due to the changed ontology, defining Acid Grassland in terms soil acidity.

Movement in this direction therefore indicates spectral attribute anomaly.

6.3.3.3 Movement SW: Decreased E, Decreased U

Movement in this direction was expected to indicate increased uncertainty about the parcel because the LCM2000 attributes would be contributing less to the Expected and Unexpected scores than the LCM1990 pixels. The parcels that were identified in this category can be placed into two groups: those with empty spectral attribution fields and those containing Spectral subclass not described in the information that accompanied the release of the LCM2000 data. Vectors moving in this direction indicated systematic artefacts in the parcel attributes.

6.3.3.4 Movement NW: Increased E, Decreased U

Parcels with vectors moving in this direction were thought to have increased certainty about parcel classification, relative to the way the parcel area was classified in 1990 and therefore indicative of land cover change. Parcels with vectors moving in this direction indicated either land cover change or misclassification in 1990 or 2000.

6.3.3.5 Summary

In summary the four directions of movement are attributable to the following:

NE — (spectral) attribution was inconsistent with the parcel land cover class and there was a failure by the expert to understand the relationships between the 1990 and 2000 ontologies;
SE — attribute inconsistency (heterogeneity);
SW — spectral heterogeneity attribute is empty *or* there are subclasses not described in the Spectral LUT;
NW — either there has been actual change on the ground *or* an error in one dataset.

6.4 DISCUSSION

The major finding of this work is the extent to which systematic analytical and data artefacts can be identified and used to generate filters for further analyses. Here, the data selected for analysis was not filtered in any way — all parcels were included if the vector was in the subset of the 100 largest vectors for their class, regardless of the origins of that vector. The analysis of results show that vector direction is a useful indicator of different artefactual trends in the data: specific anomalies or artefacts are indicated by specific vector directions. From these a set of filters or rules to eliminate parcels with specific characteristics from analysis can be constructed. The following artefacts with associated actions were identified for omission from future analyses:

- parcels with empty PPL fields;
- parcels with PPL attributes not described in the metadata accompanying the release of the data;
- parcels identified from Single Scene Imagery;
- parcels where the "knowledge-based" correction was inconsistent with the PPL attributes.

A further important finding was the extent to which the field visits revealed inconsistencies in expert understanding of how the LCM2000 class of Acid Grassland related to various 1990 classes. The results show that the semantic LUT was not consistent with observed phenomena in the field for a specific class. The expert used in this study did not fully understand the impact of the changed ontology of Acid Grassland. This was confirmed when the results were taken back to the expert and the expert's understanding of the data concepts was revisited. The expert was more familiar with lowland vegetation and ecosystems and thus when focusing on Acid Grassland it may be preferable to use another expert.

Future work will be directed in a number of areas. First, exploration of how different expressions of expert opinion may be used. Here, a single expert was used under the scenario of their *idealised mapping of how the semantics relate* — a semantic LUT. Other experts and scenarios for mapping relations between the classification schemes are available. For instance, a "Change LUT" of the expected transitions between land cover classes and a "Technical LUT" of how different land cover class concepts relate based on heuristic knowledge of where spectral confusion may occur. Evidence from these might support or be contrary to the evidence from the semantic LUT, and in turn may allow stronger or weaker inferences to be made about change for specific parcels. Second, multiple pieces of evidence from different experts under different scenarios would be suitable for combination using uncertainty formalisms such as Dempster-Shafer or Rough Sets. Third, as yet current and potential users of LCM2000 have little information about the metadata might be used for specific applications. One of the motivations for this work is to develop a "cook book" that would help users link their application specific understandings of land cover concepts with those of LCM2000 through the LCM2000 metadata.

Rules will be developed from these results and further analyses of other field data to filter data and attribute artefacts, and to identify some of the analytical limits of the data and metadata. There is an issue about the size of the parcel and the use of parcel vectors generated using percentage or area data. It may be that parcels larger than a certain area have to be eliminated from analyses, and that above a particular threshold (to be determined) vectors based on percentage change need to be applied to assess large, noisy parcels. Similarly for small parcels, there may be a minimum size below which nothing useful can be said about the probability of change.

6.5 CONCLUSIONS

The problem of detecting changes between thematic (land cover) maps was described by Fuller et al. (2003) with reference to the accuracy of LCM1990 and LCM2000. Because of the differences between the conceptualisations of the two datasets, the release of LCM2000 was accompanied with a "health warning" against comparing it to LCM1990 (Fuller et al., 2002). Fuller et al. (2003) indicated a possible way forward for detecting change between LCM1990 and LCM2000 would be to utilise the parcel structure of LCM2000 to interrogate the LCM1990 raster data, thereby providing local descriptions of LCM2000 distributions. This work has shown that change analysis is possible using the semantic statistical approach developed by Comber et al. (2003b) applied at the level of the parcel. Change areas were identified by characterising the LCM2000 parcel in terms of its spectral attribution and the intersection of LCM1990 data, and then comparing the two characterisations. Typically, change was by an increased expected score and decreased unexpected score between 1990 and 2000.

In the process of determining the extent to which change is identified reliably, certain data artefacts have also been revealed:

- Parcels that were anomalous in terms of their attribution were identified. From these rules were established that will eliminate data with specific characteristics from future analyses according to the direction of their vectors.
- Failings in the expert understanding of data concepts were revealed, which may lead to the use of different experts or revision of expert understandings.

At the time of writing there is little guidance about how the extensive LCM2000 metadata may be used and analysed. Examining the results of a field validation confirms that the semantic statistical approach provides a greater understanding of how the ontologies of LCM1990 and LCM2000 relate to each other than is provided by data documentation, specification, and class descriptions alone. This gives greater insight into the *meaning* of the LCM2000 data structures with reference to their attribution. The situation of dataset difference is endemic in natural resource inventory making it difficult to relate the information contained in one survey to that of another. Metadata commonly describes easily measurable features such as scale and accuracy when compared with some reference data. It does not describe the fundamental meaning being the data elements, nor does it report at the object level. It is hoped that the work reported here contributes to a developing area involving the analysis and use of metadata to evaluate and revise the base information. LCM2000 has allowed the identification of these issues, but they exist in many other analyses, often where there is not metadata with which to work.

ACKNOWLEDGMENTS

This chapter would not have been possible without the assistance of those involved in creating LCM2000, especially Geoff Smith. This chapter describes work done within the REVIGIS project funded by the European Commission, Project Number IST-1999-14189. We wish to thank our partners in the project, especially Andrew Frank, Robert Jeansoulin, Alfred Stein, Nic Wilson, Mike Worboys, and Barry Wyatt.

REFERENCES

Ahlqvist, O., Keukelaar, J., and Oukbir, K., 2000. Rough classification and accuracy assessment. *International Journal of Geographical Information Science*, **14**, 475–496.

Bennett, B., 2001. What is a forest? On the vagueness of certain geographic concepts. *Topoi*, **20**, 189–201.

Bishr, Y., 1998. Overcoming the semantic and other barriers to GIS interoperability. *International Journal of Geographical Information Science*, **12**, 299–314.

Brown, K., 2004. Chapter 8, this volume.

Comber, A.J., Fisher, P.F., and Wadsworth, R.A., 2002. Creating spatial information: commissioning the UK Land Cover Map 2000. In *Advances in Spatial Data*, edited by Dianne Richardson and Peter van Oosterom (Berlin: Springer-Verlag), pp. 351–362.

Comber, A.J., Fisher, P.F., and Wadsworth, R.A., 2003a. Actor network theory: a suitable framework to understand how land cover mapping projects develop? *Land Use Policy*, **20**, 299–309.

Comber, A.J., Fisher, P.F., and Wadsworth, R.A., 2003b. A semantic statistical approach for identifying change from ontologically divers land cover data. In *AGILE 2003, 5th AGILE Conference on Geographic Information Science*, edited by Michael Gould, Robert Laurini, and Stephane Coulondre, (24th–26th April, PPUR, Lausanne), pp. 123–131.

Devogele, T., Parent, C., and Spaccapietra, S., 1998. On spatial database integration. *International Journal of Geographical Information Science*, **12**, 335–352.

Frank, A.U., 2001. Tiers of ontology and consistency constraints in geographical information systems. *International Journal of Geographical Information Science*, **15**, 667–678.

Fuller, R.M., Groom, G.B., and Jones, A.R., 1994. The land cover map of Great Britain: an automated classification of Landsat Thematic Mapper data. *Photogrammetric Engineering and Remote Sensing*, **60**, 553–562.

Fuller, R.M., Smith, G.M., Sanderson, J.M., Hill, R.A., and Thomson, A.G., 2002. Land Cover Map 2000: construction of a parcel-based vector map from satellite images. *Cartographic Journal*, **39**, 15–25.

Fuller, R.M., Smith, G.M., and Devereux, B.J., 2003. The characterisation and measurement of land cover change through remote sensing: problems in operational applications? *International Journal of Applied Earth Observation and Geoinformation*, **4**, 243–253.

Guarino, N., 1995. Formal ontology, conceptual analysis and knowledge representation. *International Journal of Human-Computer Studies*, **43**, 625–640.

Kavouras, M. and Kokla, M., 2002. A method for the formalization and integration of geographical categorizations. *International Journal of Geographical Information Science*, **16**, 439–453.

Lucieer, A., Fisher, P., and Stein, A., 2004. Chapter 7, this volume.

Pundt, H. and Bishr, Y., 2002. Domain ontologies for data sharing — an example from environmental monitoring using field GIS. *Computers and Geosciences*, **28**, 95–102.

Shalan, M.A., Arora, M.K., and Elgy, J., 2004. Chapter 2, this volume.

Smith, G.M. and Fuller, R.M., 2002. Land Cover Map 2000 and meta-data at the land parcel level. In *Uncertainty in Remote Sensing and GIS*, edited by G.M. Foody and P.M. Atkinson (London: John Wiley and Sons), pp. 143–153.

Visser, U., Stuckenschmidt, H., Schuster, G., and Vogele, T., 2002. Ontologies for geographic information processing. *Computers and Geosciences*, **28**, 103–117.

CHAPTER 7

Texture-Based Segmentation of Remotely Sensed Imagery to Identify Fuzzy Coastal Objects

Arko Lucieer, Peter F. Fisher, and Alfred Stein

CONTENTS

7.1 INTRODUCTION

Recent research on remote sensing classification has focused on modelling and analysis of classification uncertainty. Both fuzzy and probabilistic approaches have been applied (Foody, 1996; Hootsmans, 1996; Canters, 1997; Fisher, 1999; Zhang and Foody, 2001). Much of this research, however, focused on uncertainty of spectral classification on a pixel-by-pixel basis, ignoring potentially useful spatial information

0-8493-2837-3/05/$0.00+$1.50

between pixels. An object-based approach instead of a pixel-based approach may be helpful in reducing classification uncertainty. Additionally, interpretation of uncertainty of real world objects may be more intuitive than interpretation of uncertainty of individual pixels. In this study, uncertainty arises from vagueness, which is characteristic for those geographical objects that are difficult to define, both thematically and in their spatial extent.

Object-oriented approaches to remotely sensed image processing have become popular with the growing amount of fine spatial resolution satellite and airborne sensor imagery. Several studies have shown that segmentation techniques can help to extract spatial objects from an image (Gorte and Stein, 1998). Segmentation differs from classification, as spatial contiguity is an explicit goal of segmentation, whereas it is only implicit in classification. Uncertainty, however, occurs in any segmented or classified image and can affect further image processing. In particular, in areas where objects with indeterminate boundaries (so-called fuzzy objects) dominate an indication of object vagueness is important.

A straightforward approach to identify fuzzy objects is to apply a (supervised) fuzzy c-means classification, or similar soft classifier (Bezdek, 1981; Foody, 1996). This classifier gives the class with the highest membership for each pixel, and possibility values of belonging to any other class. However, it does not consider spatial information, also known as pattern or texture.

Texture analysis has been addressed and usefully applied in remote sensing studies in the past. An interesting overview paper concerning texture measures is Randen and Husøy (1999). Bouman and Liu (1991) studied multiple resolution segmentation of texture images. A Markov random field (MRF) model-based segmentation approach to classification for multispectral images was carried out by Sarkar et al. (2002). For multispectral scene segmentation and anomaly detection, Hazel (2000) applied a multivariate Gaussian MRF. Recently, Ojala and his co-workers have further pursued an efficient implementation and application towards multi-scale texture-based segmentation (Ojala et al. 1996, 2002; Ojala and Pietikäinen, 1999; Pietikäinen et al., 2000). Their Local Binary Pattern (LBP) measure may be more appropriate than the traditional texture measures used in classification (Ojala et al., 1996). The LBP is a rotation invariant grey scale texture measure.

In identifying spatial objects from remotely sensed imagery, the use of texture is important. Texture reflects the spatial structure of both elevation and spectral data and is therefore valuable in classifying an area into sensible geographical units. The aim of this study is to present a supervised texture-based image segmentation technique that identifies objects from fine spatial resolution elevation (LiDAR) and multispectral airborne imagery (CASI). It is applied to a coastal area in northwest England. Information on coastal land cover and landform units is required for management of this conservation area. Since this environment is highly dynamic, (semi-) automatic and objective techniques are required to update information and maps. This chapter builds on the work of Lucieer and Stein (2002) and further explores the use of texture and the generation of an uncertainty measure to depict object vagueness in image segmentation to help the accurate identification of fuzzy objects.

7.2 STUDY AREA

The study area was the Ainsdale Sands, on the coast of northwest England. The Ainsdale Sand Dunes National Nature Reserve (NNR) totals 508 ha and forms part of the Sefton Coast. The NNR is within the coastal Special Protection Area. It is also within the Sefton Coast candidate Special Area of Conservation. The NNR contains a range of habitats, including intertidal sand flats, embryo dunes, mobile yellow dunes, fixed vegetated dunes, wet dune slacks, areas of deciduous scrub, and a predominantly pine woodland. Management of this area consists of extending the area of open dune habitat through the removal of pine plantation from the seaward edge of the NNR, maintaining and extending the area of fixed open dune by grazing and progressively creating a more diverse structure within the remaining pine plantation with associated benefits for wildlife (Ainsdale Sand Dunes NNR, 2003). Therefore, mapping of this coastal area can be useful for protection and management of the environment as a major and threatened habitat type and as a defence against coastal flooding.

In 1999, 2000, and 2001, the Environment Agency, U.K., collected fine spatial resolution digital surface models (DSM) by LiDAR, and simultaneously, acquired multispectral Compact Airborne Spectral Imager (CASI) imagery (one flight each year). The aircraft was positioned and navigated using a Global Positioning System (GPS) corrected to known ground reference points. The aircraft flew at approximately 800 m above ground level, acquiring data with a 2 m spatial resolution from the LiDAR and 1 m with the CASI. In this study, the imagery of 2001 was used. These data were also analysed in Brown (Chapter 8, this volume). These images, geometrically corrected by the Environment Agency, were spatial composites of multiple flight strips. The area covered by these images was approximately 6 km^2 (Figure 7.1).

A relevant distinction exists between land cover and landform, both characterising coastal objects. Landform properties can be extracted from digital elevation data,

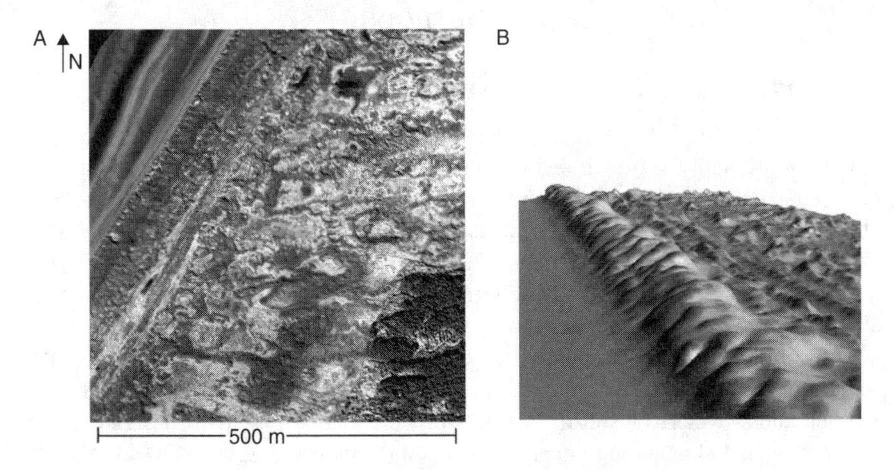

Figure 7.1 (A) Band 12 (NIR, 780 nm) of the CASI image of the Ainsdale Sands. (B) Three-dimensional view of the LiDAR DSM of the same area with the foredune clearly visible.

the LiDAR DSM of the area. Four landform classes were distinguished: beach plain, dune, dune slack, and woodland. Land cover was obtained from spectral information extracted from the CASI imagery. Four land cover classes can be distinguished: sand, marram grass, willow shrub, and woodland. Detailed mapping of these classes is required, because knowledge about the location and dynamics of these object types is important for monitoring the rare habitats in this area which also act as a defence against flooding.

Coastal objects are often characterised by fuzzy boundaries. Vagueness is the main source of uncertainty in this area as these fuzzy boundaries dominate. Therefore, information on vagueness is required to identify transition zones and to map these objects. Cheng and Molenaar (2001) proposed a fuzzy analysis of dynamic coastal landforms. They classified the beach, foreshore, and dune area as fuzzy objects based on elevation data using a semantic import model. Some classification errors, however, are likely to occur when using elevation as diagnostic information alone. For example, an area of low elevation behind the foredune was classified as beach, whereas it is almost certainly an area of sand removal by wind like a blowout or an interdune area. These types of errors can be reduced by using spatial or contextual information (e.g., by considering at morphometry or landforms). Cheng et al. (2002) and Fisher et al. (2004) proposed a multi-scale analysis for allocating fuzzy memberships to morphometric classes. This technique can be used to model objects, which are vague for scale reasons. The morphometry classes modelled at different scale levels were: channel, pass, peak, pit, plane, and ridge. Although this analysis fails to identify positions of dunes, it is possible to identify dune ridges and slacks and to monitor their changing positions. The use of textural information might aid the identification of these coastal objects.

7.3 METHODS

7.3.1 Texture

Regions with similar reflectance can be identified easily as objects in a remote sensing image. Additionally, texture is an important property of geospatial objects and should be taken into account in image analysis. In this study, texture is defined as a pattern or characteristic spatial variability of pixels over a region. The pattern may be repeated exactly, or as a set of small variations, possibly as a function of position. There is also a random aspect to texture, because size, shape, colour, and orientation of pattern elements can vary over the region.

Measures to quantify texture can be split into structural (transform-based), statistical and combination approaches. Well-known structural approaches are the Fourier and wavelet transform. Several measures can be used to describe these transforms, including entropy, energy, and inertia (Nixon and Aguado, 2002). A well-known statistical approach is the grey level co-occurrence matrix (GLCM) (Haralick et al., 1973) containing elements that are counts of the number of pixel pairs for specific brightness levels. Other texture descriptors are Gaussian Markov random fields (GMRF), Gabor filter, fractals and wavelet models. A comparative study of texture classification is given in Randen and Husøy (1999). They conclude that a direction for future research is the development of powerful texture measures that can be extracted and classified with low computational complexity. A relatively new and simple texture measure is the local binary pattern operator (LBP) (Pietikäinen et al., 2000; Ojala et al., 2002). The LBP is a theoretically simple yet efficient approach to grey scale and rotation invariant texture classification based on local binary patterns.

7.3.2 The Local Binary Pattern Operator

Ojala et al. (2002) derived the LBP by defining texture T in a local neighbourhood of a grey scale image as a function t on the grey levels of $P(P > 1)$ image pixels

$$T = t(g_c, g_0, \ldots, g_{P-1}) = t(g_c, \vec{g}_P) \tag{7.1}$$

where g_c corresponds to the value of the centre pixel (P_c) and $\vec{g}_P = (g_1, \ldots, g_P)'$ presents the values of pixels in its neighbourhood. The neighbourhood is defined by a circle of radius R with P equally spaced pixels that form a circularly symmetric neighbourhood set (Figure 7.2). The coordinates of the neighbouring pixels in a circular neighbourhood are given by

$$\{x_{c,i}, y_{c,i}\} = \left\{x_c - R\sin\left(\frac{2\pi i}{P}\right), y_c + R\cos\left(\frac{2\pi i}{P}\right)\right\} \quad \text{for } i = 0, \ldots, P-1 \tag{7.2}$$

Invariance with respect to the scaling of pixel values or illumination differences is achieved by considering the signs of the differences instead of their numerical values

$$T^* = t(\text{sign}(g_0 - g_c), \text{sign}(\vec{g}_P - g_c)) \tag{7.3}$$

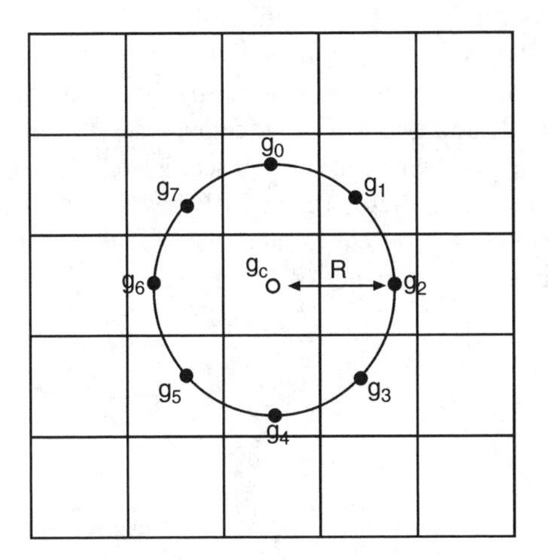

Figure 7.2 Circular pixel neighbourhood set for $P = 8$ and $R = 1$.

This results in the following operator for grey scale and rotation invariant texture description

$$\mathrm{LBP}_c = \sum_{i=0}^{P-1} \mathrm{sign}(g_i - g_c) \tag{7.4}$$

Ojala *et al.* (2002) found that not all local binary patterns are sufficiently descriptive. They introduce a uniformity measure U to define uniformity in patterns, corresponding to the number of spatial transitions or bitwise 0/1 changes in the pattern. With $g_P = g_0$, U_c is defined as

$$U_c = \sum_{i=1}^{P} |\mathrm{sign}(g_i - g_c) - \mathrm{sign}(g_{i-1} - g_c)| \tag{7.5}$$

Patterns with $U_c \leq j$ are designated as uniform. Ojala et al. (2002) found that for $j = 2$ the most suitable LBP measure was obtained for describing texture. This results in the following operator for grey scale and rotation invariant texture description

$$\mathrm{LBP}_{c,j} = \begin{cases} \sum_{i=0}^{P-1} \mathrm{sign}(g_i - g_c) & \text{if } U_c \leq j \\ P+1 & \text{otherwise} \end{cases} \tag{7.6}$$

The LBP operator thresholds the pixels in a circular neighbourhood of P equally spaced pixels on a circle of radius R, at the value of the centre pixel. It allows for the detection of uniform patterns for any quantisation of the angular space and for any spatial resolution. Non-uniform patterns are grouped under one label, $P + 1$.

7.3.3 Texture-Based Image Classification

Most approaches to supervised texture classification or segmentation assume that reference samples and unlabelled samples are identical with respect to texture scale, orientation, and grey scale properties. This is often not the case, however, as real world textures can occur at arbitrary spatial resolutions, rotations, and illumination conditions. The LBP operator is very robust in terms of grey scale variations, since the operator is by definition invariant against any monotonic transformation, and it is rotation invariant. The operator is an excellent measure of the spatial structure of local image texture, but by definition, it discards the other important property of local image texture, contrast. Therefore, the LBP measure can be further enhanced by combining it with a rotation invariant variance measure that characterises the contrast of local image texture. Local variance is defined as

$$\mathrm{VAR}_c = \frac{1}{P} \sum_{i=0}^{P-1} (g_i - \hat{\mu}_g)^2 \tag{7.7}$$

where

$$\hat{\mu}_g = \frac{1}{P} \sum_{i=0}^{P-1} g_i \tag{7.8}$$

Most approaches to texture analysis quantify texture measures by single values (e.g., mean, variance, entropy). However, much important information contained in the distributions of feature values might be lost. In this study, the final texture feature is the histogram of $LBP_{c,j}$ occurrence, computed over an image or a region of an image, or the joint distribution of the two complementary $LBP_{c,j}$ and VAR_c operators. The joint distribution of ($LBP_{c,j}$, VAR_c) is approximated by a discrete two-dimensional histogram of size $P + 2 \times b$, where P is the number of neighbours in a circular neighbourhood and b is the number of bins for VAR_c. Ojala et al. (2002) show that this is a powerful tool for rotation invariant texture classification. The number of bins used in the quantisation of the feature space plays a crucial role in the analysis. Histograms with too modest a number of bins fail to provide enough discriminative information about the distributions. If the number of entries per bin is very small (i.e., too many bins), however, histograms become sparse and unstable. In this study, following Ojala et al. (1996), the feature space was quantised by computing the total feature distribution of ($LBP_{c,j}$, VAR_c) for the whole image. This distribution was divided into 32 bins having an equal number of entries.

In texture classification, the (dis)similarity of sample and model histograms as a test of goodness-of-fit is evaluated using a nonparametric statistic, the log-likelihood ratio statistic, also known as the G-statistic (Sokal and Rohlf, 1987). Here, the sample is a histogram of the texture measure distribution of an image window. The model is a histogram of a reference image window of a particular class. By using a nonparametric test, no assumptions about the feature distributions have to be made. The value of the G-statistic indicates the probability that two sample distributions come from the same population: the higher the value, the lower the probability that the two samples are from the same population. The more alike the histograms are, the smaller is the value of G.

It should be noted that the window size should be appropriate for the computation of the texture features. However, as windows of increased size are considered, the probability that regions contain a mixture of textures is increased. This can bias the comparison, since the reference textures contain only features of individual patterns. On the other hand, if the window size is too small it is impossible to calculate a texture measure. Within this constraint, it is impossible to define an optimum size for segmenting the entire image, therefore, classifying regions of a fixed window size is inappropriate (Aguado et al., 1998). Alternatively, a top–down hierarchical segmentation process, as discussed in the next section, offers a suitable framework for classifying image regions based on texture.

7.3.4 Texture-Based Image Segmentation

Split-and-merge segmentation consists of a region-splitting phase and an agglomerative clustering (merging) phase (Horowitz and Pavlidis, 1976; Haralick and Shapiro, 1985; Lucieer and Stein, 2002). Objects derived with unsupervised segmentation have no class labels. Class labels can be assigned in a separate labelling or classification stage. In the unsupervised approach of Lucieer and Stein (2002), the image was initially considered as a block of pixel values with mean vector and covariance matrix. This block was split into four sub-blocks characterised by vectors of mean pixel values and covariance matrices. To define homogeneity, they considered a threshold for the mean and thresholds for the covariance matrix. These values were chosen in advance and kept constant during segmentation. Heterogeneous sub-blocks were split recursively until homogeneity or a minimum block size was reached. The resulting data structure was a regular quadtree. In the clustering phase, adjacent block segments were merged if the combined object was homogeneous. The homogeneity rules were applied in a similar way. However, texture was not taken into account in this approach. Recently, Ojala and Pietikäinen (1999) applied a similar unsupervised split-and-merge segmentation with splitting and merging criteria based upon the ($LBP_{c,j}$, VAR_c) texture measure.

Supervised segmentation uses explicit knowledge about the study area to train the segmentation algorithm on reference texture classes. In a supervised approach, segmentation and classification are combined and objects with class labels are obtained. Aguado et al. (1998) introduced a segmentation framework with a top–down hierarchical splitting process based on minimising uncertainty. In this study, the ($LBP_{c,j}$, VAR_c) texture measure and the segmentation/classification framework as suggested by Aguado et al. (1998) were combined. Similar to split-and-merge segmentation each square image block in the image was split into four sub-blocks forming a quadtree structure. The criterion used to determine if an image block is divided was based on a comparison between the uncertainty of the block and the uncertainty of the sub-blocks.

To obtain a segmentation, the image was divided such that classification confidence was maximised, and hence uncertainty was minimised. Uncertainty was defined as the ratio between the similarity values for the two most likely reference textures. This measure is similar to the confusion index (CI) described by Burrough (1998). Reference textures were described by histograms of ($LBP_{c,j}$, VAR_c) of characteristic regions in the image. To test for similarity between an image block texture and a reference texture the G-statistic was applied. In this study, the CI (depicting vagueness) was defined as

$$\mathrm{CI} = \frac{1 - G_2}{1 - G_1} \tag{7.9}$$

where G_1 was the lowest G value of all classes (highest similarity) and G_2 was the second lowest G value. CI contained values between 0.0 and 1.0. CI was close to 1.0, if G_1 and G_2 were similar. In this case, the decision of classifying the region was vague. The uncertainty in classification decreased, if the difference between these two texture similarities increased. The subdivision of each image block was based on this uncertainty criterion. An image block was split into four sub-blocks, if

$$\mathrm{CI_B} > \tfrac{1}{4}(\mathrm{CI_{SB1}} + \mathrm{CI_{SB2}} + \mathrm{CI_{SB3}} + \mathrm{CI_{SB4}}) \tag{7.10}$$

where the left side of Equation (7.10) defined vagueness when the sub-blocks were classified according to the class obtained by considering the whole block (B). The right side of Equation (7.10) defined vagueness obtained if the sub-blocks (SB1, SB2, SB3, and SB4) were classified into the classes obtained by the subdivision. Thus, the basic idea was to subdivide an image block only if it was composed of several textures.

Additionally, classification was always uncertain at the boundaries of textures because the image block contained a mixture of textures. Accordingly, blocks that had at least one neighbouring region of a different class were subdivided until a minimum block size was reached (Aguado et al., 1998). Finally, a partition of the image with objects labelled with reference texture class labels was obtained.

The building blocks of each of the objects gave information about object uncertainty. The measure CI_B was used to depict the vagueness with which an object sub-block was assigned a class label. It provided information about the thematic uncertainty of the building blocks. The spatial distribution of building block uncertainty within an object provided information about spatial uncertainty. Therefore, high uncertainty values were expected in object boundary blocks, caused by mixed textures or transition zones.

7.3.5 Example

To illustrate the problem of classifying regions of different texture an image (512 by 512 pixels) with a composition of photographs of five different textures was used (Figure 7.3A). Each of the classes was unique in terms of its texture. Figure 7.3 shows that the human visual system not only can distinguish image regions based on grey scale or colour, but also on pattern. Five classes could be distinguished in Figure 7.3A, labelled class NW (granite), class NE (fabric), class SW (grass), class SE (stone), and class Centre (reed mat). A pixel-based classifier does not take into account texture or spatial information. This is shown in Figure 7.3B, which gives the result of a (pixel based) supervised fuzzy c-means classifier applied using a Mahalanobis distance measure and a fuzziness value of 2.0 (Bezdek, 1981). Five regions of 40 by 40 pixels were selected in the centre of the five regions to train

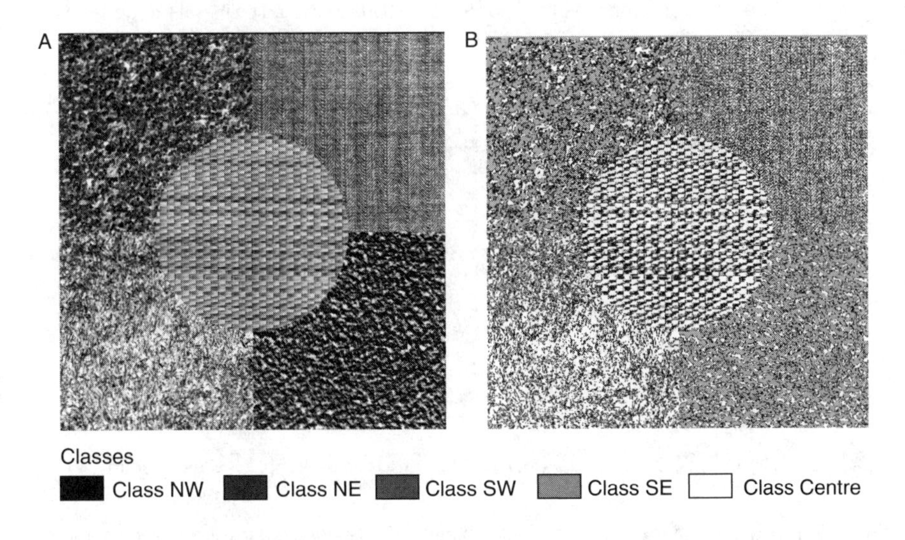

Figure 7.3 Textures image and results (A) Artificial composition of five different natural textures. (B) Result of a pixel based supervised fuzzy c-means classifier.

the classifier. Figure 7.3B shows that, although the patterns are still visible, no clear spatial partition of classes was found.

Figure 7.4 gives the results of two segmentations of Figure 7.3A. Figure 7.4A shows that a split-and-merge segmentation without texture characterisation cannot identify regions of homogeneous texture. It should be noted that this approach was unsupervised (Lucieer and Stein, 2002). Random grey values were used to depict different objects. Figure 7.4B shows a segmentation result from an unsupervised split-and-merge segmentation algorithm with the $LBP_{c,j}$, VAR_c histograms to model texture (Ojala and Pietikäinen, 1999). The spatial partition derived corresponded closely to the five different texture classes contained in the texture composite.

Figure 7.5 gives the results of a supervised texture-based segmentation of Figure 7.3A. The uncertainty criteria proposed by Aguado et al. (1998) were

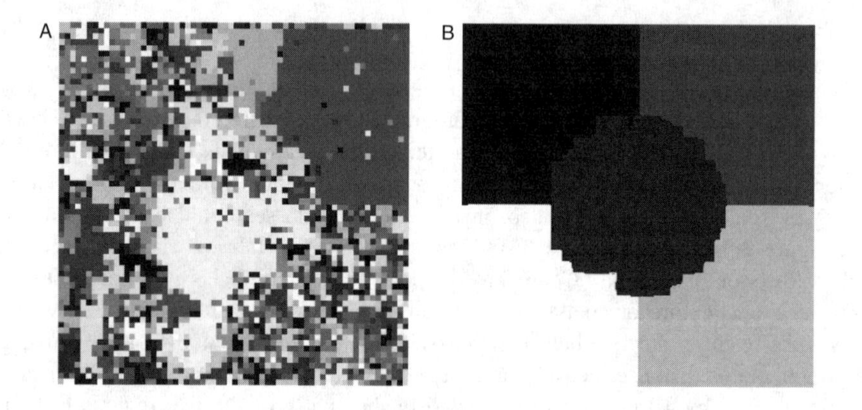

Figure 7.4 Segmentation results (A) Unsupervised split-and-merge segmentation of Figure 7.3A based on mean and variance. (B) Unsupervised split-and-merge segmentation based on texture distributions.

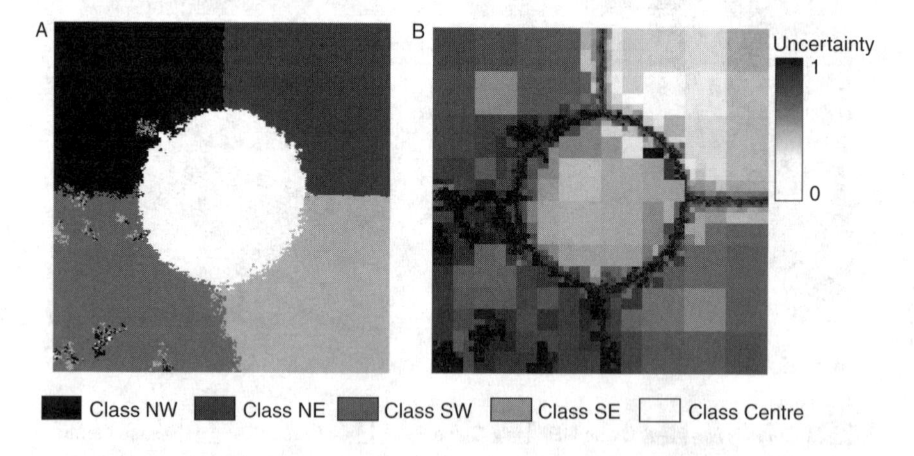

Figure 7.5 Supervised texture-based segmentations (A) Supervised texture-based segmentation of Figure 7.3A with five reference classes. (B) Related uncertainty for all object building blocks.

applied to obtain this result. Five reference regions were selected in the image, corresponding to the five different texture classes (similar to the supervised fuzzy c-means classification). Values for P and R were 8 and 1, respectively (corresponding to the eight adjacent neighbours). Figure 7.5A shows the segmented objects with their corresponding class label. In Figure 7.5B uncertainty values (CI) for each of the objects' building blocks are given. Class NE was classified with lowest uncertainty values, between 0.3 and 0.4. The centre class was classified with uncertainty values between 0.4 and 0.5. Class SE was classified correctly, but with higher uncertainty values, between 0.5 and 0.7. Confusion of this class occurred with class SW. Class NW was classified correctly, but with high uncertainty values between 0.5 and 0.75. In class NW a cluster of small objects was classified as class SW. The building blocks of these objects showed uncertainty values of 0.95 and higher, meaning that the classification confusion in these areas was very high. Confusion of this class occurred with class SE. The main area of Class SW was classified correctly. In this class, small objects were classified as class NW, SE, and Centre, however, block uncertainty values were higher than 0.94 for these objects. This type of texture, however, was very irregular (i.e., its pattern was not repetitive and the reference area did not fully represent the whole texture area). In addition, all small blocks at the boundaries of textures showed high confusion values (>0.9), because they contained mixtures of different textures.

7.4 RESULTS

7.4.1 Segmentation of the LiDAR DSM

Figure 7.6 shows the result of a supervised segmentation of a 512 by 512 pixel subset of the LiDAR DSM of the study area (Figure 7.1B). Four reference areas of

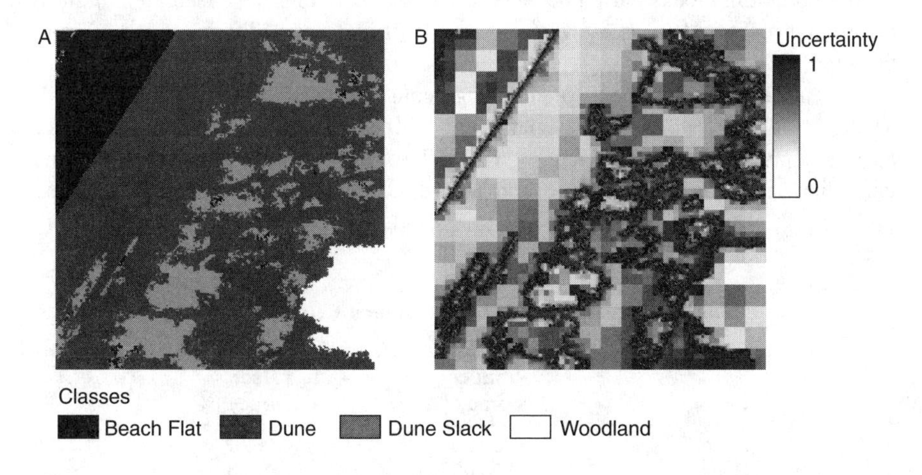

Figure 7.6 Supervised segmentation outputs (A) Supervised texture-based segmentation of the LiDAR DSM with four reference landform classes. (B) Related uncertainty for all object building blocks.

50 by 50 pixels were selected for training. These areas represented the following landform classes: Beach, dune, dune slack, and woodland. Values for P and R were 8 and 1, respectively. Figure 7.6A shows the segmented objects with class labels and Figure 7.6B shows the corresponding uncertainty values (CI). Woodland was used as one of the landform classes, as it showed a characteristic texture in the LiDAR DSM. The woodland area was classified correctly, with low uncertainty values ranging from 0.02 to 0.35. Uncertainty values increased at the border of the woodland area. Fieldwork showed that zones of willow trees occurred at the border of the main pine woodland area, which explained the higher uncertainty because of their slightly different texture. Dune slacks and blowouts are very similar in form. Blowouts are active, however, and not vegetated. Dune slacks are often stable, because they are vegetated. These texture differences could not be observed in the LiDAR DSM. Therefore, these units were classified as a single class type, called dune slacks. The core of these areas was classified correctly, with uncertainty values between 0.2 and 0.5. The boundaries of these objects, however, showed uncertainty values of 0.8 and higher. These high values can be explained by the transition zones from dune slacks to dune. No crisp boundary could be observed between these object types. Furthermore, Figure 7.6 shows that no distinction could be made between the foredune and the inland dune field. These areas have similar textures and, therefore, were classified as one class. The (steep) foredune showed, as expected, a short transition zone to the beach, depicted by high uncertainty values (>0.8), depicted by a thin black line in the upper left corner of Figure 7.6B. The dune area was classified with low uncertainty values (<0.4), except for the transition zones with the dune slacks. In the southwest and centre part of the image, small objects (with uncertainty values of 0.95) were incorrectly classified as beach. This can be explained by observations in the field showing that this area is an active flat and bare sand area, with similar texture to the beach. The beach flat was classified early in the segmentation process, as can be concluded from the large building blocks. Uncertainty related to the classification of these building blocks varied between 0.1 and 0.5. Within the beach area, highest uncertainty occurred in areas where sand is wet and has a different texture from dry sand.

An accuracy assessment of the segmentation results provided an overall accuracy of 86% and a Kappa coefficient of 0.81. Per class producer and user accuracy percentages are given in Table 7.1. It can be concluded from this table that small areas of both beach and dune slack were incorrectly labelled as dune.

Table 7.1 Producer and User Accuracy Percentages for the Landform Classes

Class	Producer Accuracy (%)	User Accuracy (%)
Beach	50.00	96.87
Dune	100.00	97.51
Dune slack	95.20	100.00
Woodland	100.00	100.00

7.4.2 Segmentation of the CASI Image

In Figure 7.7, the results of the segmentation of the CASI image (Figure 7.1A) are shown. The image was resampled to a spatial resolution of 2 m to match the spatial resolution of the LiDAR DSM. Again, a subset of 512 by 512 pixels was used for segmentation. Band 12 at 780 nm (NIR) was chosen for this study since it is suitable for discrimination of land cover types. Figure 7.7A shows the segmentation result for four land cover types: Sand, marram grass, willow shrub, woodland. Four reference areas of 50 by 50 pixels were selected to train the algorithm. Values for P and R are 8 and 1, respectively. The woodland area in the southeast corner of the image was correctly classified with uncertainty values between 0.1 and 0.5 (Figure 7.7B). The northeastern corner of the image and several small objects in the northern part of the image were also segmented as woodland. However, fieldwork showed that no woodland occurred in this area. The area was characterised by a chaotic pattern of dune slacks and dune ridges with a mixture of vegetation types. No homogeneous textures could be found, therefore this area was characterised by high uncertainty values (>0.7) in the segmentation result. The main part of the dune field was classified as willow shrub land. Fieldwork showed that marram grass was mainly found on the foredune and on the highest parts of the dune ridges in the dune field. Only a few small patches of marram grass occur in Figure 7.7A in the foredune area. Willow shrub was found all over the dune field, but mainly in the dune slacks. Image texture for these two classes, however, was very similar. Marram grass fields were characterised by a mixture of grass and sand; willow shrub areas were characterised by a mixture of baby willow shrubs and sand or low grass. High uncertainty values (higher than 0.7 in the dune field and higher than 0.95 in the foredune and dune ridge areas) in Figure 7.7B confirmed the confusion between these two classes. The sand cover on the beach was correctly segmented, because of its characteristic texture. Uncertainty values were lower than 0.2. Again, Figure 7.7B shows a short transition zone from the foredune

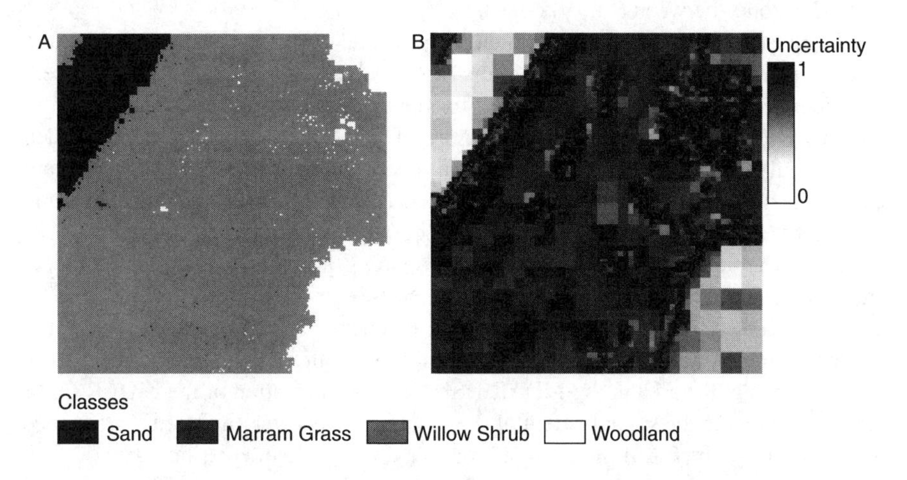

Figure 7.7 Results of the segmentation of the CASI image (A) Supervised texture-based segmentation of band 12 of the CASI image with four reference land cover classes. (B) Related uncertainty for all object building blocks.

Table 7.2 Producer and User Accuracy Percentages for the Land Cover Classes

Class	Producer Accuracy (%)	User Accuracy (%)
Sand	39.15	100.00
Marram grass	5.18	100.00
Willow shrub	87.13	55.17
Woodland	100.00	86.37

to the beach with a decreasing marram grass coverage (narrow zone with uncertainty values of 0.95 and higher southeast of the sand area).

An accuracy assessment of the segmentation results provided an overall accuracy of 64% and a Kappa coefficient of 0.51. Per class producer and user accuracy percentages are given in Table 7.2. It can be concluded from this table that major marram grass areas were incorrectly classified as willow shrub.

7.5 DISCUSSION AND CONCLUSIONS

In this chapter, a texture-based supervised segmentation algorithm derived labelled objects from remotely sensed imagery. Texture was modelled with the joint distribution of LBP and local variance. The segmentation algorithm was a hierarchical splitting technique, based on reducing classification confusion at the level of the image blocks that were obtained. By applying this segmentation technique, one does not only obtain an object-based classification, but also an indication of classification uncertainty for all the object's building blocks. The spatial distribution of uncertainty values provided information about the location and width of transition zones. This study showed that object uncertainty values provide important information to identify transition zones between (fuzzy) objects.

To illustrate the algorithm for mapping coastal objects, a LiDAR DSM and CASI image of a coastal area on the northwest coast of England was used. An accurate segmentation was obtained for the extraction of landform objects from the LiDAR DSM, with an overall accuracy of 86%. Uncertainty values provided meaningful information about transition zones between the different landforms. Land cover objects derived from the CASI image showed large uncertainty values and many incorrectly labelled objects. The overall accuracy was 64%. The woodland area showed a characteristic texture in both data sources, however, the woodland object showed a different spatial extent and area in both segmentation results. This difference was caused by the occurrence of small patches of willow trees in, and on the border of, the woodland area. The texture of these willow trees was different from the pine trees in the area in the LiDAR DSM. The segmentation of the LiDAR DSM correctly depicted the spatial extent of the pine area. However, the texture difference did not occur in the CASI image, resulting in a different segmentation result.

Errors in segmentation can possibly be prevented by taking into account spectral information from more than one spectral band. The combination of textural and spectral information from all 14 CASI bands could greatly increase the accuracy of segmentation. This combination could be useful for mapping other land cover types in the area, like grasses, herbaceous plants, mosses, scrub. Additionally, the resolution of the neighbourhood set of the LBP measure affects the segmentation result. In this study, a neighbourhood set of the nearest eight neighbouring pixels ($P = 8$, $R = 1$) was used. A multi-resolution approach with different combinations of P and R might enhance texture description. In future research, the focus will be on the assessment of the effect of different neighbourhood sets on segmentation. Additionally, a multispectral approach with a multivariate $\text{LBP}_{c,j}$ texture measure will be the focus of future research.

REFERENCES

Ainsdale Sand Dunes NNR, English Nature National Natural Reserves, 2003, URL: http://www.english-nature.org.uk. Last accessed: 24 January 2003.

Aguado, A.S., Montiel, E., and Nixon, M.S., 1998, Fuzzy image segmentation via texture density histograms. EU project Nr. ENV4-CT96-0305 — Fuzzy Land Information from Environmental Remote Sensing (FLIERS) Final Report.

Bezdek, J., 1981, *Pattern Recognition with Fuzzy Objective Function Algorithms* (New York: Plenum Press).

Burrough, P. and McDonnell, R., 1998, *Principles of Geographical Information Systems. Spatial Information Systems and Geostatistics* (Oxford: Oxford University Press).

Bouman, C. and Liu, B., 1991, Multiple resolution segmentation of textured images. *IEEE Transactions on Pattern Analysis and Machine Intelligence*, **13**, 99–113.

Brown, K., 2004, Chapter 8, this volume.

Canters, F., 1997, Evaluating the uncertainty of area estimates derived from fuzzy land cover classification. *Photogrammetric Engineering and Remote Sensing*, **63**, 403–414.

Cheng, T., Fisher, P.F., and Rogers, P., 2002, Fuzziness in multi-scale fuzzy assignment of duneness. In *Proceedings of Accuracy 2002 — International Symposium on Spatial Accuracy Assessment in Natural Resources and Environmental Sciences*, eds. G. Hunter and K. Lowell, pp. 154–159.

Cheng, T. and Molenaar, M., 2001, Formalizing fuzzy objects from uncertain classification results. *International Journal of Geographical Information Science*, **15**, 27–42.

Fisher, P.F., 1999, Models of uncertainty in spatial data. In *Geographical Information Systems: Vol. 1 Principles and Technical Issues, Vol. 2 Management Issues and Applications*, 2nd edition (New York: Wiley and Sons), pp. 191–205.

Fisher, P.F., Cheng, T., and Wood, J., 2004, Where is Helvellyn? Multiscale morphometry and the mountains of the English Lake District. *Transactions of the Institute of British Geographers*, **29**, 106–128.

Foody, G.M., 1996, Approaches for the production and evaluation of fuzzy land cover classifications from remotely sensed data. *International Journal of Remote Sensing*, **17**, 1317–1340.

Gorte, B.H.H. and Stein, A., 1998, Bayesian classification and class area estimation of satellite images using stratification. *IEEE Transactions on Geosciences and Remote Sensing*, **36**, 803–812.

Haralick, R.M., Shanmugam, K., and Dinstein, I., 1973, Textural features for image classification. *IEEE Transactions on Systems, Man and Cybernetics*, **2**, 610–621.

Haralick, R.M. and Shapiro, L.G., 1985, Image segmentation techniques. *Computer Vision, Graphics and Image Processing*, **29**, 100–132.

Hazel, G.G., 2000, Multivariate Gaussian MRF for multispectral scene segmentation and anomaly detection. *IEEE Transactions on Geoscience and Remote Sensing*, **38**, 1199–1211.

Hootsmans, R.M., 1996, *Fuzzy Sets and Series Analysis for Visual Decision Support in Spatial Data Exploration*. Ph.D. thesis, Utrecht University.

Horowitz, S.L. and Pavlidis, T., 1976, Picture segmentation by a tree traversal algorithm. *Journal of the Association for Computing Machinery*, **23**, 368–388.

Lucieer, A. and Stein, A., 2002, Existential uncertainty of spatial objects segmented from remotely sensed imagery. *IEEE Transactions on Geoscience and Remote Sensing*, **40**, 2518–2521.

Nixon, M.S. and Aguado, A.S., 2002, *Feature Extraction and Image Processing*. Butterworth-Heinemann.

Ojala, T., Pietikäinen, M., and Harwood, D., 1996, A comparative study of texture measures with classification based on feature distributions. *Pattern Recognition*, **29**, 51–59.

Ojala, T. and Pietikäinen, M., 1999, Unsupervised texture segmentation using feature distributions. *Pattern Recognition*, **32**, 477–486.

Ojala, T., Pietikäinen, M., and Mäenpää, T., 2002, Multiresolution grey-scale and rotation invariant texture classification with local binary patterns. *IEEE Transactions on Pattern Analysis and Machine Intelligence*, **24**, 971–987.

Pietikäinen, M., Ojala, T., and Xu, Z., 2000, Rotation-invariant texture classification using feature distributions. *Pattern Recognition*, **33**, 43–52.

Randen, T. and Husøy, J.H., 1999, Filtering for texture classification: a comparative study. *IEEE Transactions on Pattern Analysis and Machine Intelligence*, **21**, 291–310.

Sarkar, A., Biswas, M.K., Kartikeyan, B., Kumar, V., Majumder, K.L., and Pal, D.K., 2002, MRF model-based segmentation approach to classification for multispectral imagery. *IEEE Transactions on Geoscience and Remote Sensing*, **40**, 1102–1113.

Sokal, R.R. and Rohlf, F.J., 1987, *Introduction to Biostatistics*, 2nd edition (New York: W.H. Freeman).

Texture Analysis, University of Oulu, Finland, 2003, URL: http://www.ee.oulu.fi/research/imag/texture/. Last accessed: 14 February 2003.

Zhang, J. and Foody, G.M., 2001, Fully-fuzzy supervised classification of sub-urban land cover from remotely sensed imagery: Statistical and artificial neural network approaches. *International Journal of Remote Sensing*, **22**, 615–628.

CHAPTER 8

Per-Pixel Uncertainty for Change Detection Using Airborne Sensor Data

Kyle Brown

CONTENTS

8.1 INTRODUCTION

The 1992 European Council Habitats Directive (92/43/EEC) requires that the extent and condition of a variety of ecologically important habitats be reported on a six-yearly basis. Remote sensing provides one approach by which these requirements may be met and could also provide monitoring to assess the impact of management practices

0-8493-2837-3/05/$0.00+$1.50

on protected sites. An essential part of any monitoring program is to determine where land cover change is taking place. However, operational methods of carrying out this monitoring using remote sensing are currently not in place. There is, therefore, a need to refine aspects of remote sensing techniques, particularly in the field of change detection.

8.1.1 Change Detection

There are a number of approaches for change detection, but one of the most widely used for detecting thematic change is post-classification change analysis (Mas, 1999; de Bruin, 2000; de Bruin and Gorte, 2000). There are limitations to this approach, as errors within either classification may result in errors in change detection. An acceptable classification accuracy limit of 85% has been suggested (Wright and Morrice, 1997), but in some cases this may not be achievable. Assuming perfect co-registration, the maximum theoretical error within the final change layer is the sum of the errors of the two classifications. Therefore, even with low thematic errors in classifications, the error of the change detection layer may be relatively high.

The change detection process is subject to a number of errors at each stage of the data gathering and classification process and these errors may be subject to complex interactions as the change process is modelled. As errors are passed from source to derived data, the errors may be modified such that it may be amplified or suppressed (Veregin, 1996). For example, geometric error in two images used for change detection would result in a low misregistration error if the geometric error vectors were of a similar magnitude and direction. Errors within the final change surface could be as a result of errors from the sensors, ground data, co-registration, or due to a lack of spectral separability of classes. The classifiers used will also affect error. Though there are a number of sources of error within the change detection process, these errors manifest themselves in two major forms: thematic and misregistration errors. To estimate the confidence limits of the change detection process, the errors in each stage must be quantified and the propagation of errors through the change detection process modelled (Goodchild et al., 1992).

8.1.2 Error and Uncertainty

The most commonly used methods of quantifying error within remote sensing data are global methods. Global methods provide a single measure of classification accuracy such as overall accuracy (P_0), the Kappa coefficient of agreement (Cohen, 1960; Stehman, 1996), or by using the confusion matrix to derive per-class accuracy measures (Janssen and van der Wel, 1994; Campbell, 1996; Shalan et al., Chapter 2, this volume). Geometric and misregistration error is generally quantified using a measure such as the root mean square error (RMSE) (Janssen and van der Wel, 1994). However, these global approaches are spatially unreferenced and so error is assumed to apply uniformly across an image. Therefore, the position of likely change detection errors cannot be identified using global methods. If information is required on where change detection errors occur, local error measures need to be derived. One method of identifying where these errors occur in a pixel-based classification approach is to estimate both thematic and misregistration errors on a per-pixel basis. Deriving per-pixel error metrics has the potential to provide more accurate identification of where change is taking place.

It is impractical to quantify the actual error on a per-pixel basis, and so the focus of any error study must be on the probability of error or the uncertainty inherent in the stages involved in the change detection process. Therefore, the uncertainty in each stage must be quantified and the propagation of errors through the change detection process modelled.

The consideration of uncertainty in change detection is essential, as errors within classification layers may be magnified when they are merged for change detection. de Bruin (2000) demonstrated that when per-pixel thematic probabilities are considered in post classification change detection, the results might be very different to simple direct comparison of two classifications.

Several studies (Shi and Ehlers, 1996; de Bruin, 2000; de Bruin and Gorte, 2000) use per-pixel thematic uncertainty, but ignore the effect of geometric errors. However, even sub-pixel misregistration errors can result in large change errors (Townshend et al., 1992; Dai and Khorram, 1998; Roy, 2000).

If a pixel is incorrectly georeferenced, a thematic error may occur. This would occur if the pixel that should have occupied the position was of a different class. Geometric errors may result in thematic errors particularly at the boundaries between classes (Carmel et al., 2001). This means that misregistration results in a greater probability of change detection error at the boundary rather than the centre of classes (Dai and Khorram, 1998; Carmel et al., 2001).

Though uncertainty in change detection has been considered in several studies, there are few that consider the effect of both thematic and misregistration at the pixel level, even though these errors vary spatially. Pixels that are not represented by the training data, are mixed or belong to classes that overlap spectrally are more likely to contain error. Misregistration will be spatially complex, as it will be a combination of geometric errors within the two layers being co-registered. It is therefore essential that the spatial function of these errors is also considered (Arbia et al., 1998).

8.1.3 Thematic Uncertainty

A number of studies have generated per-pixel thematic uncertainty using the maximum likelihood classifier (Shi and Ehlers, 1996; Ediriwickrema and Khorram, 1997; de Bruin and Gorte, 2000), neural networks (Gong et al., 1996; Foody, 2000) and boosting, a machine learning technique (McIver and Friedl, 2001).

Statistical approaches to deriving probabilities have limitations as they assume a data model such as Gaussian distribution (Benediktsson et al., 1990). Machine learning techniques, such as neural networks, are dependent only on the data not on the data models and therefore do not require data to be normally distributed and uncorrelated (Benediktsson et al., 1993; Atkinson and Tatnall, 1997; Zhou, 1999). As well as being distribution-free, neural networks are importance-free (Benediktsson et al., 1990; Zhou, 1999), meaning that the network will model the relative importance of the input data surfaces during the training process without requiring operator input. This characteristic is particularly critical when considering multisource data, as *a priori* knowledge of the level of importance of data layers is not required. A neural network will set weightings to account for a data layer's importance during the training process (Zhou, 1999).

The most commonly used network in remote sensing is the multilayer perceptron (MLP). This network has previously been used to generate per-pixel thematic uncertainties. Per-pixel thematic uncertainty may also be generated by directly estimating posterior probabilities using the nonparametric probabilistic neural network (PNN) proposed by Specht (1990).

8.1.3.1 Multilayer Perceptron

Atkinson and Tatnall (1997) give a description of the MLP and its use in remote sensing classification. The basic unit of the MLP is the node, which mimics a biological neurone. The node sums the inputs and performs a function on the summed input. The MLP consists of three types of layers; input, hidden, and output (Figure 8.1). The input layer has as many nodes as there are input data layers. There may be one or more hidden layers with the number of layers and nodes specified by the user. The output layer contains as many nodes as there are output classes.

Every node in the hidden and output layers is connected to all nodes in the previous layer. As the signal passes between nodes it is modified by weights specific to each node–node connection.

Input signals are passed through the MLP, being modified by the weights associated with the connection between nodes and the functions of each node. The movement of input signals and their modification through the network from input to output is the "feedforward" stage of the MLP. The outputs of the MLP are activation levels at each output node. These activation levels may be linked to a biophysical property or land cover class.

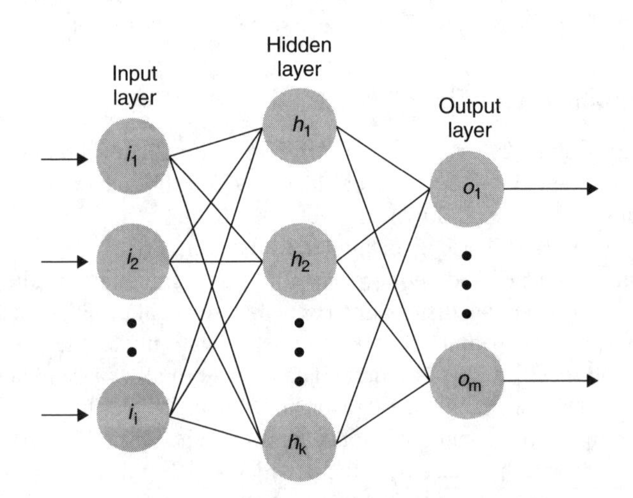

Figure 8.1 Structure of the multilayer perceptron neural network.

Training data are entered into the MLP and the activation level of each of the output nodes is compared with the validation values and an error function is calculated. A learning algorithm is applied that alters the weightings within the network to minimise the error. The whole process is then repeated until a specified number of iterations have taken place, or the error is minimised or reduced below a predetermined level (Atkinson and Tatnall, 1997; Kanellopoulos and Wilkinson, 1997).

The MLP provides an output or activation level for every class on a pixel-by-pixel basis. In a hard classification, the pixel is allocated the class with the highest activation level. However, the activation levels for all classes may be used to provide additional information for each pixel. MLP activation levels have been used as indicators of membership on a per-pixel basis (Foody, 2000). Normalised MLP activation levels have also been used as per-pixel indicators of correct allocation, where a pixel with a high normalised activation is assumed to have a high probability of correct class allocation (Gong et al., 1996).

8.1.3.2 Probabilistic Neural Networks

The PNN proposed by Specht (1990) may also be used to classify imagery and generate per-pixel thematic uncertainty measures. The PNN is a feedforward network with a similar structure to a MLP. The pattern layer corresponds to a single hidden layer in the MLP and has as many nodes as there are training pixels. Each node models a distribution function or kernel, based on the point represented in feature space by the training pixel (Figure 8.2). The output layer contains as many nodes as there are classes. Each node in the pattern layer is only connected with the class output node associated with the training data and sums the inputs from the pattern layer to generate a probability density function (PDF).

When the allocation stage of classification is carried out, the probability of membership to each of the radial nodes is calculated and these are summed for

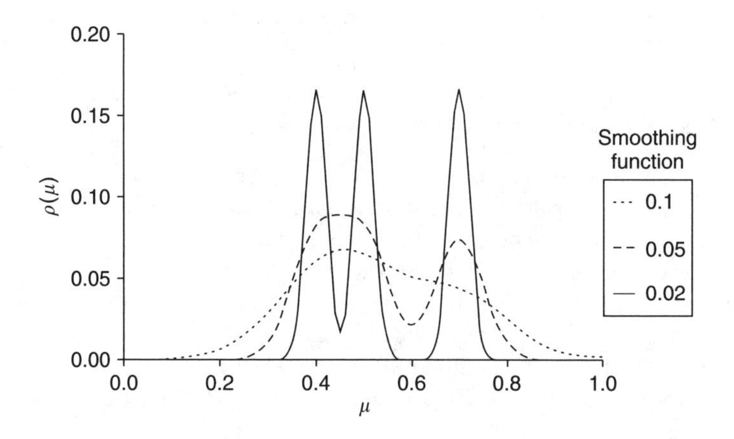

Figure 8.2 Example of the kernel approach to density estimation and the effect of smoothing function. This example uses one-dimensional Gaussian kernels and training samples at 0.4, 0.5, and 0.7.

each class and normalised to give a posterior probability of membership to each class (Figure 8.2). The pixel is then allocated the class it has the highest posterior probability of membership to.

The width of the kernel distribution is determined by the operator and is known as the smoothing function, h. When the smoothing function is too large, the estimated PDF is over smoothed, resulting in inaccurate classification (Bishop, 1995, p. 54). For an infinite sample size, as $h \to 0$ the PDFs will approach a true representation of the density. However, for a finite sample, as $h \to 0$ the PDF will approach a series of "spikes" or delta functions representing each training sample, resulting in a noisy representation of the PDF. When $h = 0$ and assuming the training samples used for different classes represent unique points in feature space, the classification accuracy of the training data will be 100%. However, any point in feature space not represented in the training data will not be classified, resulting in an inability of the network to generalise. This characteristic of PNNs means that care must be taken when determining the smoothing function to be used. The training error may give a very inaccurate indication of the ability of the PNN to correctly classify non-training data. For this reason it is essential that the optimal smoothing function should be determined by testing the PNNs with independent data.

PNNs have advantages over networks that are trained iteratively, as training requires only the generation of kernels for each of the training samples and the only variable that needs to be determined is the smoothing function. However, each training point is represented by a node in the radial layer and so the allocation process can be very intensive computationally, especially if a large number of training samples are used.

The generation of the PDFs for each class by PNNs mean that the outputs may be interpreted directly as posterior probabilities and may therefore be used to derive per-pixel thematic uncertainties.

PNNs have been used for classification of magnetometry data for the detection of buried unexploded ordnance (Hart et al., 2001); texture classification (Raghu and Yegnanarayana, 1998) and cloud classification (Tian and Azimi-Sadjadi, 2001).

8.1.4 Geometric and Misregistration Uncertainty

During the change detection process, georeferencing errors in either input classification is highly likely to result in misregistration errors in the final change detection layer. Misregistration is likely to be complex due to the interaction of spatially dependent geometric errors in either input layer (Carmel, Chapter 3, this volume). If misregistration errors are reduced, the ability to detect change accurately is increased (Dai and Khorram, 1998; Stow, 1999; Roy, 2000). The point at which misregistration results in change detection errors can be when the misregistration between images is sub-pixel in magnitude (Townshend et al., 1992; Dai and Khorram, 1998; Roy, 2000).

One approach to determine the spatial variation of misregistration would be to estimate misregistration at points across a scene and interpolate. This approach would be valid in a relatively static habitat or where known areas remain the same. However, in an environment where change is taking place, positional errors may be difficult to estimate if reference points are not known to be static. These errors may be increased in natural or seminatural environments, where there are likely to be few easily identifiable fixed points. In coastal environments this problem is likely to be increased as in many cases the greatest change occurs at the seaward side of the habitat, the area where there are least reference points. In an area where reference points cannot be found at one edge of a scene, extrapolation of the misregistration surface would be required, resulting in an increased probability of error.

In the area used in this study, there are no obvious features that may be used for geocorrecting or assessing the accuracy of geocorrection (Figure 8.3). In this case, a sensor dependent alternative approach that does not require interpolation could be used. The first stage is to georeference the imagery automatically using navigational data from instrumentation onboard the platform. Estimates are then made of the uncertainties associated with the navigational data. A geometric error model may then be generated using the navigational uncertainty and a digital elevation model (DEM) or digital surface model (DSM) to provide measures of geometric uncertainty. This negates the need for fixed reference points on the ground for geocorrection and geometric error assessment.

If the automated approach to georeferencing is taken, geometric errors within the final layer may be due to either system or orthometric errors. Instrument system errors are due to navigational data or calibration errors. These errors may be quantified prior to data gathering either using knowledge of the component parts of the system, or by testing. Orthometric errors are due to differences between the actual and assumed terrain heights when georeferencing is carried out. Orthometric errors are a function of height errors and the viewing angles of the system. At the edge of an image swath, where the nadir angle is greatest, the potential for orthometric errors is largest.

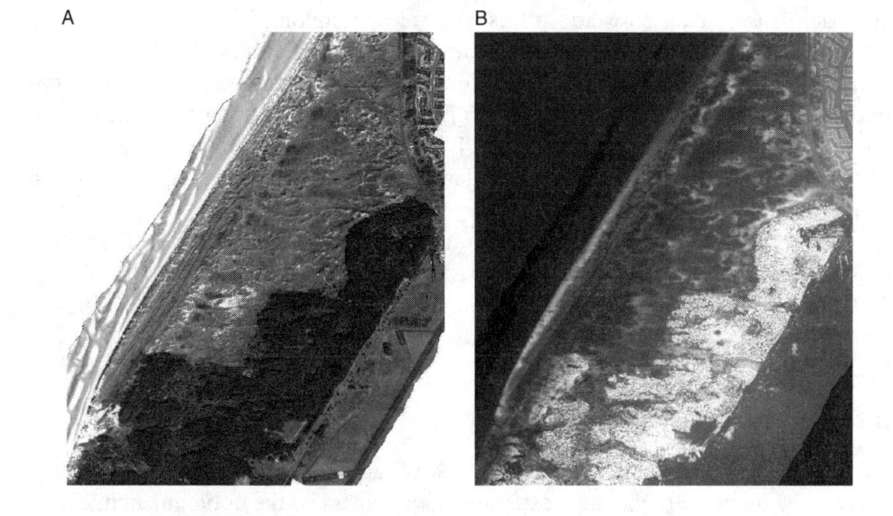

Figure 8.3 2002 Ainsdale Sand Dunes test site data (A) CASI; (B) LiDAR DSM.

8.1.5 Aims

As discussed in Section 8.1.4, there is a need for per-pixel measures of thematic and misregistration errors for change detection. The aims of this chapter are to:

1. Develop methods of deriving per-pixel thematic uncertainty from the output activations of neural network classifiers.
2. Develop a model of the misregistration uncertainty associated with co-registration of CASI imagery.

8.2 METHOD AND RESULTS

8.2.1 Data

Remotely sensed data were gathered over Ainsdale Nature Reserve, Southport, U.K. in 2001 and 2002 (Figure 8.3). Multispectral data were gathered using the ITRES Compact Airborne Spectrographic Imager (CASI) with a spatial resolution of 1 m. DSM data were gathered using Optech 2033 Light Detection and Ranging (LiDAR). The LiDAR x, y, z point data were resampled to the same 1 m raster grid as the CASI data.

The CASI data were geocorrected using ITRES automated geocorrection software using post processed differential global positioning system (dGPS), Applanix POS AV inertial measurement unit (IMU) attitude data and the LiDAR DSM.

For both flights, ground data for eight classes for classification training and accuracy assessment were also collected within 3 weeks. The classes were: water; sand; marram; grasses/mosses/herbaceous vegetation; creeping willow; reeds/rushes; sea buckthorn; and woodland/scrub. Within this 3 week time period of data acquisition, it is unlikely that there would have been change in vegetation type. Point accuracy assessment data were gathered using a geographic stratified random sampling approach. The strata used were squares with sides 60 m long. The size of the square was selected by estimating the maximum number of data points that could be collected within the study area in the time available. The accuracy assessment data were split into two sets. One set consisted of one in four points, which was used to determine the optimum heuristics for the MLP and PNN. The remaining point data were used to determine the accuracy of the final classifications and the relationship between activations and thematic uncertainty.

In order to assess geometric uncertainties of the CASI automated geocorrection system, multiple images were flown over a test site at Coventry Airport, U.K. during 2001. Easily recognisable points on the site were surveyed using dGPS. These points were used to test geocorrection error across the imagery.

8.2.2 Thematic Uncertainty

The 2002 CASI data were classified using MLP and PNN classifiers. A brief study was carried out prior to the main experimentation to assess the optimum architecture of the MLP and PNN smoothing function to classify these data. MLP networks with

5, 10, 15, 20, 25, and 30 nodes in a single hidden layer were trained using 250 iteration intervals between 250 and 3000 iterations. PNNs were trained using a smoothing function between 0.01 and 0.1, with intervals of 0.01. All networks were tested using data independent of the training and accuracy assessment data and the most accurate MLP and PNN were selected. The final MLP contained 20 nodes in a single hidden layer and was trained for 2000 iterations using a learning rate of 0.1 and a momentum of 0.3. The smoothing function of the final PNN was 0.03.

The global accuracy of the classification was tested using the Modified Kappa or Tau (Foody, 1992; Ma and Redmond, 1995; Shalan et al., Chapter 2, this volume). The classification of the Ainsdale sand dune habitat was only 0.9% more accurate using the PNN (Tau = 0.793) than the MLP (Tau = 0.784), and using Tau variance this difference was found to be not significant at 95% confidence.

MLP activations were normalised by dividing the activation for each output node by the sum of the activations for each pixel. The activation levels were tested for suitability as thematic uncertainty measures, assuming that network activations provide a direct measure of the likelihood of a class being correct.

The output activations of all the accuracy assessment pixels for all eight classes were binned according to activation level, with ten even bins between 0 and 1. The proportion of times each of the activations resulted in the correct class and the mean activation were calculated for each bin. A statistically significant relationship between mean output activation and the proportion of correctly classified pixels was found for the MLP ($r^2 = \cdot 920$, $F = 104\cdot 5$, DF $= 8$, $p < \cdot 001$) and the PNN ($r^2 = \cdot 979$, $F = 426\cdot 7$, DF $= 8$, $p < \cdot 001$) (Figure 8.4; Table 8.1).

8.2.3 Geometric and Misregistration Errors

The CASI flown over the Coventry test site data were geocorrected as described above. The DSM used was derived by resampling the dGPS surveyed points used

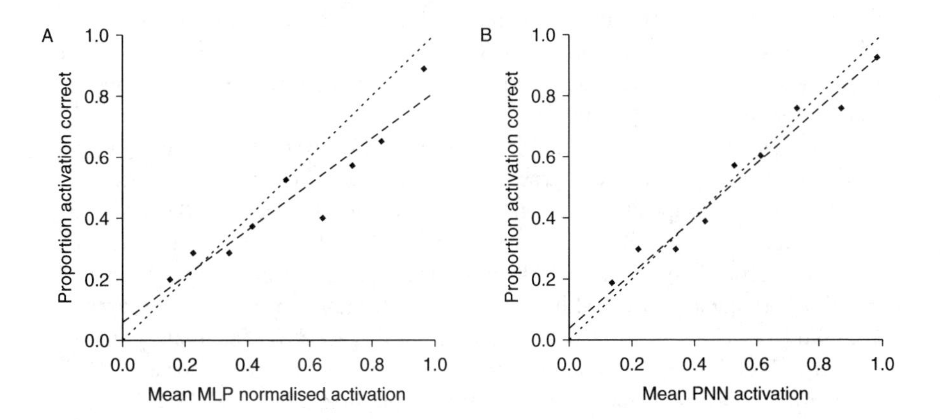

Figure 8.4 Proportion of times a given activation resulted in the correct class as a function of neural network activation. The dashed line represents the result of the linear regression. The dotted line represents the assumption that activation equals predicted proportion of times a given activation resulted in the correct class. (A) MLP and (B) PNN.

Table 8.1 Testing for Correlation between Neural Network Activation and Thematic Uncertainty Using Linear Regression

Classifier	Adjusted r^2	RMSE	F-value	Degrees Freedom	Significance
MLP	.920	0.116	104.5	8	$p < \cdot 001$
PNN	.979	0.047	426.7	8	$p < \cdot 001$

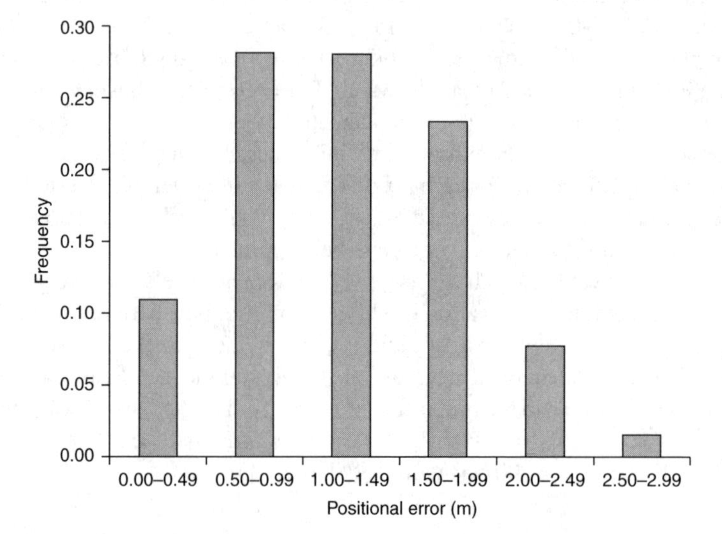

Figure 8.5 Frequency of CASI horizontal errors as a function of error distance.

to test the horizontal accuracy of the CASI data to a 10 m raster grid using nearest neighbour resampling. This provided precise elevation data at the points used to test horizontal accuracy, minimising orthometric errors in the geocorrection process over the survey points. As the orthometric error was minimised, the major error component was therefore due to the system.

The difference between GPS and CASI positions of the surveyed points was used to derive error measures. Linear regressions were carried out to test whether there were relationships between positional error and two variables derived from the navigation data: viewing angle and rate of attitude change. In neither case was a statistically significant relationship found. As a local measure of instrument error could not be derived, the frequency of CASI horizontal errors (Figure 8.5) was used to derive a global model of CASI instrument geometric error (Figure 8.6). The model consisted of a matrix based on 29 possible instrument error points or pixels. For each CASI pixel the model assumes that there are 29 possible pixels that could be the actual position on the ground and gives a probability of this occurrence.

For each square of the matrix the distance between the centre of the matrix and the centre of the square was calculated. The number of squares in the matrix that occurred at each of the bins in Figure 8.5 was calculated. The frequency within each

			0.003			
	0.003	0.011	0.061	0.011	0.003	
	0.011	0.072	0.061	0.072	0.011	
0.003	0.061	0.061	0.111	0.061	0.061	0.003
	0.011	0.072	0.061	0.072	0.011	
	0.003	0.011	0.061	0.011	0.003	
			0.003			

Figure 8.6 Pixel-based matrix of probabilities of CASI instrument geometric error assuming 1 m spatial resolution.

bin was divided by the number of squares to provide a per-square probability of error. This value was applied to the matrix.

Nadir and azimuth angles from the CASI instrument were derived from dGPS and used to estimate the potential orthometric error using:

$$o = \tan(\nu)(z_{\text{ortho}} - z_{\text{LiDAR}}) \tag{8.1}$$

where o is the orthometric error towards the CASI instrument; ν is the off nadir angle of a given pixel; z_{ortho} is the elevation used in the automated geocorrection; z_{LiDAR} is the LiDAR elevation of the instrument error point. o is then used to estimate the orthometric error in x and y:

$$x_{\text{ortho}} = \sin(\alpha)^{*}o \tag{8.2}$$

$$y_{\text{ortho}} = \cos(\alpha)^{*}o \tag{8.3}$$

where α is the azimuth angle of a given pixel.

The global model of instrument error was combined with the local model of orthometric errors to provide an overall measure of local geometric uncertainty. Testing the geometric error model was difficult due to the requirement to survey many easily recognisable points on a variety of slopes. A more practical method was to test the misregistration between two images.

The geometric uncertainty model was used to generate a model of the misregistration between CASI images. The potential error within each pixel in the matrix for 2001 was combined with the potential error within each pixel in the matrix for 2002.

The combined probabilities were used to predict a mean misregistration error.

$$x_{\text{error}} = (x_{\text{ortho},n} - x_{\text{ortho},m}) + (x_{\text{pixel},n} - x_{\text{pixel},m}) \tag{8.4}$$

$$y_{\text{error}} = (y_{\text{ortho},n} - y_{\text{ortho},m}) + (y_{\text{pixel},n} - y_{\text{pixel},m}) \tag{8.5}$$

where x_{ortho} is the error in x due to orthometric effects; x_{pixel} is the x offset from the central pixel of the matrix (Figure 8.6); n is a pixel in matrix 1; m is a pixel in matrix 2.

The misregistration between two pixels in matrix 1 and matrix 2 ($\mu_{n,m}$) is given by:

$$\mu_{n,m} = (x_{\text{error}}^2 + y_{\text{error}}^2)^{0.5} \tag{8.6}$$

and overall misregistration μ is given by:

$$\mu = \sum_{n=1}^{p} \sum_{m=1}^{p} \mu_{n,m} \rho_n \rho_m \tag{8.7}$$

where p is the number of pixels in the CASI instrument geometric uncertainty matrix (29 in this study); ρ_n and ρ_m are the probability of geometric error for matrix 1 and matrix 2, respectively. They are derived from Figure 8.6.

This model was tested using an urban area next to the Ainsdale sand dune test site. The positions of 100 easily identifiable points on the 2001 and 2002 data sets were estimated and the actual misregistration between the images was compared to that estimated by the model.

To compare the actual and predicted misregistration, predicted misregistration was sorted by magnitude of error for every point and binned in groups of ten to generate average predicted and actual misregistration. A linear regression was carried out on the averaged values and a statistically significant relationship was found between predicted misregistration and actual misregistration ($r^2 = \cdot 530$, $F = 11\cdot 1$, DF $= 8$, $p < \cdot 05$) (Figure 8.7). The relationship between predicted and actual misregistration is given by:

$$\mu_{\text{actual}} = 1\cdot 27\mu_{\text{predicted}} - 0\cdot 39 \tag{8.8}$$

Though there was correlation between predicted and actual misregistration, the slope of the relationship is not close to one (Equation (8.8)) and therefore more work is required on the model.

8.3 DISCUSSION AND CONCLUSIONS

The PNN and MLP classifier had very similar global thematic accuracies, with Tau coefficients of 0.793 and 0.784, respectively. However, when estimating per-pixel thematic uncertainty the PNN approach had a 59% lower RMS error than the MLP (Table 8.1). This indicates that the PNN was more suitable for deriving local thematic uncertainty than the MLP for the data in this study. The increased accuracy in

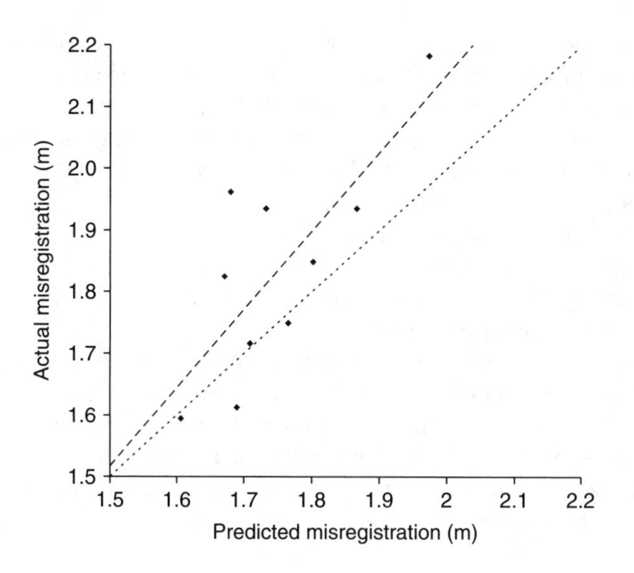

Figure 8.7 Actual misregistration as a function of values predicted in model. Each point on the graph is the average of ten ground points. The dashed line represents the result of a linear regression. The dotted line represents the assumption that predicted misregistration equals actual misregistration.

determining thematic uncertainty may have been due to the ability of the PNN to directly output posterior probabilities rather than the indirect approach of the MLP.

Although the PNN was more accurate in estimating local thematic uncertainty measures from airborne remote sensing data, both approaches resulted in significant relationships between per-pixel output and thematic uncertainty (Table 8.1). This builds on previous studies that have derived thematic uncertainty using a variety of classifiers (Gong et al., 1996; Shi and Ehlers, 1996; de Bruin and Gorte, 2000; Foody, 2000; McIver and Friedl, 2001). The measures derived will allow mapping of the potential uncertainty associated with every class on a per-pixel basis.

Local measures of geometric uncertainty may be derived using a global model of CASI instrumentation uncertainty and a fine spatial resolution DSM to provide measures of potential orthometric error. Though it was possible to derive a local model of instrument error, more complex models involving interaction between GPS and IMU and aircraft movement should be developed. This would allow uncertainty to be modelled more precisely on a per-pixel basis by accounting for the local variation in error due to geometric correction errors. The accuracy of the misregistration model also has the potential to be increased by modelling the uncertainty associated with the DSM used in georeferencing image data.

The measures derived in this chapter have the potential to be used as inputs for change detection, allowing the spatial context of change to be modelled on a per-pixel basis in a probabilistic framework, quantifying inaccuracies in the post classification change detection process. The misregistration metric in this chapter was a measure of the potential misregistration between two pixels, effectively an averaged value (Equation (8.7)). However, one method of combining thematic and misregistration models to provide a per-pixel model of change would be to use the geometric error model. The geometric error model consisted of a series of geometric error vectors, each of which had a probability associated with them. Due to geometric error, every pixel could actually occupy a series of alternative positions. The geometric error model estimates the probability associated with each of those positions. During the change detection process, the thematic uncertainty for every pixel derived from the classification would be modified by the thematic uncertainty of the pixels in these alternative positions. The magnitude of the modification would be determined by the geometric model probability. Misregistration would therefore be considered in a probabilistic framework using thematic uncertainty.

This chapter focuses on the use of uncertainty for hard land cover change detection. However, there are a number of issues that have not been addressed, particularly representing a continuous surface as a series of discrete pixels, each given a single land cover category (Fisher, 1997). Further work needs to be carried out to determine whether the thematic uncertainty measures derived above describe the continuous nature of the surfaces being examined as well as error.

ACKNOWLEDGEMENTS

Many thanks to Giles M. Foody and Pete Atkinson of the University of Southampton for inputs at various stages of this study and to Rob Wolstenholme of English Nature, the warden of Ainsdale Sand Dunes Nature Reserve. Also thanks to the anonymous reviewers who provided many helpful comments on the original manuscript.

REFERENCES

Arbia, G., Griffith, D., and Haining, R., 1998, Error propagation modelling in raster GIS: overlay operations. *International Journal of Geographical Information Science*, **12**, 145–167.

Atkinson, P.M. and Tatnall, A.R.L., 1997, Neural networks in remote sensing. *International Journal of Remote Sensing*, **18**, 699–709.

Benediktsson, J.A., Swain, P.H., and Ersoy, O.K., 1990, Neural network approaches versus statistical methods in classification of multisource remote sensing data. *IEEE Transactions on Geoscience and Remote Sensing*, **28**, 540–551.

Benediktsson, J.A., Swain, P.H., and Ersoy, O.K., 1993, Conjugate-gradient neural networks in classification of multisource and very-high-dimensional remote sensing data. *International Journal of Remote Sensing*, **14**, 2883–2903.

Bishop, C.M., 1995, *Neural Networks for Pattern Recognition*. Oxford University Press, Oxford.

Campbell, J.B., 1996, *Introduction to Remote Sensing* (2nd ed.). Guildford Publications, New York.

Carmel, Y., Chapter 3, this volume.

Carmel, Y., Dean, D.J., and Flather, C.H., 2001, Combining location and classification error sources for estimating multi-temporal database accuracy. *Photogrammetric Engineering and Remote Sensing*, **67**, 865–872.

Cohen, J., 1960, A coefficient of agreement for nominal scales. *Educational and Psychological Measurement*, **20**, 37–46.

Dai, X. and Khorram, S., 1998, The effects of image misregistration on the accuracy of remotely sensed change detection. *IEEE Transactions on Geoscience and Remote Sensing*, **36**, 1566–1577.

De Bruin, S., 2000, Querying probabilistic land cover data using fuzzy set theory. *International Journal of Geographical Information Sciences*, **14**, 359–372.

De Bruin, S. and Gorte, B.G.H., 2000, Probabilistic image classification using geological map units applied to land-cover change detection. *International Journal of Remote Sensing*, **21**, 2389–2402.

Ediriwickrema, J. and Khorram, S., 1997, Hierarchical maximum-likelihood classification for improved accuracies. *IEEE Transactions on Geoscience and Remote Sensing*, **35**, 810–816.

Fisher, P., 1997, The pixel: a snare and delusion. *International Journal of Remote Sensing*, **18**, 679–685.

Foody, G.M., 1992, On the compensation for chance agreement in image classification accuracy assessment. *Photogrammetric Engineering and Remote Sensing*, **58**, 1459–1460.

Foody, G.M., 2000, Mapping land cover from remotely sensed data with a softened feedforward neural network classification. *Journal of Intelligent and Robotic Systems*, **29**, 433–449.

Gong, P., Pu, R., and Chen, J., 1996, Mapping ecological land systems and classification uncertainties from digital elevation and forest-cover data using neural networks. *Photogrammetric Engineering and Remote Sensing*, **62**, 1249–1260.

Goodchild, M.F., Guoqing, S., and Shiren, Y., 1992, Development and test of an error model for categorical data. *International Journal of Geographical Information Sciences*, **6**, 87–104.

Hart, S.J., Shaffer, R.E., Rose-pehrsson, S.L., and McDonald, J.R., 2001, Using physics-based modeler outputs to train probabilistic neural networks for unexploded ordnance (UXO) classification in magnetometry surveys. *IEEE Transactions on Geoscience and Remote Sensing*, **39**, 797–804.

Janssen, L.L.F. and van der wel, F.J.M., 1994, Accuracy assessment of satellite derived land-cover data: A review. *Photogrammetric Engineering and Remote Sensing*, **60**, 419–426.

Kanellopoulos, I. and Wilkinson, G.G., 1997, Strategies and best practice for neural network image classification. *International Journal of Remote Sensing*, **18**, 711–725.

Ma, Z., and Redmond, R.L., 1995, Tau coefficients for accuracy assessment of classification of remote sensing data. *Photogrammetric Engineering and Remote Sensing*, **61**, 435–439.

Mas, J.F., 1999, Monitoring land-cover changes: a comparison of change detection techniques. *International Journal of Remote Sensing*, **20**, 139–152.

McIver, D.K. and Friedl, M.A., 2001, Estimating pixel-scale land cover classification confidence using non-parametric machine learning methods. *IEEE Transactions on Geoscience and Remote Sensing*, **39**, 1959–1968.

Raghu, P.P. and Yegnanarayana, B., 1998, Supervised texture classification using a probabilistic neural network and constraint satisfaction model. *IEEE Transactions on Neural Networks*, **9**, 516–522.

Roy, D.P., 2000, The impact of misregistration upon composited wide field of view satellite data and implications for change detection. *IEEE Transactions on Geoscience and Remote Sensing*, **38**, 2017–2032.

Shalan, M.A., Arora, M.J., and Elgy, J., Chapter 2, this volume.

Shi, W.Z. and Ehlers, M., 1996, Determining uncertainties and their propagation in dynamic change detection based on classified remotely-sensed images. *International Journal of Remote Sensing*, **17**, 2729–2741.

Specht, D.F., 1990, Probabilistic neural networks. *Neural Networks*, **3**, 109–118.

Stehman, S.V., 1996, Estimating the Kappa coefficient and its variance under stratified random sampling. *Photogrammetric Engineering and Remote Sensing*, **62**, 401–407.

Stow, D.A., 1999, Reducing the effects of misregistration on pixel-level change detection. *International Journal of Remote Sensing*, **20**, 2477–2483.

Tian, B. and Azimi-Sadjadi, M.R., 2001, Comparison of two different PNN training approaches for satellite cloud data classification. *IEEE Transactions on Neural Networks*, **12**, 164–168.

Townshend, J.R.G., Justice, C.O., Gurney, C., and McManus, J., 1992, The impact of misregistration on change detection. *IEEE Transactions on Geoscience and Remote Sensing*, **30**, 1054–1060.

Veregin, H., 1996, Error propagation through the buffer operation for probability surfaces. *Photogrammetric Engineering and Remote Sensing*, **62**, 419–428.

Wright, G.G. and Morrice, J.G., 1997, Landsat TM spectral information to enhance the land cover of Scotland 1988 dataset. *International Journal of Remote Sensing*, **18**, 3811–3834.

Zhou, W., 1999, Verification of the nonparametric characteristics of backpropagation neural networks for image classification. *IEEE Transactions on Geoscience and Remote Sensing*, **37**, 771–779.

Part II

Physical Processes

CHAPTER 9

Introduction — Spatially Distributed Dynamic Modelling

Stephen E. Darby, Fulong Wu, Peter M. Atkinson, and Giles M. Foody

It is perhaps no coincidence that an emerging interest in the use of spatially distributed dynamic modelling (hereafter SDDM) to simulate geographical phenomena has coincided with a shift in the nature of the "big" questions facing the Earth and environmental science communities. Until comparatively recently most of these "big" questions have tended to focus on the identification and characterisation of land cover and climate change processes — an area of research in which remote sensing has made numerous important contributions. However, the net results of these studies have served to focus the attention of policy makers on the fact that global change processes are real, and that they have tangible, large-scale, environmental, and socio-economic impacts. At the same time, a scientific consensus has now essentially been reached regarding the significance and nature of future climate change. Accordingly, it can be argued that the core research agenda is now moving away from identification towards impacts assessment. SDDM has therefore emerged as an important discipline in its own right because it represents a means of predicting the response of a range (examples in this section comprise hydrological, geomorphological, and ecological applications) of environmental systems to environmental change felt from the local to the global scale.

Particularly in the key context of impacts assessment, it can even be argued that SDDM is the *only* viable means of predicting the response of complex, spatially distributed, environmental systems to global change drivers. The nature of SDDM

0-8493-2837-3/05/$0.00+$1.50

is that it employs a spatially explicit approach with any reasonable number of process models representing various facets (e.g., tree growth, seed dispersal, etc.) of the environmental system of interest. However, due to spatial feedbacks and interactions between process models, very complex model behaviour can emerge even if the basic spatial data structure and sub-models employed are apparently relatively simple. What this implies is that the value of SDDM may lie primarily in its use as a tool for predicting system response that would otherwise be unforeseen. By combining large-scale spatial data sets with process models, SDDM offers the advantage of being able to simulate dynamic phenomena across a cascade of scales ranging from the local and rapid through to the large and slow. It should also be clear that the spatially distributed data sets required for parameterising and validating SDDM models are married to the forms of data acquisition associated with the remote sensing technologies that are a core concern of this book. Put another way, the chapters in this book suggest that there is now a synergistic relationship between remote sensing and SDDM.

Turning now to the specific content of the contributions in this section, two chapters focus on simulating ecosystem dynamics. First, Svoray and Nathan (Chapter 10, this volume) present a process-based model of tree (Aleppo pine) population spread, while Malanson and Zeng (Chapter 11, this volume) use cellular automata to simulate the interaction between forest and tundra ecosystems as a basis for understanding the controls on the shifting location of treelines. An interesting feature of both these (and many other SDDM) studies is that a combination of relatively simple process rules, together with the use of spatially explicit data sets to parameterise the models, results in the emergence of complex non-linearities and feedbacks, thereby producing complex interactions that could not otherwise have been foreseen. The unique ability to identify this type of behaviour serves to highlight one of the key advantages of SDDM approaches.

The two chapters that focus on geomorphological response are similarly characterised by the emergence of complex model behaviour as a result of interactions between apparently simple process rules and spatial feedbacks. Wichmann and Becht's (Chapter 12, this volume) contribution is concerned with predicting the sediment yield derived from steep, alpine environments. They highlight the importance of representing both the physical mechanics of the governing processes (in this case, triggering mechanisms for rapid mass movements) and their spatial relation (in this case, the pathways along which sediment is routed after initiation of motion). Along similar lines, Schmidt (Chapter 13, this volume) is concerned with predicting the onset of landsliding. In this case, the key control that influences the spatial and, as a result of climate change, temporal distribution of shallow landslides is the spatial distribution of sedimentary properties combined with the temporal evolution of pore water pressures predicted through the use of a groundwater model.

The focus of the first four chapters in this section on process representation suggests that while progress has already been made in SDDM applications, there are fundamental challenges still to address. Indeed, it is possible to argue that errors and uncertainties in our ability to model geodynamic processes might be derived primarily from a lack of knowledge about these governing processes. However, this argument would overlook the difficulties that are associated with the use of the spatial data sets that are used to parameterise geodynamic models. The last chapter of Section II, by Wilson and Atkinson (Chapter 14, this volume), focuses on this issue. They explore the extent to which uncertainties in the creation of digital elevation models (DEMs) of floodplain topography influence the prediction uncertainty of flood inundation models that use the DEM data.

The chapters in this Section, comprising as they do case studies drawn from such apparently diverse fields as ecosystem dynamics, drainage basin geomorphology, and flood modelling, serve to remind us that the issues involved in SDDM cut across the traditional boundaries of subject disciplines. As with the other sections in this book, taken together these contributions highlight the concept that "GeoDynamics" in general, and SDDM in particular, have much to offer, while simultaneously identifying the issues that remain and signposting avenues for fruitful further research.

REFERENCES

Malanson, G.P. and Zeng, Y., Chapter 11, this volume.
Schmidt, J., Chapter 13, this volume.
Svoray, T. and Nathan, R., Chapter 10, this volume.
Wichmann, V. and Becht, M., Chapter 12, this volume.
Wilson, M. D. and Atkinson, P.M., Chapter 14, this volume.

CHAPTER 10

Dynamic Modelling of the Effects of Water, Temperature, and Light on Tree Population Spread

Tal Svoray and R. Nathan

CONTENTS

10.1 INTRODUCTION

The recent rise of spatial ecology emphasises the critical importance of the spatial context at which ecological processes take place (Levin, 1992; Tilman and Kareiva, 1997; Clark et al., 1998; Silvertown and Antonovics, 2001). For example, there has been a growing recognition that seed dispersal — one of the most critical processes in plant spatial dynamics (Harper, 1977; Schupp and Fuentes, 1995;

0-8493-2837-3/05/$0.00+$1.50

Nathan and Muller-Landau, 2000) — should be incorporated in a spatially realistic manner in models of plant population dynamics, because different distributions of dispersal distances can give rise to entirely different dynamics (Levin et al., 2003). In addition, several features of the environment, such as water availability and soil surface temperature, which are of critical importance to plant recruitment dynamics, typically exhibit pronounced variation in space and time.

It is therefore surprising that plant population models have not yet incorporated spatially and temporally realistic descriptions of key environmental factors that shape seed dispersal and plant recruitment processes. This can be explained by the great complexity of natural habitats, and the difficulty in identifying key factors and measuring the variation over large spatial and long temporal scales. Recent advances in remote sensing, GIS, geocomputation, and field measurements, along with increased knowledge of recruitment processes, may now allow us to cope with this challenge.

Adult plants produce seeds with considerable spatiotemporal variation among individuals inhabiting different sites. This generates a non-random spatial structure in the annual (or seasonal) seed output within the population (Nathan and Muller-Landau, 2000). Seed release is followed by seed dispersal, the major (or only) stage during which individual plants move in space. The process of dispersal is affected by multiple factors (Chambers and MacMahon, 1994) that can also vary substantially in space and time (Schupp and Fuentes, 1995; Nathan and Muller-Landau, 2000). Seed predators can also be affected by environmental heterogeneity and can drastically alter the spatial structure of dispersed seeds (Hulme, 1993). Seed dormancy can induce an additional component of temporal variation in the availability of seeds for germination (Andersson and Milberg, 1998). Spatial patterns of seedlings can be different from those of seeds, due to spatial heterogeneity in the distribution of suitable micro-habitats for germination. Similarly, spatial variation in seedling survival, generated, for example, by differential herbivory, may further alter the pattern of seedlings. Overall, successive stages of early recruitment generally show low concordance (Schupp and Fuentes, 1995); hence, modelling plant population dynamics requires detailed descriptions of spatiotemporal variation in environmental conditions and their effects on different recruitment processes.

This research proposes a spatially explicit dynamic model to predict long-term tree population spread in a heterogeneous environment. The model is formulated based on mechanistic principles, which describe specific conditions during seed dispersal, germination, seedling survival, and tree establishment. We apply the model to predict the dynamics of pine population spread during two consecutive generations and evaluate the predictions using field data, including visual interpretation of aerial photography and tree age measurements from annual growth rings.

10.2 SPECIES AND STUDY AREA

The Aleppo pine (*Pinus halepensis* Miller) is the most widely distributed pine of the Mediterranean Basin (Quézel, 2000). It is also common in plantations within and outside its natural range, spreading rapidly from plantations to nearby natural habitats (Richardson, 2000). The species, considered one of the most invasive pines (Rejmánek and Richardson, 1996), threatens native biodiversity in various habitats and has even caused severe financial losses, especially across the Southern Hemisphere (Richardson, 2000). Early recruitment processes play a key role in determining Aleppo pine spatial dynamics (Nathan and Ne'eman, 2004); successful management of natural, planted, and invasive populations should therefore be guided by models that incorporate spatial heterogeneity and its effects on early recruitment.

Mt. Pithulim at the Judean Hills of Israel holds a native Aleppo pine population that was well isolated for a long time from any neighbouring population, with no evidence for any planting, cutting, or fire. This population has expanded from five trees at the early 20th century, to the thousands of trees that inhabit the site today. The history of this spatial spread, and of major influencing factors, has been reconstructed in exceptionally fine detail, providing a uniquely detailed long-term perspective into the dynamics of Aleppo Pine populations. The study site on Mt. Pithulim and its surroundings covers an area of roughly 4 km^2, and contains several tens of thousand pine trees, including recent plantations. To concentrate our efforts on the population of interest, a subset area of 60 ha (750×800 m^2) was selected. This 60 ha plot was selected to (a) include the core of the old stand and a buffer of at least 150 m around it; (b) to avoid very steep terrain (>35°) in which fieldwork is extremely difficult; and (c) to represent the major topographical and edaphic units of the area where current tree population spread processes occur.

10.3 THE MODEL

10.3.1 Overall

Our model incorporates the spatial and temporal variation of tree recruitment, focusing on factors operating on early stages of seed dispersal and germination. The temporal resolution is a single month for rainfall and temperature and the spatial resolution is 5×5 m. To simulate seed dispersal, the model draws on a previous dispersal model. Then post-dispersal recruitment is simulated in three steps: (1) evaluation of the conditions for seed survival and germination, on a monthly basis; (2) evaluation of the conditions for seedling-to-adult survival, on a yearly basis; (3) selection of cells that have met a set of expert-defined threshold values for each recruitment factor for each grid cell, followed by selection of cells that have met an expert-defined threshold value for the joint (all factors combined) suitability for each grid cell. The overall conceptual model with its three stages is described in Figure 10.1.

10.3.2 Assumptions

Six assumptions are made regarding the processes that dictate tree population spread:

Seed dispersal

1. Seed dispersal can be simulated from data on a few physical and biological factors, as implemented in the dispersal simulator WINDISPER (see below).

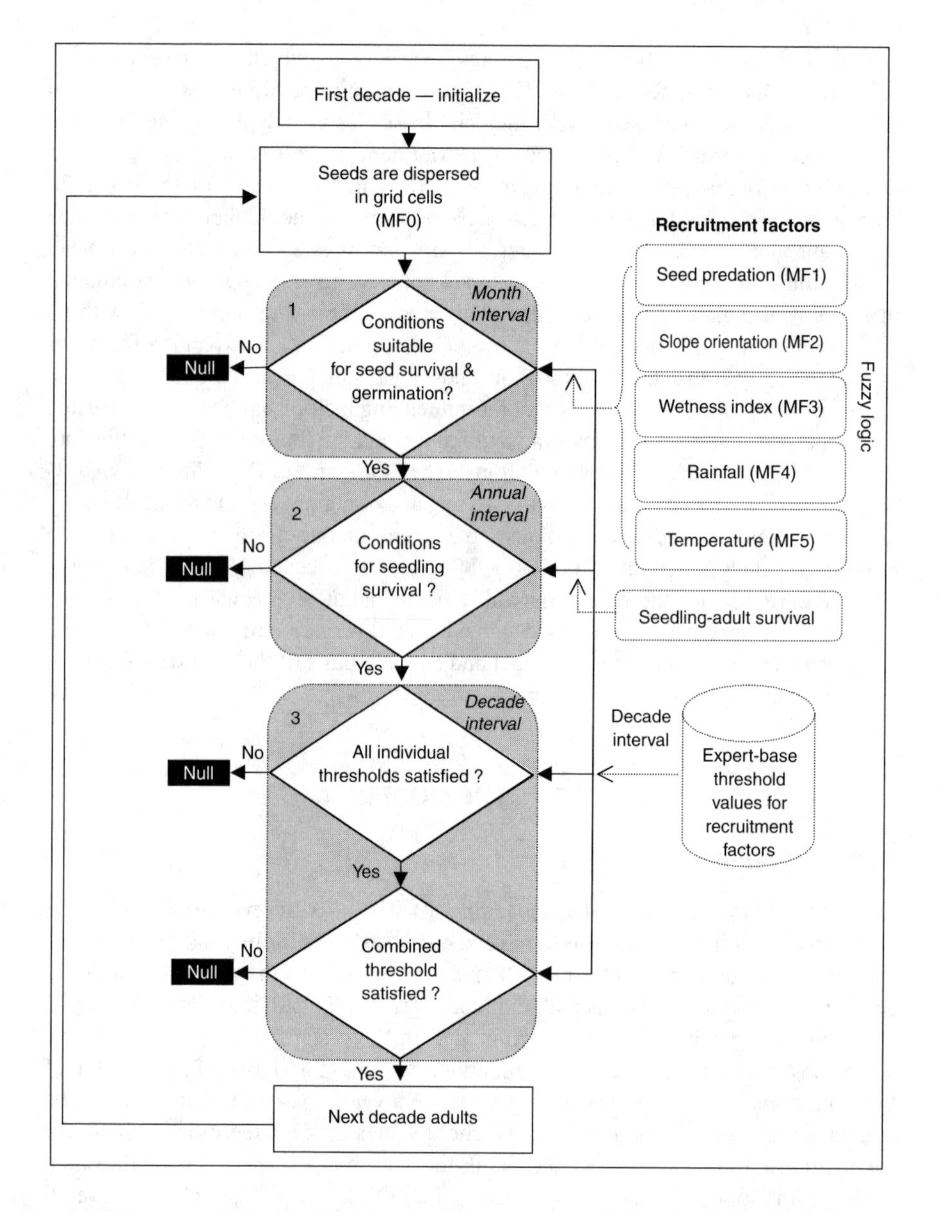

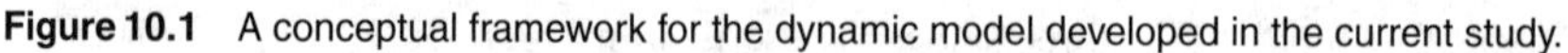
Figure 10.1 A conceptual framework for the dynamic model developed in the current study.

Seed survival and germination

2. Seed survival increases with distance from adult trees due to the attraction of seed predators to the vicinity of adult trees (Nathan and Ne'eman, 2004).
3. Seeds germination is restricted to the winter and spring (between November and April inclusive); seeds that did not germinate do not survive for the next year (Izhaki and Ne'eman, 2000).
4. Seed germination is affected primarily by surface temperature and water availability (Thanos, 2000).

Establishment

5. Seedling survival increases with distance from adult trees due to shading, competition for water with adults, sibling competition and the attraction of seedling herbivores to the vicinity of adult trees (Nathan et al., 2000).
6. Tree mortality occurs mostly during early establishment (Nathan and Muller-Landau, 2000); hence individuals that survive the seed and seedling stages will become adults in the model (Nathan and Ne'eman, 2004).

10.3.3 Seed Dispersal Model

Seed dispersal was simulated using the mechanistic simulator WINDISPER (Nathan et al., 2001). WINDISPER assumes a log-normal distribution of horizontal windspeed and normal distribution of vertical windspeed (truncated to exclude net upward movements), height of seed release and seed terminal velocity. The model assumes a logarithmic vertical profile of the horizontal windspeed. It has been tested against extensive seed trap data collected in two native Aleppo pine populations: (i) Mt. Pithulim (within the focal area of the current study site); and (ii) Mt. Carmel (Israel). The results show close agreement between predictions and observations. As with these previous applications of this simulator, we assume here that the seed output is a linear function of the tree canopy projection, and the distance travelled by each individual seed was calculated after random selection of parameter values, based on their measured empirical distribution.

10.3.3.1 Spatial and Temporal Variation of Recruitment Factors

We assemble a set of parameters that reflect different expressions of the key relationships assumed above. Seed survival, represented by its distance-dependent effects (assumption 2), is translated by assigning three levels of survival for grid cells of different proximity to adult-containing cells. Adult-containing cells were assigned zero probability of survival, adjacent (first neighbouring) cells were assigned an intermediate level, while all other cells were assigned the highest level. The role of water availability and surface temperature (assumption 4) were incorporated in a more complex manner, by distinguishing the effects of topographic and climatic factors. Spatial variation in soil water content is highly dependent on topographic conditions through surface and subsurface runoff convergence and dispersion. It is expected that lower areas in the catchment are moister due to the accumulated water reaching these plots through upper and lower runoff flows (Beven and Kirkby, 1979). Among the key topographic parameters affecting surface hydrology, the local slope and the catchment area determine the hydraulic gradient and the potential water flux to a given area (Barling et al., 1994). Following Beven and Kirkby (1979) and Burrough et al. (1992), we used these two parameters in conjunction as a wetness index (Equation (10.1)):

$$\mathrm{WI_i} = \mathrm{Ln}\left[\frac{As_\mathrm{i}}{\tan\theta_\mathrm{i}}\right] \tag{10.1}$$

where As_i is the specific catchment area (the upslope contributing area) and θ_i is slope angle of the surface. The parameter As_i was calculated using the ArcGIS 8.2

flow accumulation algorithm, while the local slope (degrees) was calculated using the ERDAS Imagine Terrain Analysis algorithm.

We also incorporated the effects of slope orientation on water availability. Slopes of different orientation differ in the amount of solar radiation they receive, hence in evapotranspiration and water loss rates. Consequently, south-facing slopes are less humid than slopes oriented to the north, east, or west (Kutiel, 1992). The aspect layer was calculated from the DEM using the algorithms of the Terrain Analysis module of the ERDAS IMAGINE software. The output value range of the aspect characteristics (0–360°) was categorised into three sectors: north-facing slopes (315–45°), south-facing slopes (135–225°), and a sector of united east- and west-facing slopes (45–135° and 225–315°).

The above topographic parameters change in space over the simulated landscape but not in time. The climatic parameters, however, do the reverse: they change in time (month) but are assumed to be the same over the entire simulated landscape. Monthly statistics of surface temperature and rainfall are from measurements taken at the Israel Meteorological Service (IMS) station at Beit-Jimal, 16 km west of the study site, between 1945 and 2003. These data show high temporal variance in both rainfall (ranging between 1.5 and 460 mm per month) and temperature (ranging between maximum temperature of 12 and 34°C).

The biotic conditions for seedling survival are based also on the distance from the trees that was set for the nearest neighbour cell of each tree. Thus, cells that are already occupied by trees will not allow the development of a new tree and the adjacent cells are of less favoured conditions from more distant cells. Consequently, the second neighbourhood and onward provide the best conditions for seed and seedling survival.

10.3.4 Fuzzy Model

Fuzzy logic is used here to assess the suitability of grid cells for tree establishment. Several membership functions (MFs) are used to determine the degree of membership of each individual recruitment factor to the set A (Table 10.1). The set A represents a group with sufficient conditions for the establishment of a new Aleppo pine tree.

The weights are determined based on our understanding of the tree population spread processes: seed dispersal, survival and germination, and tree establishment. Their weights are assumed to be of equal importance; thus, MF0, MF1, and MF6 that represent processes 1, 2, and 4, respectively, are set to a weight value of 0.25. Process 3 of seed germination is represented here by the combined effect of wetness index, slope orientation, rainfall, and temperature. The factors that affect this process (MF2 to MF5) were also assumed to have equal importance, and therefore, their weights are set to 0.0625.

To perform the first step in the overall model (Figure 10.1), the membership functions are joint (JMF) with their respective weights ($\lambda_{11...6}$; Table 10.1) to provide the monthly assessment as in Equation (10.2):

$$\mathrm{JMF(month)} = \lambda_0\mathrm{MF0} + \lambda_1\mathrm{MF1} + \cdots + \lambda_5\mathrm{MF5} \qquad (10.2)$$

Table 10.1 A Summary of the MFs Used in Our Model

Attribute	Type	Id	MF type	MF	Weights λ
Available seeds	Biotic	MF0	Linear	$\gamma - x/\gamma - \alpha$	0·25
Seed predation	Biotic	MF1	S^+	$1 - (x - \gamma/\gamma - \alpha)^2$	0·25
Slope orientation	Topography	MF2	Linear	$\gamma - x/\gamma - \alpha$	0·0625
Wetness index	Topography	MF3	Linear	$\gamma - x/\gamma - \alpha$	0·0625
Temperature	Climate	MF4	Combined linear and S^+	$1 - (x - \gamma/\gamma - \alpha)^2$; $\gamma - x/\gamma - \alpha$	0.0625
Rainfall	Climate	MF5	S^+	$1 - (x - \gamma/\gamma - \alpha)^2$	0·0625
Seedling mortality	Biotic	MF6	S^+	$1 - (x - \gamma/\gamma - \alpha)^2$	0·25

Note: α is the minimum point; γ is the maximum point; and x is the grid cell value. All λ are the weights while $0 < \lambda < 1$ and $\sum \lambda = 1$. S^+ is S-shape MF with a right open shoulder often used in fuzzy logic according the conditions described in Robinson (2003).

This JMF model is well known as the convex combination JMF where the membership of a cell in the new fuzzy set A and it is determined based on the weighted sum of the membership functions MF0, ..., MF5.

This approach could be very useful to represent complexities in tree population spread processes. For example: in cases where MF2 is low due to a south-facing slope location, MF3 can compensate with a high membership value due to footslope location. However, in cases where no seeds reach their destination to a given cell, and consequently MF0 is equal to zero, other MF values may falsely compensate and the cell might be attributed to the set A. To overcome this problem, we added to the model a rule that adjusts the entire JMF (month) to zero when MF0 is equal to zero.

The second step introduces the annual timescale by adding seedling mortality as in Equation (10.3):

$$\mathrm{JMF(Year)} = \frac{\sum \mathrm{JMF_{Nov}} + \cdots + \mathrm{JMF_{April}}}{6} + \lambda_6 \mathrm{MF6} \tag{10.3}$$

The sum of monthly JMFs is divided by the number of months and thus the annual joint membership function of the recruit factors is the average monthly JMF. This step is applied to normalise the data and thus the total JMF for the next step (decade JMF) will also sum to one in the best-case scenario.

The third step introduces the temporal scale of a decade, simply as the sum of annual JMFs — Equation (10.4):

$$\mathrm{JMF(Decade)} = \sum \mathrm{JMF_{1,\,\ldots\,,\,10}} \tag{10.4}$$

This third and last step of the model is followed by two tests to determine which cells are expected to be established by trees in the next generation. This is done using simulated threshold values that represent degrees of membership from which the establishment of a tree is expected (defuzzification). The threshold values are determined based on mechanistic assumptions that are used to populate Equations (10.2) to (10.4). The tests include threshold values for the suitability of the final JMF and for the suitability of each of the recruitment factors separately. The predicted suitability of all grid cells within the simulated landscape to tree establishment is then tested against visual interpretation of historical air photos and tree ring measurements.

10.3.5 Fieldwork — Model Validation

We have gathered a database of 93 of the oldest trees in the study area. The trees were inspected from historical airphotos (1956 and 1974). The airphotos were scanned and the images were rectified using the *Orthobase Pro* tool of ERDAS IMAGINE. The reference image is a colour air photo from 1996 that was available from our previous work in the site (Nathan et al., 1999, 2001), including an orthophoto and a digital elevation model (DEM) at 0.25 m horizontal resolution. The spatial resolution of the orthophotos is 0.43 m with 13 ground control points and total RMS error of <0·6 m. The interpretation of the trees was verified by a sample of 43 trees whose age was measured using the tree ring methodology. Since tree age estimation using the tree ring methodology is very labour-demanding and therefore cannot be implemented to the entire population, the use of high accuracy, well-rectified historical remote sensing data is an important component of the model validation.

10.4 RESULTS

The observed Aleppo pine tree population of the first generation and the potential conditions of the modelled cells, that is, grades of the total joint membership function

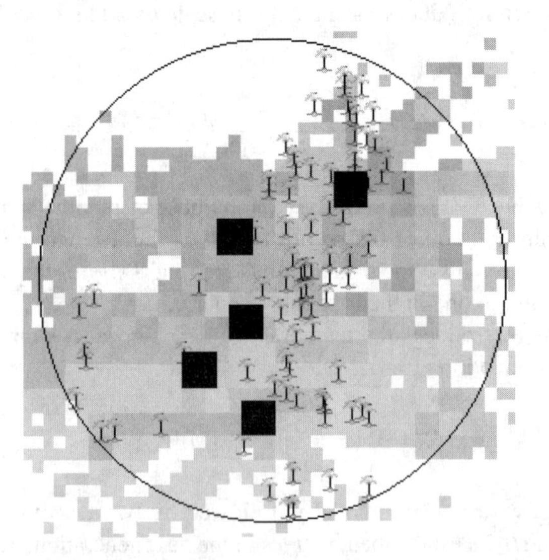

Figure 10.2 **(see color insert following page 168)** The estimated spatial distribution of the first generation of Aleppo pines (tree symbols) on the Mt. Pithulim site, within a radius of 100 m from the central location of the five older trees (■) that have inhabited the site prior to 1940. The background is a map of the total (decade) grades where lighter grey represents lower membership grades and darker grey represents higher membership grades.

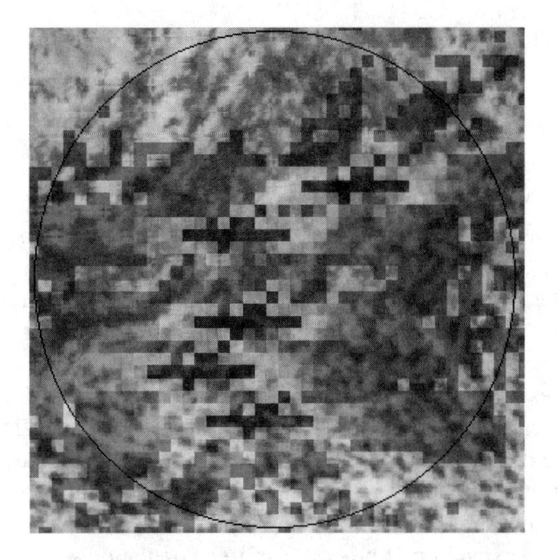

Figure 10.3 **(see color insert following page 168)** A map of the membership grades of seed dispersal where black colours represent higher values and dark grey colours represent lower values. The background is a panchromatic orthophoto acquired in 1956.

are illustrated in Figure 10.2. The results are examined within a radius of 100 m from the central location of the previous generation — the five old trees — since this distance roughly marks the limit of short-distance dispersal that can effectively be predicted by WINDISPER (Nathan et al., 2001). Figure 10.3 provides the seed dispersal membership values within this range. The 100 m limit excludes trees that could have been established through rare long-distance dispersal events. Preliminary data suggest that such events did take place during this period, but their analysis is beyond the scope of this chapter.

Only 14 of the 93 trees are located on cells with extremely low potential for tree establishment (represented by white cells; Figure 10.2). Many of the other trees, mostly those of the northern part of the study area, are located in cells with relatively high predicted suitability values while up to 20 trees, located mostly in the southern part of the studied area, are located in cells of rather low potential.

A quantitative analysis of the results in Figure 10.2 requires a test of how many trees did and how many trees did not actually establish in cells of high and low potential. We distinguished between cells of "high" and "low" suitability by setting an arbitrary threshold of total JMF >0.59 for "high" suitability. The proportion of trees

Table 10.2 **A Summary of the Four Categories that Represent the Presence and Absence of Trees within Cells of High and Low Suitability**

	Suitability Cells	
	High	**Low**
Presence of trees	51	28
Absence of trees	541	692

located in cells of high predicted suitability is significantly higher than their proportion in cells of low predicted suitability (Table 10.2; chi-square test, df = 3; $P < \cdot 0001$).

10.5 CONCLUSIONS

We have shown in this chapter a mechanistic model that represents spatial and temporal variation in water conditions, temperature, and seed and seedling predation. The model was applied to a heterogeneous Mediterranean environment of Mt. Pitulim, Israel. The results show that sites (cells) of high predicted suitability for recruitment are occupied by more trees than cells of low predicted suitability (df = 3; $P < \cdot 0001$). Furthermore, sites predicted to hold no potential for recruitment were populated by trees in very rare cases and mainly in the margins of the study area. These cases were attributed to long distance seed dispersal processes that were excluded from the current analysis. Further application of the model framework suggested here to longer dynamics in the study area and to other sites populated by Aleppo pine trees will help to increase our understanding of processes of tree population spread.

ACKNOWLEDGEMENTS

We gratefully thank Rakefet Shafran and Arnon Tsairi for their help with the preprocessing of RS and GIS databases and David Troupin and Chemdat Bannai for their help with estimating the trees age by the tree ring methodology. We also thank the Israeli Nature & Parks Authority's cooperation and permission to work in the Nachal Soreq Nature Reserve. Finally, Ran Nathan is happy to acknowledge the financial support of the German-Israeli Foundation (GIF 2006-1032.12/2000) and the Israeli Science Foundation (ISF-474/02) to his work.

REFERENCES

Andersson, L. and Milberg, P., 1998, Variation in seed dormancy among mother plants, populations and years of seed collection. *Seed Science Research*, **8**, 29–38.

Barling, R.D., Moore, I.D., and Grayson, R.B., 1994, A quasi dynamic wetness index for characterizing the spatial distribution of zones of surface saturation and soil water content. *Water Resources Research*, **30**, 1029–1044.

Beven, K. and Kirkby, M.J., 1979, A physically-based, variable contributing area model of basin hydrology. *Hydrological Sciences Bulletin*, **24**, 1–10.

Burrough, P.A., MacMillan, R.A., and Deursen, W.V., 1992, Fuzzy classification methods for determining land suitability from soil profile observations and topography. *Journal of Soil Science*, **43**, 193–210.

Chambers, J.C. and MacMahon, J.A., 1994, A day in the life of a seed: movement and fates of seeds and their implications for natural and managed systems. *Annual Review of Ecology and Systematics*, **25**, 263–292.

Clark, J.S., Fastie, C., Hurtt, G., Jackson, S.T., Johnson, C., King, G.A., Lewis, M., Lynch, J., Pacala, S., Prentice, C., Schupp, E.W., Webb, T., and Wyckoff, P., 1998, Reid's paradox of rapid plant migration — Dispersal theory and interpretation of paleoecological records. *Bioscience*, **48**, 13–24.

Harper, J.L., 1977, *Population Biology of Plants*. Academic Press, London.

Hulme, P.E., 1993, Postdispersal seed predation by small mammals. *Symposium of the Zoological Society of London*, **65**, 269–287.

Izhaki, I. and Ne'eman, G., 2000, Soil seed banks in Mediterranean pine forests. In *Ecology, Biogeography and Management of Pinus halepensis and P. brutia Forest Ecosystems in the Mediterranean Basin* (Ne'eman, G. and Trabaud, L., Eds.), pp. 167–182. Backhuys, Leiden, The Netherlands.

Kutiel, P., 1992, Slope and aspect effect on soil and vegetation in a Mediterranean ecosystem. *Israel Journal of Botany*, **41**, 243–250.

Levin, S.A., 1992, The problem of pattern and scale in ecology. *Ecology*, **73**, 1943–1967.

Levin, S.A., Muller-Landau, H.C., Nathan, R., and Chave, J., 2003, The ecology and evolution of seed dispersal: a theoretical perspective. *Annual Review of Ecology, Evolution and Systematics*, **34**, 575–604.

Nathan, R. and Muller-Landau, H.C., 2000, Spatial patterns of seed dispersal, their determinants and consequences for recruitment. *Trends in Ecology & Evolution*, **15**, 278–285.

Nathan, R. and Ne'eman, G., 2004, Spatiotemporal dynamics of recruitment in Aleppo pine (*Pinus halepensis* Miller). *Plant Ecology*, **171**, 123–127.

Nathan, R., Safriel, U.N., Noy-Meir, I., and Schiller, G., 1999, Seed release without fire in *Pinus halepensis*, a Mediterranean serotinous wind-dispersed tree. *Journal of Ecology*, **87**, 659–669.

Nathan, R., Safriel, U.N., Noy-Meir, I., and Schiller, G., 2000, Spatiotemporal variation in seed dispersal and recruitment near and far from *Pinus halepensis* trees. *Ecology*, **81**, 2156–2169.

Nathan, R., Safriel, U.N., and Noy-Meir, I., 2001, Field validation and sensitivity analysis of a mechanistic model for tree seed dispersal by wind. *Ecology*, **82**, 374–388.

Quezel, P., 2000, Taxonomy and biogeography of Mediterranean pines (*Pinus halepensis* and *P. brutia*). In *Ecology, Biogeography, and Management of Pinus halepensis and P. brutia Forest Ecosystems in the Mediterranean Basin* (Ne'eman, G. and Trabaud, L., Eds.), pp. 1–12. Backhuys, Leiden, The Netherlands.

Rejmanek, M. and Richardson, D.M., 1996, What attributes make some plant species more invasive? *Ecology*, **77**, 1655–1661.

Richardson, D.M., 2000, Mediterranean pines as invaders in the Southern Hemisphere. In *Ecology, Biogeography and Management of Pinus halepensis and P. brutia Forest Ecosystems in the Mediterranean Basin* (Ne'eman, G. and Trabaud, L., Eds.), pp. 131–142. Backhuys, Leiden, The Netherlands.

Robinson, V.B., 2003, A perspective on the fundamental of fuzzy sets and their use in geographic information systems. *Transactions in GIS*, **7**, 3–30.

Schupp, E.W. and Fuentes, M., 1995, Spatial patterns of seed dispersal and the unification of plant population ecology. *Ecoscience*, **2**, 267–275.

Silvertown, J. and Antonovics, J. (Eds.), 2001, *Integrating Ecology and Evolution in a Spatial Context*. Blackwell Science, Oxford.

Thanos, C.A., 2000, Ecophysiology of seed germination in *Pinus halepensis* and *P. brutia*. In *Ecology, Biogeography and Management of Pinus halepensis and P. brutia Forest Ecosystems in the Mediterranean Basin* (Ne'eman, G. and Trabaud, L., Eds.), pp. 13–35. Backhuys, Leiden, The Netherlands.

Tilman, D. and Kareiva, P. (Eds.), 1997, *Spatial Ecology: The Role of Space in Population Dynamics and Interspecific Interactions*. Princeton University Press, Princeton.

CHAPTER 11

Uncovering Spatial Feedbacks at the Alpine Treeline Using Spatial Metrics in Evolutionary Simulations

George P. Malanson and Yu Zeng

CONTENTS

0-8493-2837-3/05/$0.00+$1.50

11.1 INTRODUCTION

The spatial patterns of ecosystems in landscapes are important to their functioning (Turner, 1989). Exchanges of energy, matter, and individual organisms and species across landscapes are in part determined by pattern, and it is this exchange, along with vertical fluxes, that creates the pattern. Central to our understanding of pattern, spatial fluxes across landscapes, and changing pattern are ecotones, the transition zones between adjacent ecosystems (Risser, 1995). Ecotones are primarily characterized by the change in species composition, but are also seen as areas of interaction among ecosystems. The change in species is important because it indicates the limits of a species range, at least locally, and can be indicative of a species' relationship to the environment and other species. Fluxes may be higher here because of steeper gradients in the environment that coincide with the difference in species.

11.1.1 Treeline Ecotones

Treelines are the ecotones most studied. Most such studies are of treelines where forest borders grassland or tundra (rather than tree–shrub boundaries). The physical contrast has led to a number of hypotheses of treeline formation, location, and pattern. All recognize multiple causes, but they have different emphases. Stevens and Fox (1991) propose stature- and growth-related hypotheses. The stature-related hypotheses involve such problems as those a large plant encounters when resources are spatially and temporally limited, such as the inability to accumulate enough resources in its space. Other stature-related problems are those of exposure of the terminal buds and/or photosynthetic tissue damage. Growth-related hypotheses are related to the respiration costs of a large woody plant when resources are few in time (i.e., seasonally) and/or space. Analyses of carbon balance indicate that calculations of photosynthesis minus respiration are a fair indicator of treeline location (Cairns and Malanson, 1998), but physiological studies indicate that treeline carbon storage may be more flexible than previously thought (Hoch and Korner, 2003). These studies have not, however, considered alpine treeline advance and its spatial configuration from the perspective of complexity. Our recent work indicates that endogenous fractal dynamics may be the underlying mechanism (Zeng and Malanson, in review).

Important to understanding the above is the recognition of feedbacks between trees and their physical environment. Wilson and Agnew (1992) theorized that positive feedbacks occur where trees can modify their local environment in their own favor or cause environmental deterioration nearby, and that forbs and grasses can do the same. Negative feedback has not been as well theorized for treelines in general, but shading the soil is one such effect (Korner, 1998). With feedback comes the recognition of the role of spatial pattern in affecting the process. Feedbacks are local. They occur in the immediate vicinity of the trees, so that the pattern of trees determines the pattern according to which the underlying and surrounding environment is modified; this process affects the pattern in a feedback loop (Malanson, 1997).

The alpine treeline, also called the forest–alpine tundra ecotone, typifies such ecotones (Figure 11.1). Positive feedbacks for trees encroaching on tundra include

Figure 11.1 Patterns at treeline include advancing fingers, isolated patches, and openings remaining within contiguous forest.

warming the canopy by means of lower albedo, increasing nutrients and water by increasing local atmospheric deposition, and reducing transpiration and abrasion by slowing wind. The boundary is between adjacent ecosystems of very different physical characterisitics. The macro-scale controls are clearly related to climate, given the elevation gradient of treelines from the equator to high latitudes. The local elevations and patterns are not so straightforward, and so may help us understand the interaction of pattern and process.

We wish to study the relationship between the spatial pattern of trees, the feedbacks that ensue, and the potential change in the spatial pattern that results. This study is part of a project that includes other modeling approaches. Having started with a topography-based empirical model (Brown et al., 1994) and a physiologically mechanistic model that calculates photosynthesis, respiration, and the allocation of carbon (Cairns and Malanson, 1998), we added a forest dynamics model to simulate the population dynamics of trees (Bekker et al., 2001). We are now exploring cellular automata (CAs), wherein the weights given to neighboring trees in determining a cell's tree/tundra status are based on the calculation of physiological effects (Alftine and Malanson, 2004). However, there are two obstacles. First, landscape processes and their interactions with spatial patterns are complex, and thus it is difficult to identify specific mechanisms. Second, even if an expression can be used, it is difficult to evaluate any set of parameters. Here, we attempt to determine whether the weights for the effects of neighboring trees on the status of a cell in a CA can be determined using evolutionary computation (EC).

11.1.2 Evolutionary Computation

Evolutionary computation is a means of producing aspects of computer programs that achieve specified ends. EC is commonly applied to optimization problems in geography wherein the locations of facilities or land use sites are to be optimized relative to multiple criteria. It is essentially an engineering approach that models an evolutionary process to search a large parameter space; it is used because of its genetic operations and its stochastic nature under selection pressure to avoid becoming trapped in a local minimum while approximating the global maximum. The inherent parallelism in EC can explore a large solution space in an efficient way (Mitchell, 1996). Techniques based on EC have been pioneered in human geography for location-allocation (e.g., Hosage and Goodchild, 1986; Xiao et al., 2002), and for spatial modeling — including pattern recognition — by Openshaw (1992, 1995, 1998; Diplock and Openshaw, 1996; Turton and Openshaw, 1998). They have been applied in land use (Bennett et al., 1999; Matthews et al., 1999) and transportation studies (e.g., Pereira, 1996). Manson (2003) used genetic programming to determine local farmers' land use strategies. Applications related to biogeography are few (e.g., Noever et al., 1996), and those with spatial detail are rare (e.g., Hraber and Milne, 1997; Giske et al., 1998). The issue of how to make use of this optimization method for scientific simulation needs to be explored further. Here, we evaluate genetic algorithms (GAs), which are one of the methods under the umbrella of EC, to help us in this task.

11.2 ANALYSIS WITH THREE SPATIAL METRICS

11.2.1 Methods A

We apply GAs to a CA model of the alpine treeline to simulate the advance of trees into tundra. We begin with a CA in which the cells are either tree or tundra (cf. Noble, 1993; Malanson et al., 2001). We assume that the presence of trees in a neighborhood will increase the probability that a tundra cell will become a tree cell based on the positive feedback mechanism (Wilson and Agnew, 1992; and others). The number of trees in the neighborhood can be considered in several ways. Landscape ecologists have developed numerous spatial metrics for quantifying landscape pattern (McGarigal and Marks, 1995). Here, we chose three metrics that we thought should be important: number of trees, size of patches, and total number of cells in a 3×3 window. Because we do not know the optimal weights or parameters in an expression of feedback, we adopted evolutionary computation to search for the most likely ones. A genetic algorithm is essentially a goal-directed random search program, and we can take advantage of its implicit parallelism in maintaining a large set of search points at any given time. Here, the task of the GA is to find the weights that determine the effect of the chosen metrics on a cell. Another reason to adopt GA is related to the uncertainty about the exact form of feedbacks. It is possible that the feedback can be expressed in different but equivalent forms. Even though the exact forms of feedbacks are uncertain or unknown, we can allow a GA to approach the best results that can be obtained under different forms of transition rules.

11.2.1.1 General Design of the Simulation System

There are three modules in this system: the GA, the CA, and the spatial metrics calculation (Figure 11.2). Operationally, they are organized into three layers: the first layer is the GA, which creates a population of values in the parameter space expressed as sets of weights according to the genetic operations of mutation, crossover, and

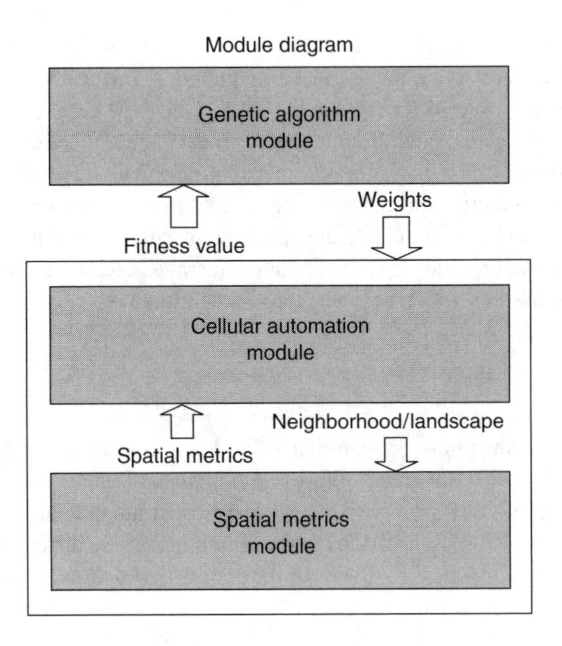

Figure 11.2 Outline of the interaction between components of the EC model for an alpine treeline.

selection. The second layer is the CA model, which accepts a set of weights into its transition rule, which in turn drives the treeline advance. The third layer is the spatial metrics module, which is called by the CA model to calculate the spatial metrics in a designated neighborhood. The resulting treeline spatial patterns are in turn fed back into the GA module for the purpose of computing a fitness value in the genetic operations and producing the next generation of parameters or sets of weights for future CA steps.

11.2.1.1.1 CA model

Our CA is a two-dimensional lattice with each cell having two possible states, tree or tundra. The lattice represents a long gradual slope of tundra, initially with trees near the bottom. Although, nonuniform conditions, random and not, are found (e.g., Malanson et al., 2002; Walsh et al., 2003), we assumed uniform conditions across the slope in order to examine what patterns could emerge from endogenous processes alone. Tundra cells can change to tree cells, but not the reverse. The lattice is initialized with a gradient of probability of a tundra cell changing into a trees. The initial probability of a cell changing into a tundra cell is modified by trees in its neighborhood. The change in the probability is calculated as a function of selected spatial metrics.

11.2.1.1.2 Spatial metrics

Our fitness function is defined as the combination of spatial metrics that best matches the observed pattern at an alpine treeline. Landscape ecology is largely founded on the notion that environmental patterns greatly influence ecological processes

(Turner, 1989). Landscape patterns can be quantified in a variety of ways depending on the type of data collected, the manner in which they are collected, and the objectives of the investigation (McGarigal and Marks, 1995). At each iteration the pattern of trees and tundra is analyzed. We are trying to determine whether the clumping of trees has a different effect from the number of trees. We construct patches of contiguous trees and quantify the number and size of patches and the density of trees. The fitness function is calculated as the sum of normalized similarities in the shape-adjusted patch number, total tree cells, and the average patch size between simulated and observed (classified ADAR image) treeline landscapes.

$$\text{Fitness value} = \Delta\text{PN} + \Delta\text{TCN} + \Delta\text{APS} \tag{11.1}$$

where ΔPN = patch number in simulated landscape − patch number in observed landscape; ΔTCN = total tree cells in simulated landscape − total tree cells in observed landscape; ΔAPS = average patch size in simulated landscape − average patch size in observed landscape. We ran the simulation using different neighborhood window sizes ranging from 3×3 to 9×9; the results based on different window sizes are indistinguishable.

11.2.1.1.3 Genetic algorithm

To find the optimal weights for the spatial metrics used to set the probabilities in the CA, we used a GA. Population-level operations included (1) roulette wheel selection. Subsequent chromosome-level operations included (2) reproduction by N-point crossover; (3) mutation; and (4) statistics of the average, highest, and lowest fitness values. We used a distance measure in the space defined by the three spatial metrics as the fitness function. Remote sensing is a primary source of input to the model. Similarity to the pattern observed in a classified ADAR image of a section of treeline on Lee Ridge, Glacier National Park, MT (Figure 11.3) was calculated. The ADAR 5500 System is a second-generation, charge-coupled device frame camera system. It operates in four channels in the visible and near-infrared wavelengths. For this study, spectral information was captured at a spatial resolution of $1 \times 1\ \text{m}^2$ in July 1999. On-board GPS technology spatially relates each acquired image frame to ground coordinates. We generated 1 m representations of trees/no-trees using a supervised, maximum likelihood classification based on field data.

11.2.1.1.4 Monte Carlo runs

Because of high variances of the resultant spatial patterns from the same set of weights in the transition rule, we adopted a Monte Carlo approach to try to increase the reliability of the fitness comparison and the next generation of weights used in the model. We tried the variance reduction technique of antithetic variables, but without success. We believe that the variance of the resultant spatial patterns is equivalent to different stages of treeline advance and can be considered as differential realization of the same system, and that it arises from self-organization within this system, as discussed below.

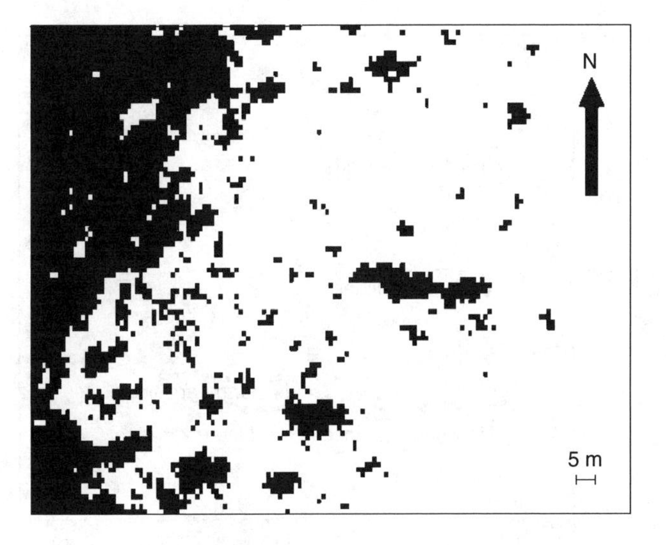

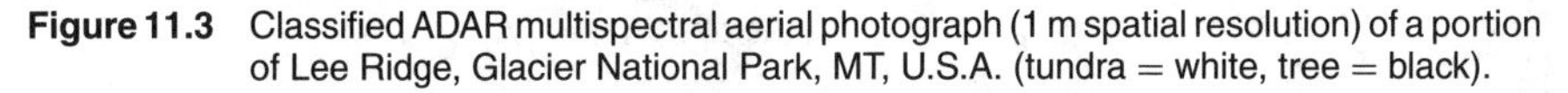

Figure 11.3 Classified ADAR multispectral aerial photograph (1 m spatial resolution) of a portion of Lee Ridge, Glacier National Park, MT, U.S.A. (tundra = white, tree = black).

11.2.1.2 Tests

We use EC to determine the weights in the CA for two conditions, a linear model:

$$P = \text{weight } 1 \times \text{patch number} + \text{weight } 2 \times \text{average patch size} \\ + \text{weight } 3 \times \text{total cell number} \tag{11.2}$$

and a nonlinear model:

$$p = \text{weight } 1 \times \text{patch number} + \text{weight } 2 \times (\text{average patch size})^2 \\ + \text{weight } 3 \times \text{total cell number} \tag{11.3}$$

Mean fitness is calculated by averaging the fitness values of one generation of chromosomes, which provides a representation of how fit this generation is and how close it is to the global maximum. The mean fitness series also conveys the progress of the GA. Elite fitness measures the best chromosome from the current generation, the one producing the highest fitness value. In this program, we use elitism to keep the best chromosome in the next generation in order to accelerate the evolutionary process of optimization.

11.2.2 Results A

The landscapes resulting from the nonlinear transition rule are much closer to the observed one than those from the linear transition rule after optimization (Figure 11.4).

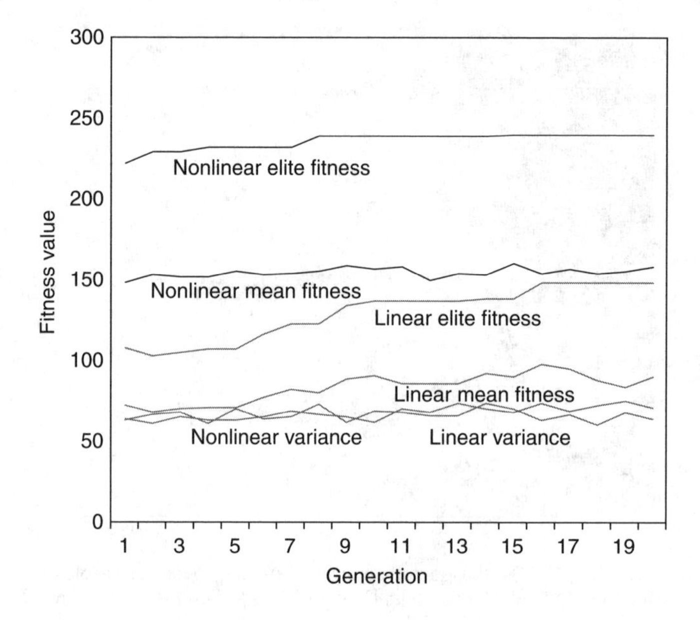

Figure 11.4 Fitness performance of the models based on spatial metrics with and without a nonlinear function and with and without the use of elite fitness.

Even the elite fitness of the linear model was not as high as the mean fitness of the nonlinear model. The overall results do not show improvement over time after the early generations. These results suggest that nonlinear positive feedback is more likely to be the driving force in creating alpine treeline patterns. They do not inform us about the usefulness of EC, but instead point to another line of research wherein EC may be useful, that is, the determination of the value for the exponent in the nonlinear function. However, the variances from the linear and nonlinear simulations are indistinguishable. This suggests that some internal variability is independent of the choice of transition rules, and thus that limiting by evolutionary optimization techniques is not satisfactory. How best to apply optimization methods for such a stochastic system needs further study. The variance may also be due to the spatial metrics chosen to describe the landscape. Perhaps these cannot exactly quantify the internal complexity of this system, which causes the "moving target" problem for optimization.

The results show that it is difficult to distinguish different transition rules in terms of different sets of weights. The variances in the results are very large, and the Monte Carlo approach cannot reduce them. The underlying cause has been explored, and it seems that self-organized complexity may be responsible for such phenomena; the nearly steady state after a few generations supports this idea. The endogenous self-organization process makes the system insensitive to quantitative changes of internal mechanisms, which is consistent with universality in self-organized complex systems. We suggest that the quantitative modification of transition rules may not be a good way of searching for appropriate mechanisms.

11.3 ANALYSIS WITH POLYNOMIAL FEEDBACK

11.3.1 Methods B

Close examination of the spatial patterns generated using the above method reveals that there are fewer big patches and more small patches than in the observed landscape. To further explore the usefulness of a GA to the analysis of a self-organizing system, we turn to a description of establishment and mortality probabilities that take into account both positive and negative feedbacks that are linear or nonlinear. Taking the Weierstrass Approximation Theorem, that is, that a continuous function on a bounded and close interval can be approximated on that interval by polynomials (Krantz, 1999), we chose polynomials of the following form to represent the establishment process:

$$\text{Probability}\ (X) = \sum_{i=1}^{N} w_i\, X^i \tag{11.4}$$

where X is the average size of patches in a neighborhood for establishment and w_i are the weights determined by the EC. The probability of tree mortality (P_{m}) is a function of the age of a tree and the number of trees in its immediate neighborhood. If the tree's age is 6 years or under,

$$P_{\text{m}} = \frac{1}{C\sqrt{n}}\left(1 - \frac{1}{1 + \lambda\, e^{-z}}\right) \tag{11.5}$$

where λ is a coefficient; here it takes the value 50. C is a parameter to adjust the strength of positive feedback in mortality, here it takes the value 3.0; n is the number of tree cells in the immediate neighborhood (3×3); z is the age of the tree. If the tree's age is over 6 years, mortality probability is 0.02.

We then rewrote the GA program so that the GA searches the function space of the polynomials by searching the parameter space. In accordance with our previous work exploring self-organized complexity (Turcotte and Rundle, 2002) in the alpine treeline (Zeng and Malanson, in review), we used the difference between the power law exponents of the frequency–size distributions of the observed and simulated landscapes as the fitness function. In this way, one can see whether more similar spatial patterns can be generated and identify the relative importance of the two processes. The rationale is that, even if one does not know the specific functions for establishment and mortality, the GA will search for the best results of the two processes, which can then be compared. This is different from, and a conceptual improvement on, the Monte Carlo approach in that we have used our theoretical understanding of the system and the optimization power of genetic algorithms to search for the best performance to see whether realistic spatial patterns can be obtained and to identify the relative importance.

We ran two simulations. One had establishment as a random process and used the mortality with feedback described above, whereas the second reversed this comparison and had mortality as a random process and establishment with feedback.

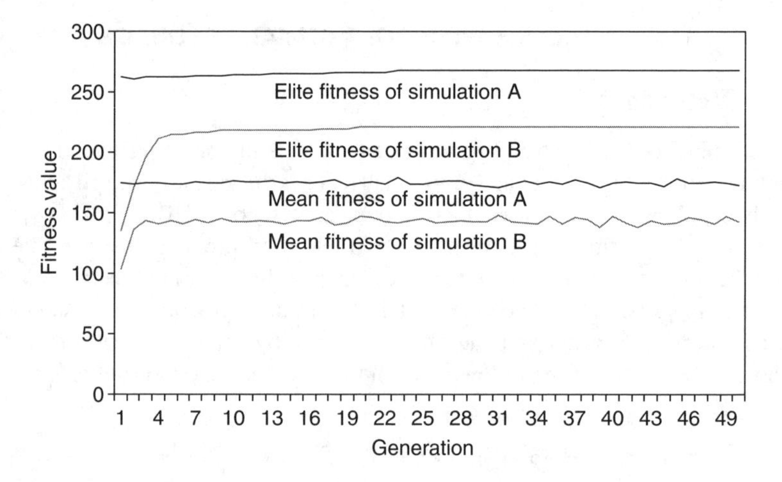

Figure 11.5 Fitness performance of the models based on polynomials with either establishment or mortality random and the other with feedback, and with and without the use of elite fitness.

11.3.2 Results B

This approach produces useful insights into the system. Figure 11.5 shows the dynamics of the mean and elite fitness values from the two simulations. It shows that both mean and elite fitness values with random establishment and spatially explicit mortality are higher than those from the reverse case. This result indicates that mortality is the more essential process, and more important than the establishment process in generating observed spatial patterns. Establishment without spatial positive feedback in mortality does not generate many large patches. The highest fitness will occur when both processes include spatially explicit feedback. The differences among the runs are, however, informative. The result indicates that ongoing mortality, tempered by the ameliorated environment of nearby trees, is the part of the process that most influences the spatial pattern of an advancing treeline.

The GA is not able to sharpen our understanding of system behavior by focusing on particular values for the coefficient in the polynomials. It does, however, set some boundaries. At lower levels of establishment, or higher levels of mortality, trees do not advance at all, whereas in the reverse case the advance does not allow the development of a spatial pattern. These results in general are trivial, but this is where quantitative specificity may be most useful.

11.4 DISCUSSION AND CONCLUSIONS

This research supports the theory that feedback is important in structuring the tree establishment pattern found at the alpine treeline. Further research is necessary to determine the relative importance of the various components of feedback, including directional forces. The overall balance between negative and positive feedbacks and the roles of establishment and mortality need additional investigation. Trees or krummholz create a negative feedback by cooling soil temperatures below and immediately adjacent to themselves (Korner, 1998), but nearby they create positive feedbacks by increasing canopy temperature and reducing wind. The balance of these two forces in creating the pattern varies with the wind and the canopy structure, which depend on each other. The two will affect establishment and seedling mortality in different directions.

As a result of these simulations, we tested a theoretical model of treeline advance (Zeng and Malanson, in review). Here, feedback modifies establishment probability as a logistic function that captures the positive neighborhood feedback or facilitation, which is reduced by negative neighborhood feedback when too many trees are nearby (and shading dominates). Tree mortality is a function of a tree's age and the number of trees in its immediate neighborhood. We analyzed the relationship between the time series of landscape potential and that of the exponent of the frequency distribution of patch size (Figure 11.6). When lagged by 5 years, the exponent of the frequency distribution of patch size is negatively correlated with landscape potential (−0.4186; $p < .001$), suggesting that spatial structure may change ahead of landscape potential and exert a positive impact on the latter.

The rules specify the nonlinear positive and negative feedbacks between pattern and process at a local scale, but these local interactions diffuse stochastically across the landscape. Global patterns and a linear correlation between global pattern and process with temporal and spatial fractal scaling properties emerge from these dispersed, localized interactions through a cross-scale self-organizing process (Figure 11.7).

Fractal dynamics driven by dispersed, localized pattern–process interactions collectively are capable of self-organizing to create long-term landscape-scale correlations. When new tree patches form or existing patches expand, localized pattern–process interactions are established across the landscape. The spatial pattern of trees increases environmental heterogeneity and establishes landscape connectivity at various scales diffusively. These interactions cause fluctuations at small and

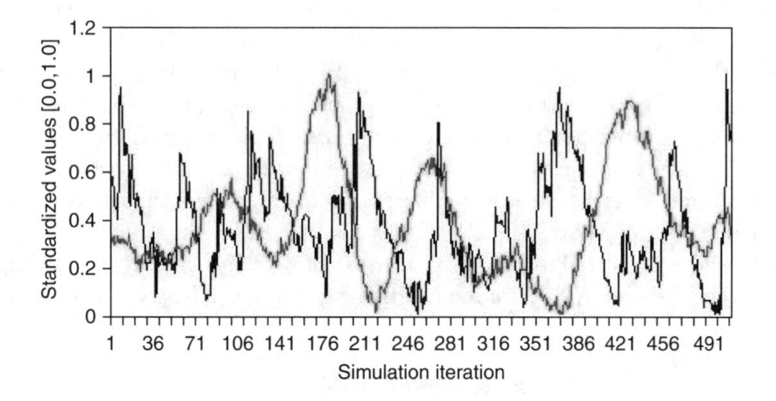

Figure 11.6 The frequency distribution of patch size is a useful indicator of the spatial structure of alpine treeline and it is correlated with the potential of the landscape to support new advance based on positive feedback.

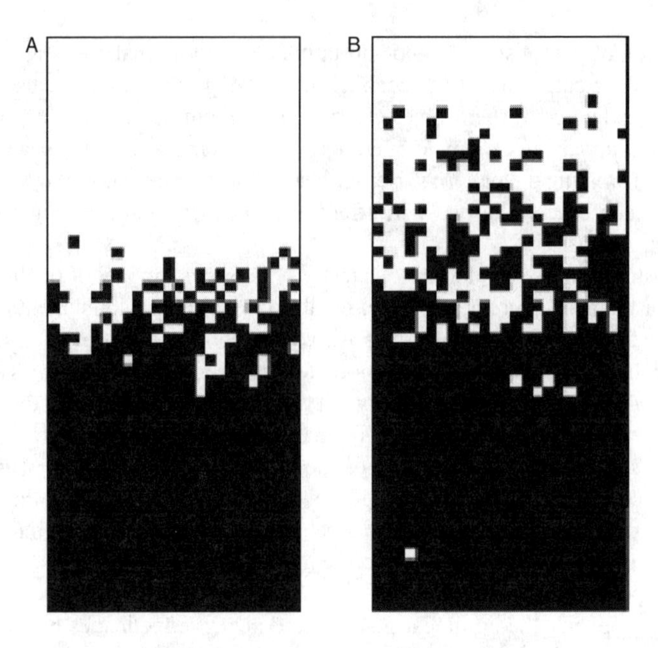

Figure 11.7 Simulated treelines for the (A) linear and (B) nonlinear transition rules (tundra = white, tree = black).

medium scales at most times, but occasionally the interactions create connectivity at the landscape scale through long-term and large-scale feedbacks in a coalescence of patches. A second-order phase transition between high and low fractal states is triggered that collapses the environmental variability, and connectivity extends across the landscape. We propose that self-organized complexity (in the sense of Turcotte and Rundle, 2002; not self-organized criticality in the sense of Bak et al., 1988) is an organizing theory that covers this pattern–process interaction.

The GA–CA simulations presented here support the interpretation of self-organization in the alpine treeline ecotone. Moreover, in self-organizing systems, EC has the potential to help us understand the broad form of the functions describing system behavior. Within the form of these functions, however, the exact values of the coefficients may not be narrowed down by EC, but they may not be so important. The nature of self-organization and the basic optimization purpose of EC are not harmonious. The particular coefficients that we use in our model are not very important, and the overall system behavior is robust across some range. Future research will be aimed at assessing the range of the coefficients, the extent to which the range can be explained in terms of biophysical processes, and, finally, whether EC can be used to determine the general form of the function if not the specific coefficient values.

ACKNOWLEDGMENTS

This work was supported by NSF Geography & Regional Science Program grant 0001738 to GPM by a cooperative agreement with the USGS Global Change Program coordinated by Dan Fagre.

REFERENCES

Alftine, K.J. and Malanson, G.P. 2004. Directional positive feedback and pattern at an alpine treeline. *Journal of Vegetation Science*, **15**, 3–12.

Bak, P., Tang, C., and Wiesenfeld, K. 1988. Self-organized criticality. *Physical Review A*, **38**, 364–374.

Bekker M.F., Malanson, G.P., Alftine, K.J., and Cairns, D.M. 2001. Feedback and pattern in computer simulations of the alpine treeline ecotone. In Millington, A.C., Walsh, S.J., and Osborne, P.E., Eds. *GIS and Remote Sensing Applications in Biogeography and Ecology*. Kluwer, Dordrecht, pp. 123–138.

Bennett, D.A., Wade, G.A., and Armstrong, M.P. 1999. Exploring the solution space of semi-structured geographical problems using genetic algorithms. *Transactions in GIS*, **3**, 89–109.

Brown, D.G., Cairns, D.M., Malanson, G.P., Walsh, S.J., and Butler, D.R. 1994. Remote sensing and GIS techniques for spatial and biophysical analyses of alpine treeline through process and empirical models. In Michener W.K., Stafford S., and Brunt J., Eds. *Environmental Information Management and Analysis: Ecosystem to Global Scales*. Taylor and Francis, Philadelphia, pp. 453–481.

Cairns, D.M. and Malanson, G.P. 1998. Environmental variables influencing carbon balance at the alpine treeline ecotone: a modeling approach. *Journal of Vegetation Science*, **9**, 679–692.

Diplock, G. 1998. Building new spatial interaction models by using genetic programming and a supercomputer. *Environment and Planning A*, **30**, 1893–1904.

Diplock, G. and Openshaw, S. 1996. Using simple genetic algorithms to calibrate spatial interaction models. *Geographical Analysis*, **28**, 262–279.

Giske, J., Huse, G., and Fiksen, O. 1998. Modelling spatial dynamics of fish. *Reviews in Fish Biology and Fisheries*, **8**, 57–91.

Hoch, G. and Korner, C. 2003. The carbon charging of pines at the climatic treeline: a global comparison. *Oecologia*, **135**, 10–21.

Hosage, C.M. and Goodchild, M.F. 1986. Discrete space location-allocation solutions from genetic algorithms. *Annals of Operations Research*, **6**, 35–46.

Hraber, P.T. and Milne, B.T. 1997. Community assembly in a model ecosystem. *Ecological Modelling*, **103**, 267–285.

Korner, C. 1998. A reassessment of high elevation treeline positions and their explanation. *Oecologia*, **115**, 445–459.

Krantz, S.G. 1999. *Handbook of Complex Variables*. Birkhäuser, Boston.

Malanson, G.P. 1997. Effects of feedbacks and seed rain on ecotone patterns. *Landscape Ecology*, **12**, 27–38.

Malanson, G.P., Xiao, N., and Alftine, K.J. 2001. A simulation test of the resource averaging hypothesis of ecotone formation. *Journal of Vegetation Science*, **12**, 743–748.

Malanson, G.P., Butler, D.R., Cairns, D.M., Welsh, T.E., and Resler, L.M. 2002. Variability in a soil depth indicator in alpine tundra. *Catena*, **49**, 203–215.

Manson, S.M. 2003. The SYPR integrative assessment model: complexity in development. In Turner II, B.L., Foster D., and Geoghegan J., Eds. *Final Frontiers: Understanding Land Change in the Southern Yucatan Peninsular Region*. Claredon Press, Oxford, pp.271–291.

Matthews, K.B., Sibaald, A.R., and Craw, S. 1999. Implementation of a spatial decision support system for rural land use planning: integrating geographic information system and environmental models with search and optimisation algorithms. *Computers and Electronics in Agriculture*, **23**, 9–26.

McGarigal, K. and Marks, B.J. 1995. *FRAGSTATS: Spatial Pattern Analysis Program for Quantifying Landscape Structure*. Oregon State University, Forest Science Department, Corvallis, OR.

Mitchell, M. 1996. *An Introduction to Genetic Algorithms*. MIT Press, Cambridge, MA.

Noble, I.R. 1993. A model of the responses of ecotones to climate change. *Ecological Applications*, **3**, 396–403.

Noever, D.A., Brittain, A., Matsos, H.C., Baskaran, S., and Obenhuber, D. 1996. The effects of variable biome distribution on global climate. *Biosystems*, **39**, 135–141.

Openshaw, S. 1992. Some suggestions concerning the development of artificial intelligence tools for spatial modeling and analysis in GIS. *Annals of Regional Science*, **26**, 35–51.

Openshaw, S. 1995. Developing automated and smart pattern exploration tools for geographical information systems applications. *Statistician*, **44**, 3–16.

Openshaw, S. 1998. Neural network, genetic, and fuzzy logic models of spatial interaction. *Environment and Planning A*, **30**, 1857–1872.

Pereira, A.G. 1996. Generating alternative routes by multicriteria evaluation and a genetic algorithm. *Environment and Planning B*, **23**, 711–720.

Risser, P.G. 1995. The status of the science examining ecotones. *BioScience*, **45**, 318–325.

Stevens, G.C. and Fox, J.F. 1991. The causes of treeline. *Annual Review Ecology and Systematics*, **22**, 177–191.

Turcotte, D.L. and Rundle, J.B. 2002. Self-organized complexity in the physical, biological, and social sciences. *Proceedings of the National Academy of Sciences, USA*, **99**, 2463–2465.

Turner, M.G. 1989. Landscape ecology — the effect of pattern on process. *Annual Review of Ecology and Systematics*, **20**, 171–197.

Turton, I. and Openshaw, S. 1998. High-performance computing and geography; development, issues, and case studies. *Environment and Planning A*, **30**, 1839–1856.

Walsh, S.J., Butler, D.R., Malanson, G.P., Crews-Meyer, K.A., Messina, J.P., and Xiao, N. 2003. Mapping, modeling, and visualization of the influences of geomorphic processes on the alpine treeline ecotone, Glacier National Park, Montana, USA. *Geomorphology*, **53**, 129–145.

Wilson, J.B. and Agnew, A.D.Q. 1992. Positive-feedback switches in plant communities. *Advances in Ecological Research*, **23**, 263–336.

Xiao, N.C., Bennett, D.A., and Armstrong, M.P. 2002. Using evolutionary algorithms to generate alternatives for multiobjective site-search problems. *Environment and Planning A*, **34**, 639–656.

Zeng, Y. and Malanson, G.P. In review. Endogenous fractal dynamics at alpine treeline ecotones.

CHAPTER 12

Modeling of Geomorphic Processes in an Alpine Catchment

Volker Wichmann and Michael Becht

CONTENTS

0-8493-2837-3/05/$0.00+$1.50

12.1 INTRODUCTION

The work presented here is part of the *sed*iment cascades in *a*lpine *g*eosystems project (SEDAG), within which a research team from five universities is studying alpine sediment transfer by various geomorphic processes. It is intended to obtain more detailed information about the sediment budget and landscape evolution of two catchment areas in the Northern Limestone Alps, Germany (Lahnenwiesgraben, Ammergebirge mountains and Reintal, Wetterstein mountains). Therefore, the spatial interaction of hill slope and channel processes — including soil erosion, rockfall, debris flows on slopes and in channels, shallow landslides and full-depth avalanches — is studied (see Heckmann et al., 2002; Keller and Moser, 2002; Schrott et al., 2002; Unbenannt, 2002). It is attempted to develop new spatially distributed modeling approaches to describe the sediment cascade. The modeling task is to identify starting zones of the processes (disposition modeling; e.g., Becht and Rieger, 1997) and to determine which areas would be affected (process modeling; e.g., Wichmann et al., 2002; Wichmann and Becht, 2004a,b). It is intended to explore and illustrate the potential effects of land management strategies and climate change on landscape evolution (see Schmidt, Chapter 13, this volume, for an analysis of the effects of climate change on landslides). Because of the detailed modeling of process path, run-out distance and erosion and deposition areas, the models can also be applied to natural hazard assessment.

For both catchments, each with a surface area of about 16 km^2, digital elevation models (DEMs) with a cell size of 5 m have been calculated from photogrammetric contour data using ARC/INFO's TOPOGRID. Raw data at a 1:10,000 scale was obtained from the Bavarian Ordnance Survey (© Bayerisches Landesvermessungsamt München, Az: VM-DLZ-LB0628). Aerial photographs and orthophotos provided a basis for land use and geomorphologic mapping. Additional data layers with geomorphological, geologic, and land use data were prepared at the same spatial resolution like the DEMs.

A new GIS, *s*ystem for an *a*utomated *g*eo-scientific *a*nalysis (SAGA, see Böhner et al., 2003), developed by the working group 'geosystem analysis' at the Department of Physical Geography, University of Goettingen, was used to build the models. SAGA is capable of processing raster and vector data of various formats and is based on a graphical user interface. Additional functionality is added by loading external run time libraries; so-called module libraries. Thus, it is possible to extend SAGA without altering the main program. The module libraries are programmed with C++ which is rather easy, because many object classes and basic functions are already provided. All models presented here are coded as such module libraries.

12.2 METHODS

12.2.1 Process Starting Zones

Different approaches have been used to determine possible process starting zones: rule-based models, multivariate statistical analysis, and more physically based approaches.

12.2.1.1 Rule-Based Models

Areas susceptible to process initiation may be derived by qualitative and quantitative analysis of significant data layers. This includes the classification and weighting of each parameter map. For example, the following procedure was used to determine possible starting points of debris flows in channels (Zimmermann et al., 1997; Buwal et al., 1998):

- Extraction of the channel network from digital elevation data using flow accumulation and plan curvature thresholds.
- Extraction of channel cells receiving enough sediment from the hill slopes to produce debris flows. Therefore, a maximum distance to the channel network and a minimum slope threshold in the uphill direction were specified to determine hill slope cells that may deliver material. This material contributing area was weighted in relation to vegetation cover and active process (e.g., a cell with bare soil will deliver more material than a cell covered with vegetation, a cell with landslide activity will deliver more material than a cell with rockfall activity).
- Extraction of possible starting points by combining empirically derived thresholds for flow accumulation, slope and potential sediment supply. The basic idea is that the triggering of a debris flow is governed by sufficient peak discharge, channel bed slope, and available sediment. On steeper channel beds a lower discharge is needed than on lower slopes.

Potential rockfall source areas, that is, mostly uncovered sheer rock faces, were derived by applying a simple slope threshold as described in Section 12.2.2.

12.2.1.2 Multivariate Statistical Analysis

A multivariate regression analysis for alphanumeric data is used to delineate possible starting zones of debris flows on slopes. A similar approach has been used by Jäger (1997) to determine potential landslide areas. The analysis is based on a binary grid containing mapped debris flows (1: presence, 0: absence of debris flow starting zones) and several (classified) grids containing relevant geofactors (e.g., slope, vegetation) in the area under study. In order to get a larger sample, the analysis was carried out on the catchment area of the observed debris flows instead of using point data. All relevant parameter combinations were examined in SAGA and written to a contingency table that is exported to an external statistical software package (SPSS, 2001). A log-linear model was used to calculate the probability of process occurrence. The result was retransformed to a probability map in SAGA. In the special case of using the catchment area instead of point data, one obtains the probability of a grid cell belonging to a catchment that may produce debris flows on slopes.

12.2.1.3 Physically Based Approaches

A more physically based approach is used to analyze the topographic influence on shallow landslide initiation (Montgomery and Dietrich, 1994; Montgomery et al., 2000). Therefore, a hydrologic model (O'Loughlin, 1986) is coupled with a slope stability model. Soil saturation is predicted in response to a steady state rainfall for each cell of the DEM. An *infinite-slope* stability model uses this relative soil saturation to analyze the stability of each topographic element. The model was extended to use spatially variable soil properties (soil thickness, effective soil cohesion including the effect of reinforcement by roots, bulk density, hydraulic conductivity, and friction angle). Thus, it was possible to calculate which elements become unstable for a given steady state rainfall, or to calculate the necessary steady state rainfall, which causes instability in an element. The latter can be seen as a measure of the relative potential of each element for shallow landsliding. Besides the stability classes stable and unstable, Montgomery and Dietrich (1994) define two further stability classes: unconditionally unstable elements are those predicted to be unstable even when dry and unconditionally stable elements are those predicted to be stable even when saturated. First results are presented in Wichmann and Becht (2004a) and show a good agreement with mapped initiation sites. Schmidt (Chapter 13, this volume) uses physically based approaches to explore the influence of climate change on landslide activity in central Germany (Bonn).

12.2.2 Process Path and Run-Out Distance

Process pathways are modeled by a combination of single and multiple flow direction algorithms. The algorithms are incorporated in a random walk model, which can be adjusted to different processes by three calibration parameters. The total process area results from Monte Carlo simulation. Run-out distances are modeled by calculating the velocity along the process path by either one or two parameter friction models.

12.2.2.1 Random Walk

The process path is modeled by a grid based random walk similar to the *dfwalk* model of Gamma (2000). All immediate neighbor cells in a 3×3 window, which have a lower elevation than the central cell, are potential flow path cells. To reduce this set **N**, two parameters are available: a slope threshold and a parameter for divergent flow. Possible flow path cells are determined by:

$$\mathbf{N} = \left\{ n_i \middle| \begin{cases} \gamma_i \geq (\gamma_{\max})^a & \text{if } 0 < \gamma_{\max} \leq 1, \\ \gamma_i = \gamma_{\max} & \text{if } \quad > \gamma_{\max} > 1, \end{cases} \quad i \in \{1, 2, \ldots, 8\}, a \geq 1 \right\} \tag{12.1}$$

and

$$\gamma_i = \frac{\tan \beta_i}{\tan \beta_{\text{thres}}}, \quad \beta_i \geq 0, \ i \in \{1, 2, \ldots, 8\} \tag{12.2}$$

where $\gamma_{\max}$ is the $\max(\gamma_i)$, β_i is the slope to neighbor i, β_{thres} is the slope threshold, and a is a parameter for divergent flow. **N** is reduced to the neighbor of steepest descent if the slope to the neighbor is greater than the slope threshold. This results in a single flow direction algorithm like D8 (Jenson and Domingue, 1988). Otherwise Equation (12.1) provides a set of potential flow path cells. The probability for each cell to be selected from this set as flow path is given by

$$p_i = \begin{cases} \dfrac{\tan \cdot \beta_i \cdot p}{\sum_j \tan \beta_j} & \text{if } i' \in \mathbf{N}, \\ \dfrac{\tan \beta_i}{\sum_j \tan \beta_j} & \text{if } i' \notin \mathbf{N}, \end{cases} \quad i, j \in \mathbf{N} \tag{12.3}$$

If the set contains the previous flow direction i', abrupt changes in direction can be reduced by a higher weighting of i'. Therefore the persistence factor p is introduced, which is also contained in the calculation of the sum. The calculated transition probabilities are scaled to accumulated values between 0 and 1, and a random number generator is used to select one flow path cell from the set.

For each starting point, several random walks are calculated (Monte Carlo simulation). Each run results in a slightly different process path. A high enough number of iterations assures that the whole process area is reproduced.

The approach offers the following properties (for more details see Gamma, 2000):

- The slope threshold allows the model to adjust to different relief. In steep passages near the threshold, only steep neighbors are allowed in addition to the steepest descent. In flat regions, almost all lower neighbors are possible flow path cells. The tendency for divergent flow is increased. Above the slope threshold a single flow direction algorithm is used.
- The degree of divergent flow is controlled by parameter a.
- Abrupt changes in flow direction are reduced by a higher persistence factor.
- A tendency towards the steepest descent is achieved as the transition probabilities are weighted by slope.

With these properties, it is possible to calibrate the model in order to match the behavior of different geomorphic processes. A higher persistence factor implies a greater fixation in the direction of movement (accounting for inertia) as may be observed by debris flows or wet snow avalanches. A process like rockfall may be modeled with no persistence and a higher degree of divergence.

12.2.2.2 1-Parameter Friction Model

A general method for defining the run-out distance of rockfall was developed by Scheidegger (1975) and extended by van Dijke and van Westen (1990) and Meissl (1998). With this method, the velocity of a rock particle is calculated along a profile line that is divided into a number of triangles. We adapted this method to grid-based modeling (see Figure 12.1). The velocity on the processed grid cell depends on the velocity on the previous cell of the process path. It is updated as soon as a new cell is delineated as a flow path by the random walk model. The area between the rockfall source and the point at which the velocity becomes zero is considered to be the potential process area.

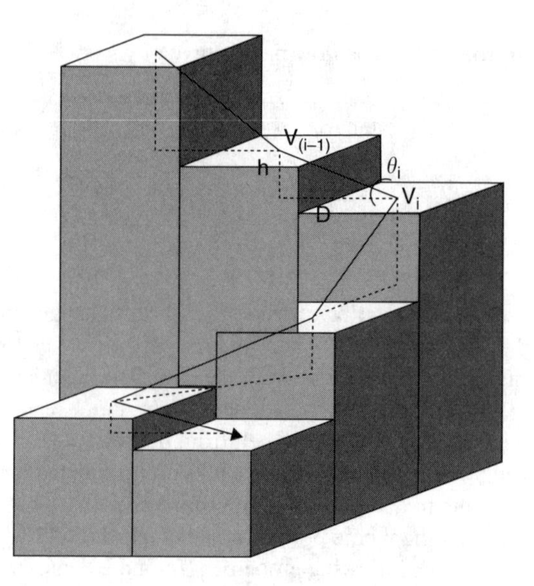

Figure 12.1 Grid-based approach to separate the flow path into triangles. Labels refer to 1- and 2-parameter friction model.

After a block is detached from the rock face, it is falling in free air (Equation (12.4)). The impact on the talus slope is accounted by reducing the velocity to a specified amount (75% after Broilli (1974), Equation (12.5)). Instead of using a separate grid with coded impact areas as done by Meissl (1998), a slope threshold is used to verify if the talus slope is reached. After the impact, the block is modeled as either sliding or rolling (Equations (12.6) and (12.7)).

$$\text{Falling}: \quad v_i = \sqrt{2gh_f} \tag{12.4}$$

$$\text{Impact}: \quad v_i = \sqrt{2gh_f} - r\sqrt{2gh_f} \tag{12.5}$$

$$\text{Sliding}: \quad v_i = \sqrt{v_{(i-1)}^2 + 2g(h - \mu_\mathrm{s} D)} \tag{12.6}$$

$$\text{Rolling}: \quad v_i = \sqrt{v_{(i-1)}^2 + \tfrac{10}{7} g(h - \mu_\mathrm{r} D)} \tag{12.7}$$

where v is the velocity [m/s], g is the acceleration due to gravity [m/s^2], h_f is the height difference between start point and element i [m], r is the energy loss, h is the height difference between adjacent elements [m], D is the horizontal difference between adjacent elements [m], μ_s is the sliding friction coefficient, and μ_r is the rolling friction coefficient.

The model is calibrated by three parameters: a slope threshold to determine if the block is in free fall (due to the grid data structure, a slope near or over 90° does not exist); a reduction parameter to account for the energy losses because of the impact on the talus slope; and a friction parameter. It is possible to use spatially distributed friction parameters to account for different geological materials and the effect of vegetation. The module is producing several output grids to facilitate the calibration process (including a grid with all cells in which free fall occurred and a grid with modeled maximum velocities).

12.2.2.3 2-Parameter Friction Model

A 2-parameter friction model (Perla et al., 1980) was used to calculate the run-out distance of snow avalanches and debris flows. Originally developed for snow avalanches, the model has also been applied to debris flows more recently (Rickenmann, 1990; Zimmermann et al., 1997; Gamma, 2000). The process is assumed to have a finite mass and the position in space or time of the center of mass is calculated. It is assumed that the motion is mainly governed by a sliding friction coefficient (μ) and a mass-to-drag ratio (M/D). M/D has a higher influence on velocity in steeper parts of the track, whereas the velocity in the run-out area is dominated by μ. Again, the process path is divided into segments of constant slope and an iterative solution is used to calculate the velocity along the path. The velocity on the processed grid cell depends on the velocity of the previous cell and is calculated by:

$$v_i = \sqrt{\alpha_i * (M/D)_i (1 - \exp^{\beta_i}) + (v_{(i-1)})^2 \exp^{\beta_i}} \tag{12.8}$$

and

$$\alpha_i = g(\sin\theta_i - \mu_i \cos\theta_i) \tag{12.9}$$

$$\beta_i = \frac{-2L_i}{(M/D)_i} \tag{12.10}$$

where v is the velocity [m/s], g is the acceleration due to gravity [m/s^2], θ is the local slope, L is the slope length between adjacent elements [m], μ_s is the sliding friction coefficient, and (M/D) is the mass-to-drag ratio [m]. At concave transitions in slope, Perla et al. (1980) assume the following velocity correction for $v_{(i-1)}$ before v_i is calculated with Equation (12.8):

$$v^*_{(i-1)} = \begin{cases} v_{(i-1)} \cos(\theta_{(i-1)} - \theta_i) & \text{if } \theta_{(i-1)} \geq \theta_i \\ v_{(i-1)} & \text{if } \theta_{(i-1)} < \theta_i \end{cases} \tag{12.11}$$

The correction is based on the conservation of linear momentum. For the case $\theta_{(i-1)} < \theta_i$ the authors expect that velocity decrease due to momentum change is compensated to a larger extent by velocity increase due to the reduced friction as the process tends to lift off the slope. If the process stops at a mid-segment position, the shortened segment length s may be calculated by:

$$s = \frac{(M/D)_i}{2} \ln\left(1 - \frac{(v_{(i-1)})^2}{\alpha_i (M/D)_i}\right) \tag{12.12}$$

In our grid-based approach, we circumvent solving Equation (12.12) and the process stops as soon as the square root in Equation (12.8) becomes undefined.

The model is calibrated by the two parameters μ and M/D. To overcome the problem of mathematical redundancy of a two parameter model (different combinations of μ and M/D can result in the same run-out distance), the parameter M/D is taken to be constant along the process path in the case of debris flow modeling (Zimmermann et al., 1997; Gamma, 2000). It is calibrated only once to obtain realistic velocity ranges. The parameter μ is calculated from the catchment area of each grid cell by empirically derived estimating functions (Gamma, 2000) as described in Section 12.3.2.

12.2.3 Erosion and Deposition Modeling

To model the sediment transfer throughout the catchment, erosion and deposition sites need to be identified. We use simple methods to classify the process area accordingly and calculate relative erosion and deposition heights assuming transport-limited conditions. The latter may be used to scale event-based measurements of sediment yield along the process path. Another possible field of application are stochastic-driven models that estimate erosion volumes from probability functions.

In the case of rockfall, a specified amount of material is subtracted from the DEM at the starting location in each model run. This amount is added to the DEM at the last cell of the process path (i.e., where the process stops).

A different approach is used for debris flows. We use simple threshold functions of slope and modeled velocity to delineate relative erosion and deposition heights along the process path. The threshold functions for erosion and deposition are combined in such a way that artifacts resulting from the usage of one threshold alone are minimized. For example, no material is deposited in flat parts of the profile if the velocity is still high. And no material is deposited in steep passages even if the velocity is low. A more detailed description of the method is given in Wichmann and Becht (2004b). In a slightly different version of the model, a specified amount of material is eroded and deposited during each model iteration and thus influences the following runs. Deposited material may produce sinks in the DEM and a special algorithm is used to fill sinks as soon as they are detected by a following run. Thus the filling of sinks and barriers and the plugging of the channel can be simulated. The algorithm fills the sink and, if needed, further cells of the process path upslope with available material up to heights that assure valid flow directions for the next runs.

12.3 FIRST RESULTS

This section provides first results obtained for rockfall and debris flows on slopes and in channels. Up to now it is not possible to calibrate the models to a full extent, since the data obtained from the individual working groups have not yet been fully

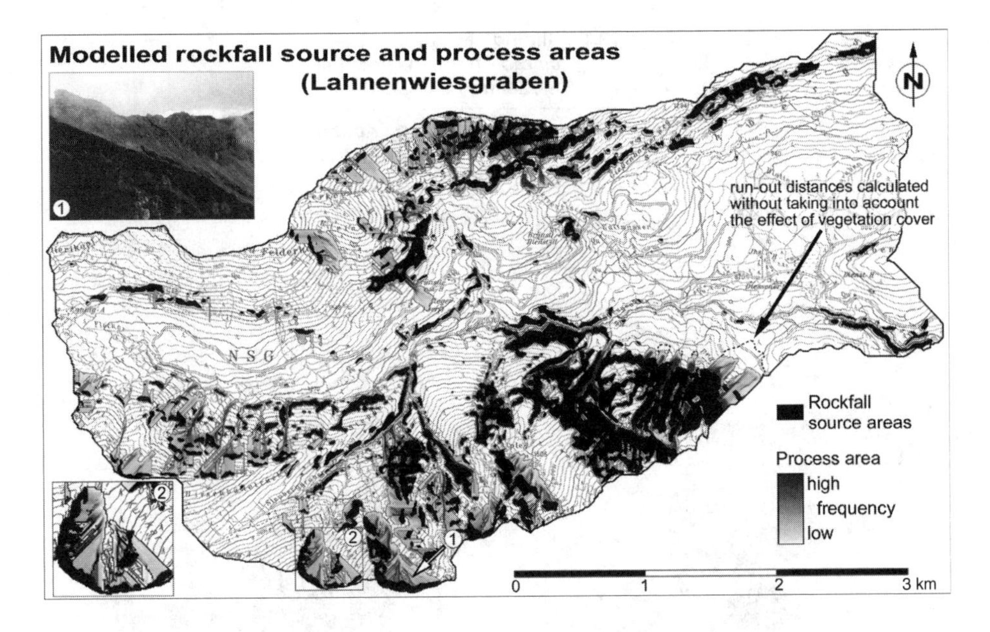

Figure 12.2 Results of rockfall modeling in the Lahnenwiesgraben catchment area.

analyzed. Nevertheless it is possible to show the potential of the models to describe spatially distributed sediment transfers in the Lahnenwiesgraben catchment area.

12.3.1 Rockfall Modeling

The results of rockfall modeling in the Lahnenwiesgraben catchment area are shown in Figure 12.2. Rockfall source areas are derived from the slope map including the effect of vegetation. The slope map was reclassified to slopes >40° (a value also used by Dorren and Seijmonsbergen, 2003) and then the true surface area of each grid cell was calculated (cell area/*cos*(slope)). The land use map was weighted by the fraction of free ground surface (100% uncovered, full of gaps (50%), and 25% uncovered). Potential rockfall source areas including relative process intensities result from the multiplication of the two maps. This method results in 13.9% of the Lahnenwiesgraben catchment area to be classified as rockfall producing area. Besides sheer rock walls, very steep slopes are also included. Talus slopes, in contrast, exhibit lower gradients than the threshold (20 to 40°, see photo (1) and magnified map (2) insets in Figure 12.2).

Process paths and run-out distances are modeled with a combination of random walk and a 1-parameter friction model. A slope threshold of 30°, a divergence factor of 2 and a persistence factor of 1 are used in the random walk model. The 1-parameter model is calibrated to fit the observed run-out distances on different slope materials by using spatially distributed friction coefficients (see Table 12.1). Free fall occurs as long as the slope is steeper than 60° and the velocity reduction by impact is 75%. On

Table 12.1 Sliding Friction Coefficients Used in Rockfall modeling

Material/Vegetation Cover	Sliding Friction Coefficient (μ)
Marl	0.4
Fluvial materials	0.5
Glacial deposits	0.6
Dolomite	0.7
Limestone	0.8
Forest	1.2

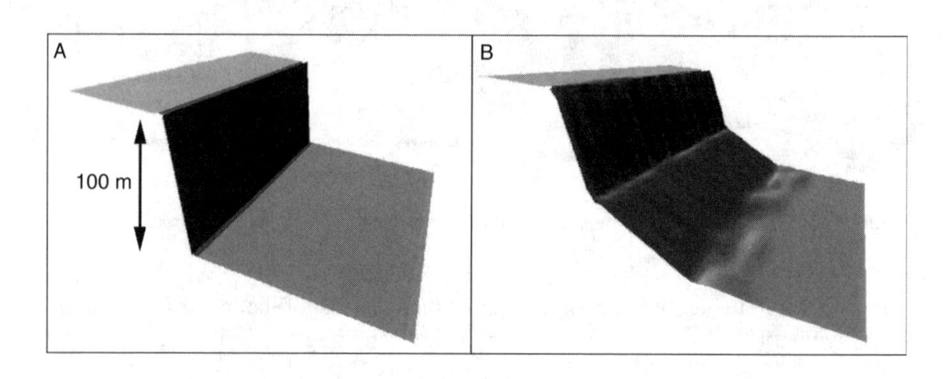

Figure 12.3 Three-dimensional view of (A) the initial rock face and (B) after 375,000 m^3 of material have been eroded and deposited.

the scree slope a sliding motion is modeled (Kirkby and Statham, 1975; Scheidegger, 1975).

Modeled run-out distances match the observed deposits. The effect of forest cover must be included in the friction coefficient, otherwise the run-out distances are overestimated (see dashed lines in Figure 12.2, SE part of the catchment). This shows the importance of dense forests for rockfall hazard mitigation.

The rockfall module was extended to test the model for the simulation of long-term landscape evolution. Instead of using a grid with coded process initiation cells, a slope threshold is used to define a set of potential starting cells. Start cells are selected from this set by random and the process path, run-out distance, and erosion and deposition are modeled. The set of potential starting cells is regularly updated to account for changes in elevation. Figure 12.3A shows an artificially produced rock face with a height of 100 m. The result after 1,500,000 random walks with 0.01 m erosion in each iteration is shown in Figure 12.3B. The rock face retreated and scree slopes developed with an inclination corresponding to the selected sliding friction coefficient. In the future, it is intended to test if the model is capable of reproducing the rockfall deposits (sediment storages) observed in the Reintal study area.

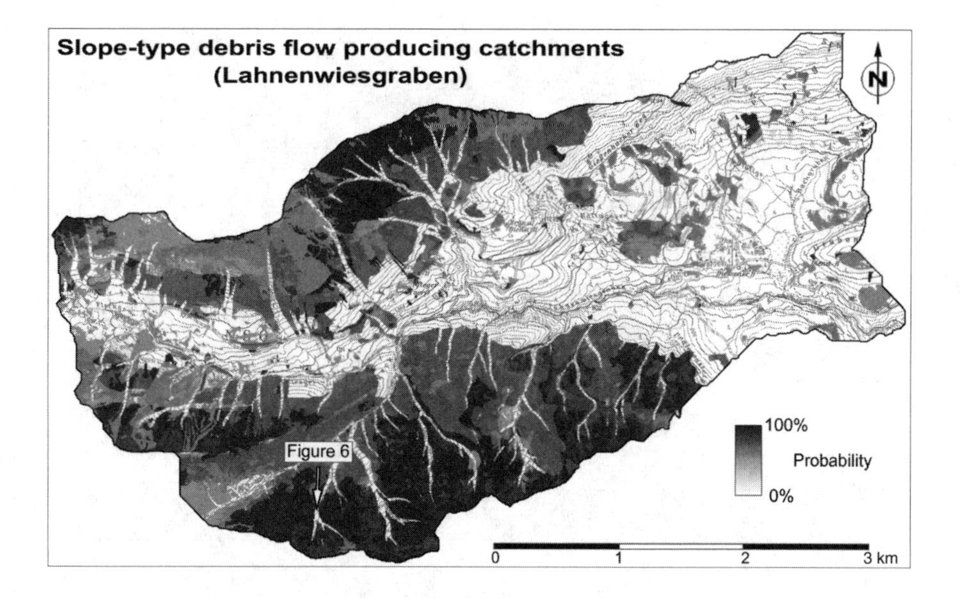

Figure 12.4 Map of potential debris flow producing grid cells in the Lahnenwiesgraben catchment area.

12.3.2 Debris Flow Modeling on Slopes

Process initiation cells for debris flows on slopes in the Lahnenwiesgraben catchment area are derived by multivariate statistical analysis as described in Section 12.2.1.2. The analysis is based on reclassified maps of slope, vegetation, infiltration capacity, and flow accumulation. Although material properties are important for the occurrence of debris flows, it was not possible to incorporate geological information into the analysis, because nearly all of the observed debris flows occur on talus slopes composed of dolomite. The results of the statistical analysis show that very low gradients and gradients $>60^\circ$ reduce the probability of debris flow initiation. Dense vegetation cover prevents high probabilities, debris flow initiation occurs mainly in areas with sparse or no vegetation. Higher infiltration rates favor the saturation of slope debris and thus debris flow initiation. High flow accumulation values, which are associated with higher discharges, do not allow enough material to accumulate for debris flow initiation and thus reduce the probability. Unfortunately, the small sample size of observed debris flows prevents a significant statistical analysis and makes further investigations in other catchments necessary. Nevertheless, modeling results show a relatively good agreement with observed debris flow catchments in the Lahnenwiesgraben. Within susceptible areas, the spatial distribution of high probabilities lacks some detail (Figure 12.4).

To derive potential starting points from the calculated probabilities to belong to a debris flow producing catchment, a separate module was used to accumulate probabilities in the downslope direction. Potential starting cells exceed a user-specified threshold of accumulated 'disposition' and must have slopes in a range typical for debris flows (20 to 45°). The user-threshold was set to values slightly higher than

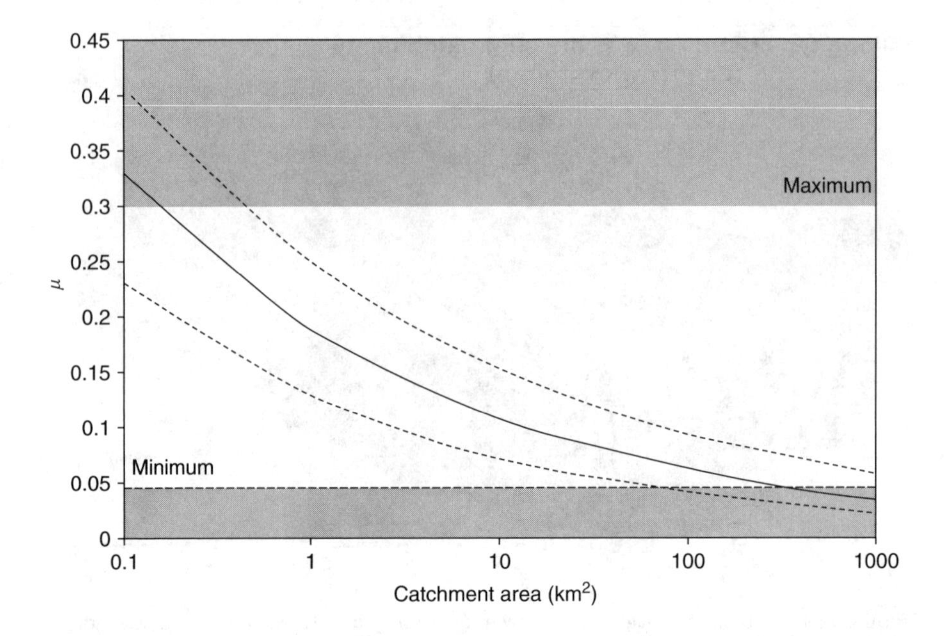

Figure 12.5 Relation between friction coefficient μ and catchment area *a*. A minimum threshold is set to 0.045 and a maximum threshold to 0.3. Estimator functions represent minimum (upper dashed line), likely and maximum (lower dashed line) run-out distance.

those calculated for observed initiation sites. The catchment area of each starting cell is delineated simultaneously and may be weighted by its mean probability to obtain a relative measure of debris flow release potential.

Process paths and run-out distances were modeled with a combination of random walk and a 2-parameter friction model. A slope threshold of 20°, a divergence factor of 1.3 and a persistence factor of 1.5 are used in the random walk model. Run-out distance is calculated with a constant mass-to-drag ratio of 75 m and spatially distributed friction coefficients (μ). The friction coefficient of each cell is calculated in relation to its catchment area a. The estimating functions (least squares method) in Figure 12.5 are empirically derived from mapped debris flows in Switzerland by Gamma (2000). The relationship is based on the observation that the sliding friction coefficients tend towards lower values with increasing catchment area. This is attributed to changing rheology with higher discharges along the process path. A minimum threshold is set to 0.045 and a maximum threshold to 0.3. Between the thresholds, μ is decreased with increasing catchment area upstream of the position of the debris flow in accordance with one of the estimating functions. The functions represent different scenarios as indicated in Figure 12.5. As slope type debris flows normally exhibit small catchment areas, they are mostly modeled with a constant maximum μ of 0.3.

Figure 12.6 shows a three-dimensional-view of the "Kuhkar" cirque (looking south). Erosion and deposition are modeled with arbitrary quantities of material to support visualization. Large amounts of erosion occur in steep parts of the process path and are followed by a zone of slight erosion before deposition sets in. Material

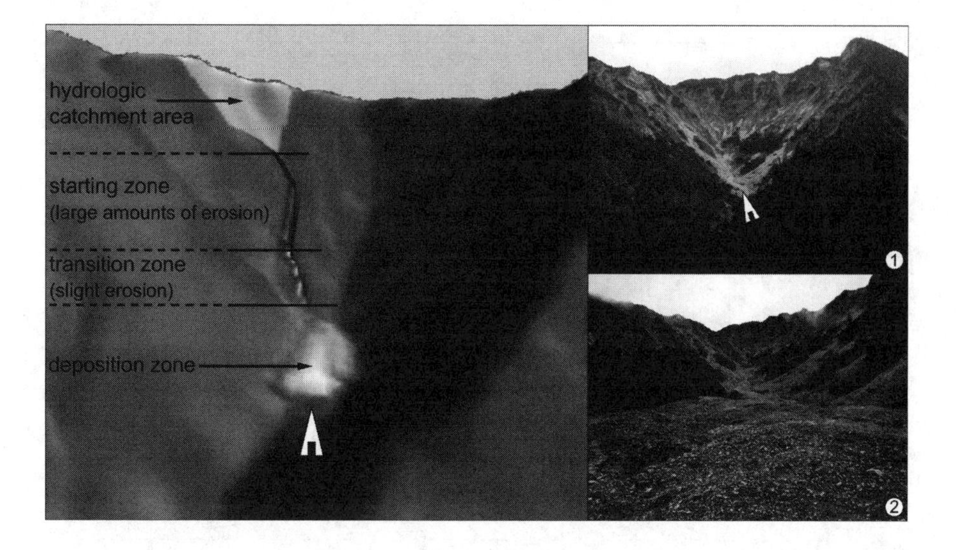

Figure 12.6 Three-dimensional-view of debris flow modeling on slopes in the "Kuhkar" cirque, Lahnenwiesgraben catchment area. Photo (1) shows a view of the cirque and white arrows indicate the line of sight of photo (2).

is eroded and deposited during every model iteration. Because of the sink filling algorithm used, the appearance of the resulting deposit is very similar to observed debris flow deposits. The deposits exhibit a steep front and decreasing deposition heights upslope.

12.3.3 Debris Flow Modeling in Channels

Process initiation cells for debris flow in channels are derived by qualitative and quantitative analysis as described in Section 12.2.1.1. Those cells which have a flow accumulation higher than 2500 m^2 and a concave plan curvature are classified as channel cells. All cells within a maximum distance of 250 m to the channel network and with a slope (in the direction of flow) steeper than 20° are selected as material contributing area. Each cell may be weighted in relation to its intensity of material supply, but up to now not enough information is available to do so. The relation between channel slope and catchment area in Figure 12.7 was derived empirically by Zimmermann et al. (1997) from debris flows in Switzerland. Potential starting cells exceed this threshold and have a minimum material contributing area of 10,000 m^2.

The resulting grid contains possible process initiation cells. A filter along the channel network is used to reduce the number of start cells since it is unnecessary to calculate the process path for each cell of the grid. If a starting cell is detected by the filtering algorithm, all lower starting cells in a specified distance along the channel reach (here: 500 m) are eliminated. Process path and run-out distance are modeled with a combination of random walk and 2-parameter friction model. A slope threshold of 20°, a divergence factor of 1.3, and a persistence factor of 1.5 are used in the random

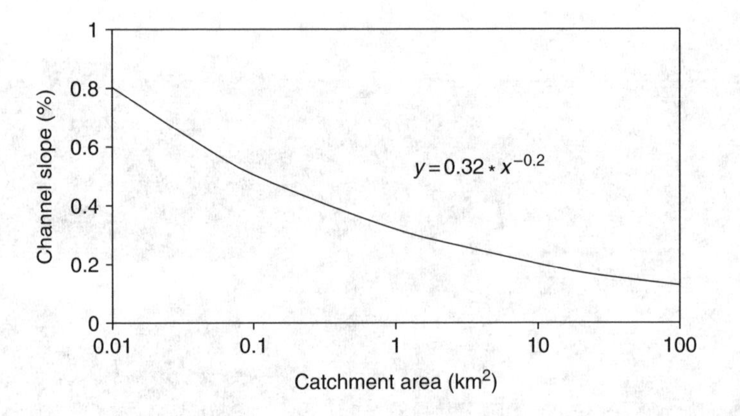

Figure 12.7 Relation between channel slope and catchment area. The threshold function is used to derive potential debris flow initiation cells.

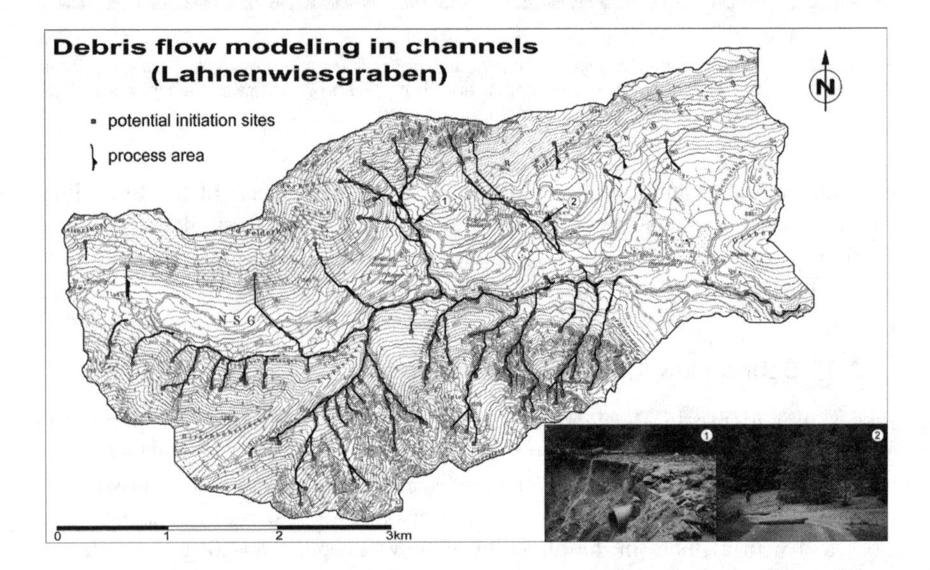

Figure 12.8 Debris flow modeling in channels in the Lahnenwiesgraben catchment area. Arrows indicate the line of sight of photo (1) and (2).

walk model. Run-out distance is calculated with a constant mass-to-drag ratio of 75 m and spatially distributed friction coefficients (μ) as described in Section 12.3.2. As the catchment area a of channel-type debris flows increases significantly downstream, the sliding friction coefficient is gradually reduced by one of the estimating functions. In Figure 12.8, the maximum run-out distances were calculated with the estimator function $\mu = 0.13 * a^{-0.25}$ (Gamma, 2000).

Although the material contributing area was not weighted, the results match well with observed debris flows in the catchment area. Material is eroded and deposited along the process path following the rules stated in Section 12.2.3. Most debris flows stop as soon as they reach the main channel with lower slopes. Further transportation of the deposited material is then accomplished by high discharges. In the year 2002, a high magnitude rainstorm event triggered several debris flows in some of the steeper torrents of the Lahnenwiesgraben. Photos (1) and (2) in Figure 12.8 show some of the damages done to the forest road. Wichmann and Becht (2004b) provide more details about the event in 2002 and the corresponding modeling results.

12.4 CONCLUSION

The first results presented here look promising and it should be possible to combine the models to describe parts of the sediment cascade in alpine catchments. Although the used models were originally developed for natural hazard zonation, they are applicable to geomorphologic problems too. By overlaying the model outputs it is possible to do a detailed terrain classification with respect of the spatial distribution of different geomorphic processes. As the calculated process areas are further subdivided into sections of erosion and deposition, it becomes possible to analyze the spatial relocation of material. Further research and data is needed to couple the models with measured sediment yield. Process rates may be obtained from historical data (e.g., dendrochronology, sediment accumulation dating by radiocarbon, geophysical methods) or from measurements of recent process intensities. In the future, we intend to incorporate more physically based approaches for erosion and deposition modeling. But in the meantime, the simple approaches to subdivide the process area are useable. Although the calculation of relative erosion and deposition heights lacks a true physical background, it is possible to obtain realistic results when measured sediment yield is scaled accordingly.

A validation of the modeling results is difficult, because it is impossible to map the complete spatial distribution of geomorphic processes. Mapping during the field trips is done on orthophotos, but a spatial displacement of the orthophotos with respect to the DEM makes the calculation of model accuracy difficult. A visual comparison of our own maps and available topographic maps with the modeled process areas reveals a high degree of conformance. In some cases, we discovered old deposits already covered with vegetation that had not been mapped yet.

Especially the validation of disposition modeling is difficult, as the model output consists of potential initiation sites. The most deficient results are those of the multivariate statistical analysis used for slope-type debris flows. The model needs to be improved, otherwise minor differences in the spatial occurrence of debris flows are not reproduced. We obtained more convenient results with conditional analysis as used by Clerici et al. (2002) for landslide susceptibility zonation. Satisfactory results are produced by the disposition models for rockfall and torrent bed-type debris flows. The empirical functions used in the latter case seem to be transferable to the natural conditions found in the Lahnenwiesgraben catchment area.

The random walk model in conjunction with a Monte Carlo simulation is capable of reproducing the observed process paths. The three calibration parameters allow the model to be applied to different processes. Some problems arise because of inaccuracies of the DEM. Our grid-based approach of the friction models used to calculate the run-out distances yields also satisfactory results. We intend to extend the models to reproduce different event magnitudes by using adequate friction coefficients. Further models, for example, slope wash and channel erosion, are under development.

ACKNOWLEDGMENTS

This research was funded by the German Research Foundation (DFG, Bonn), which is gratefully acknowledged by the authors. The comments of Stephen E. Darby and Peter M. Atkinson on the paper are highly appreciated.

REFERENCES

Becht, M. and Rieger, D. (1997): Spatial and temporal distribution of debris-flow occurrence on slopes in the Eastern Alps, In CHEN, C. [Ed.]: *Debris-Flow Hazard Mitigation: Mechanics, Prediction, and Assessment*, pp. 516–529.

Böhner, J., Conrad, O., Köthe, R., and Ringeler, A. (2003): System for an automated geographical analysis. http://134.76.76.30

Broilli, L. (1974): Ein Felssturz im Großversuch. *Rock Mechanics*, Suppl. **3**, 69–78 (in German).

Buwal, Bundesamt Für Umwelt, Wald, und Landschaft (1998): Methoden zur Analyse und Bewertung von Naturgefahren. *Umwelt-Materialien*, **85**, Naturgefahren, 247 (in German).

Clerici, A., Perego, S., Tellini, C., and Vescovi, P. (2002): A procedure for landslide susceptibility zonation by the conditional analysis method. *Geomorphology*, **48**, 349–364.

Dorren, L. and Seijmonsbergen, A. (2003): Comparison of three GIS-based models for predicting rockfall runout zones at a regional scale. *Geomorphology*, **56**, 49–64.

Gamma, P. (2000): Dfwalk — Ein Murgang-Simulationsprogramm zur Gefahrenzonierung. *Geographica Bernensia*, **G66**, 144 (in German).

Heckmann, T., Wichmann, V., and Becht, M. (2002): Quantifying sediment transport by avalanches in the Bavarian Alps — first results. *Zeitschrift für Geomorphologie N. F.*, Suppl. **127**, 137–152.

Jäger, S. (1997): Fallstudien zur Bewertung von Massenbewegungen als geomorphologische Naturgefahr. *Heidelberger Geographische Arbeiten*, **108**, 151 (in German).

Jenson, S.K. and Domingue, J.O. (1988): Extracting topographic structure from digital elevation data for Geographic Information System analysis. *Photogrammetric Engineering and Remote Sensing*, **54/11**, 1593–1600.

Keller, D. and Moser, M. (2002): Assessments of field methods for rock fall and soil slip modelling. *Zeitschrift für Geomorphologie N. F.*, Suppl. **127**, 127–135.

Kirkby, M.J. and Statham, I. (1975): Surface stone movement and scree formation. *Journal of Geology*, **83**, 349–362.

Meissl, G. (1998): Modellierung der Reichweite von Felsstürzen. *Innsbrucker Geographische Studien*, **28**, 249 (in German).

Montgomery, D.R. and Dietrich, W.E. (1994): A physically based model for the topographic control on shallow landsliding. *Water Resources Research*, **30**, 1153–1171.

Montgomery, D.R., Sullivan, K., and Greenberg, M. (2000): Regional test of a model for shallow landsliding, In Gurnell, A.M. and Montgomery, D.R. [Eds.]: *Hydrological applications of GIS*, pp. 123–135.

O'Loughlin, E.M. (1986): Prediction of surface saturation zones in natural catchments by topographic analysis. *Water Resources Research*, **22**, 794–804.

Perla, R., Cheng, T.T., and McClung, D.M. (1980): A two-parameter model of snow-avalanche motion. *Journal of Glaciology*, **26/94**, 197–207.

Rickenmann, D. (1990): Debris flows 1987 in Switzerland: modelling and fluvial sediment transport. *IAHS Publications*, **194**, 371–378.

Scheidegger, A.E. (1975): *Physical Aspects of Natural Catastrophes*, 289.

Schmidt, J. (2004): Chapter 13, this volume.

Schrott, L., Niederheide, A., Hankammer, M. Hufschmidt, G., and Dikau, R. (2002): Sediment storage in a mountain catchment: geomorphic coupling and temporal variability (Reintal, Bavarian Alps, Germany). *Zeitschrift für Geomorphologie N. F.*, Suppl. **127**, 175–196.

SPSS (2001): *SPSS for Windows*. Release 11.0.1, SPSS Inc.

Unbenannt, M. (2002): Fluvial sediment transport dynamics in small alpine rivers — first results from two upper Bavarian catchments. *Zeitschrift für Geomorphologie N. F.*, Suppl. **127**, 197–212.

van Dijke, J.J. and van Westen, C.J. (1990): Rockfall hazard: a geomorphologic application of neighbourhood analysis with ILWIS. *ITC Journal*, **1990/1**, 40–44.

Wichmann, V., Mittelsten Scheid, T., and Becht, M. (2002): Gefahrenpotential durch Muren: Möglichkeiten und Grenzen einer Quantifizierung. *Trierer Geographische Studien*, **25**, 131–142 (in German).

Wichmann, V. and Becht, M. (2004a): Modellierung geomorphologischer Prozesse zur Abschätzung von Gefahrenpotentialen. *Zeitschrift für Geomorphologie N. F.*, Suppl. **135**, 147–165.

Wichmann, V. and Becht, M. (2004b): Spatial modelling of debris flows in an alpine drainage basin. *IAHS Publications*, **288**, 370–376.

Zimmermann, M., Mani, P., Gamma, P., Gsteiger, P., Heininger, O., and Hunziker, G. (1997): *Murganggefahr und Klimaänderung — ein GIS-basierter Ansatz*. 161 (in German).

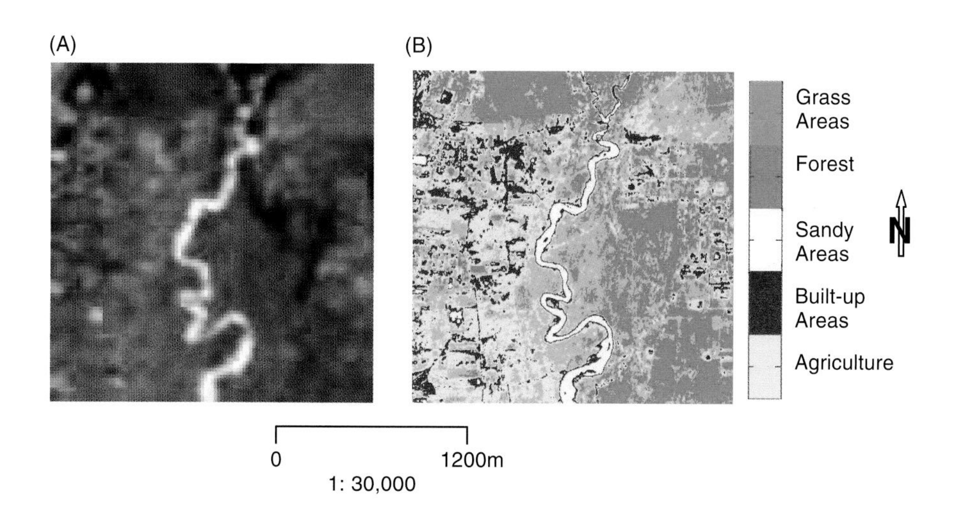

Figure 2.1 (A) IRS 1C LISS III FCC (Red: band 4; Blue: band 2; Green: band 1). (B) Classified PAN Image used as reference Data.

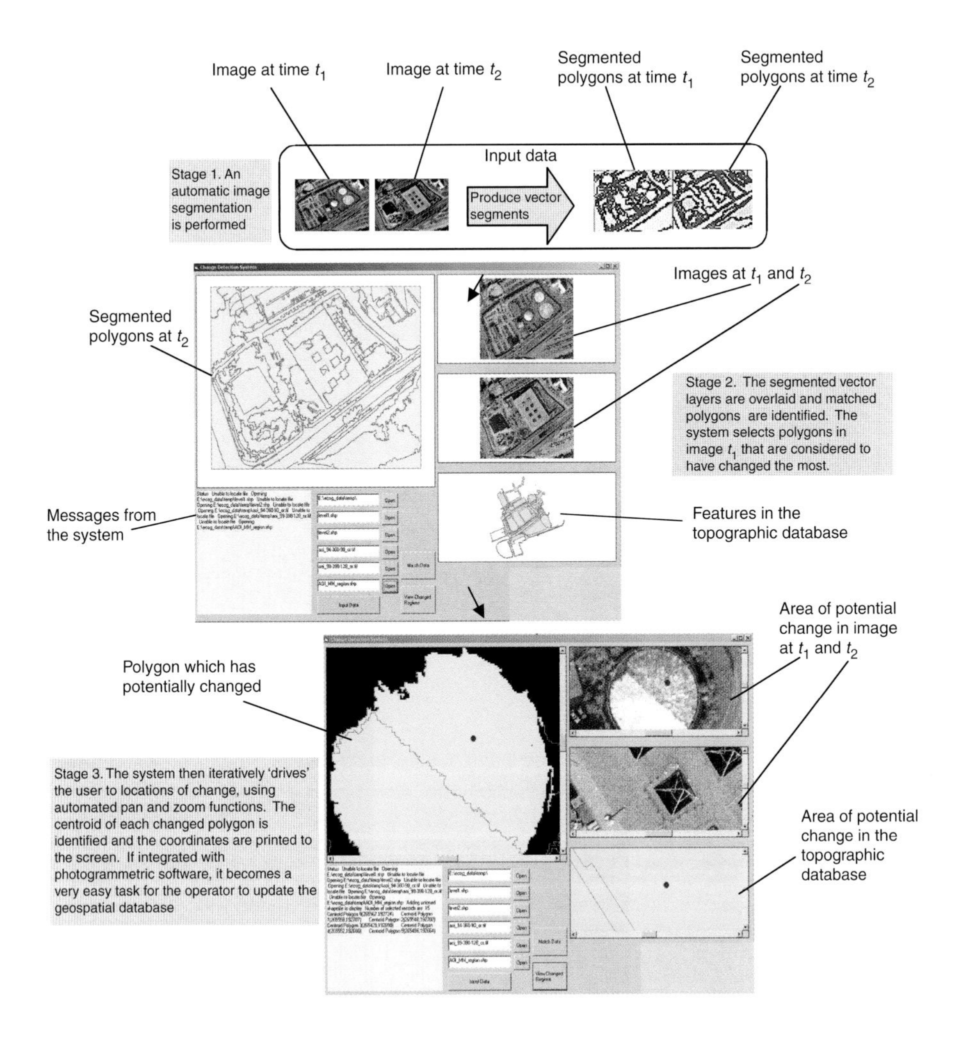

Figure 4.3 Process flowline through the prototype change detection system.

LCM2000 class, 1990 to 2000 change process, *1990 classes*	**Geo-Context**	**Picture**
Broadleaved Woodland, Maturation, from *Grass Heath*	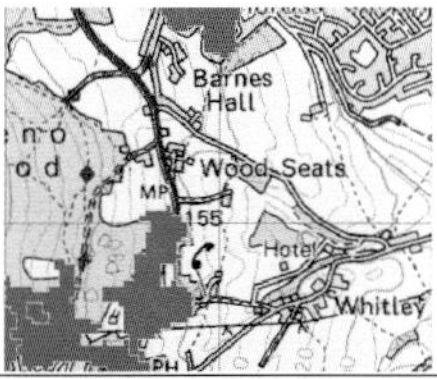 Barnes Hall Wood Seats MP 156 Hotel Whitley PH	
Broadleaved Woodland, Maturation, from *Bracken* and *Moorland Grass*	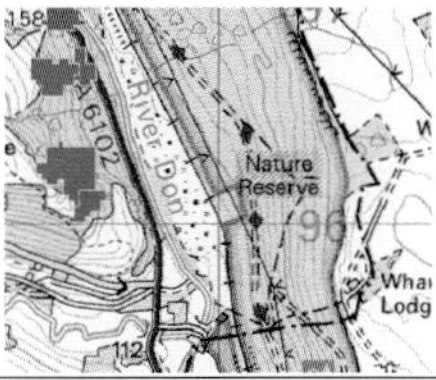 158 A6102 River Don Nature Reserve Wha Lodg 112	
Acid Grassland, degraded land quality by overstocking, from *Mown* and *Grazed Turf* and *Pasture*	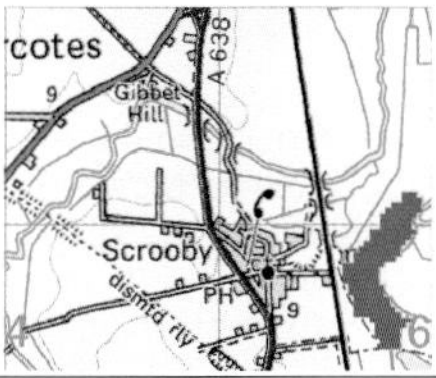 cotes 9 Gibbet Hill A 638 Scrooby dismtd rly PH 9	
Acid Grassland, changes in land quality, let go, degraded from *Tilled Land*	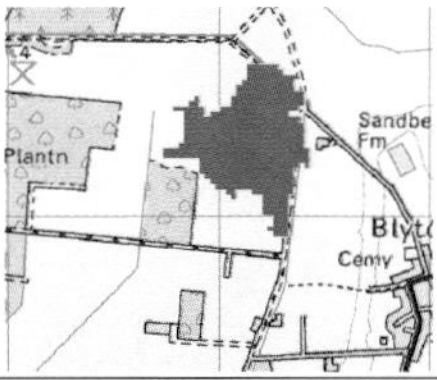 Plantn Sandbe Fm Blyt Cemy	
Suburban and Rural Development, new housing development, from *Tilled Land*	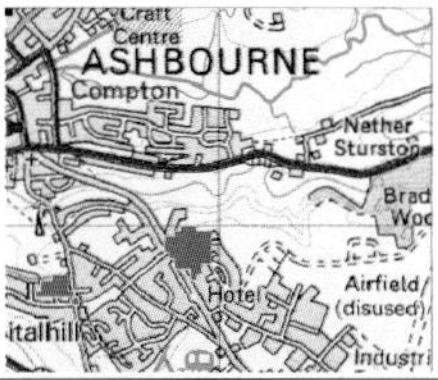 Craft Centre ASHBOURNE Compton Nether Sturston Brad Woo Hotel Airfield (disused) italhill Industri	
Suburban and Rural Development, new housing development, from *Mown* and *Grazed Turf* and *Pasture*	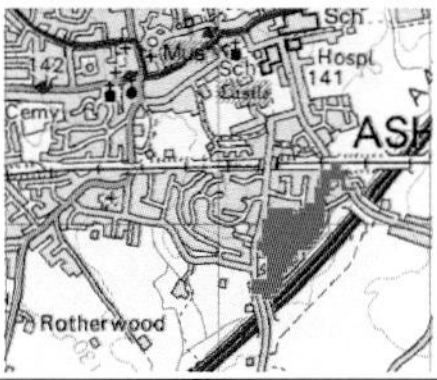 Sch 142 Mus Sch Castle Hospl 141 Cemy ASH Rotherwood	

Figure 6.3 Six example change parcels, a brief description of the nature of the change, the original (1990) class, and some context from OS 1:50000 Raster scanned maps (© Crown Copyright Ordnance Survey. An EDINA Digimap/JISC supplied service).

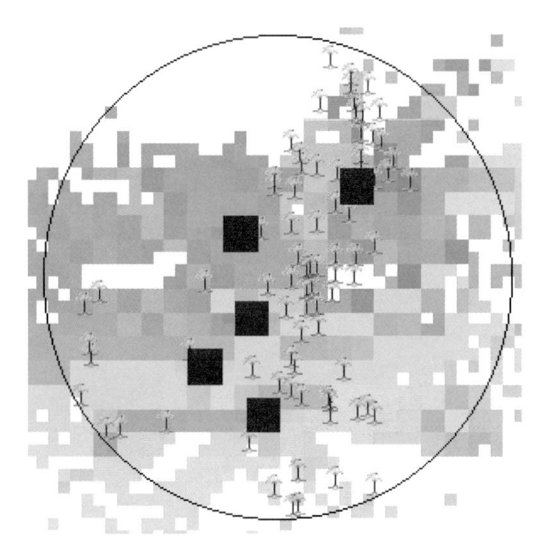

Figure 10.2 The estimated spatial distribution of the first generation of Aleppo pines (tree symbols) on the Mt. Pithulim site, within a radius of 100 m from the central location of the five older trees (red squares) that have inhabited the site prior to 1940. The background is a map of the total JMF (decade) grades where reddish colours represent lower membership grades and greenish colours represent higher membership grades.

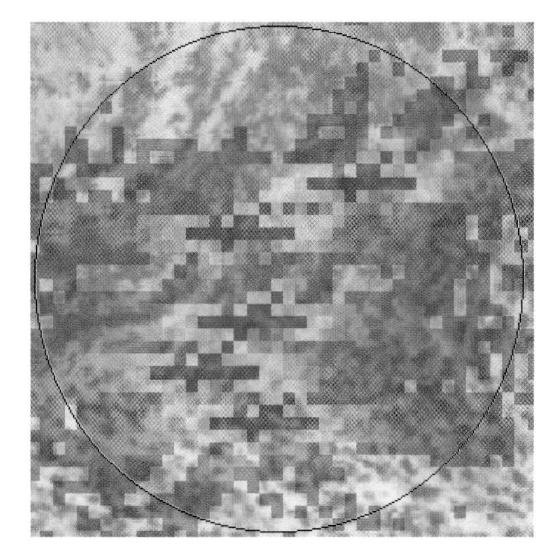

Figure 10.3 A map of the membership grades of seed dispersal where bluish colours represent higher values and reddish colours represent lower values. The background is a panchromatic orthophoto acquired in 1956.

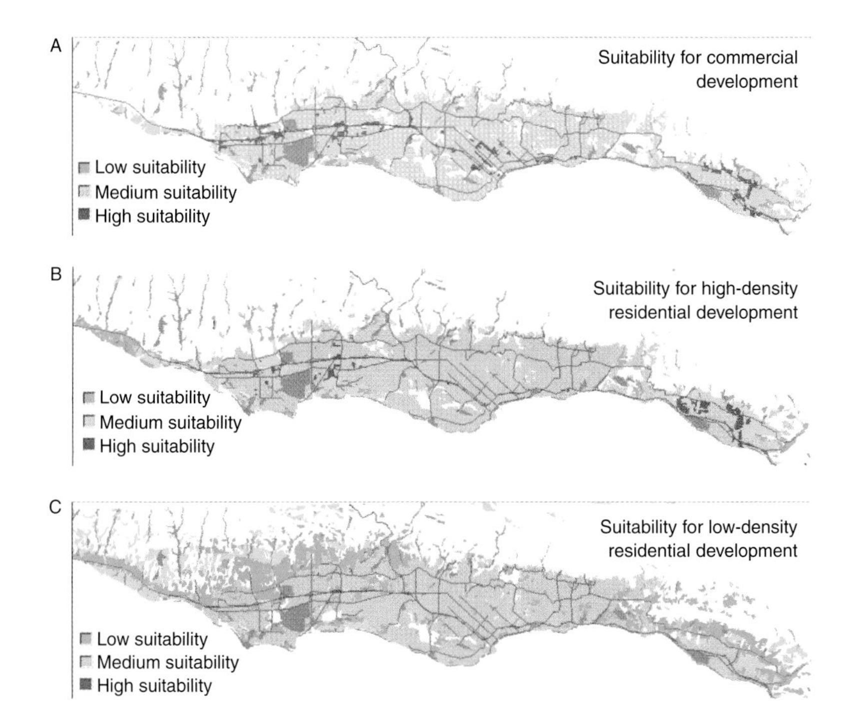

Figure 16.3 What if? land suitability assessments for Santa Barbara.

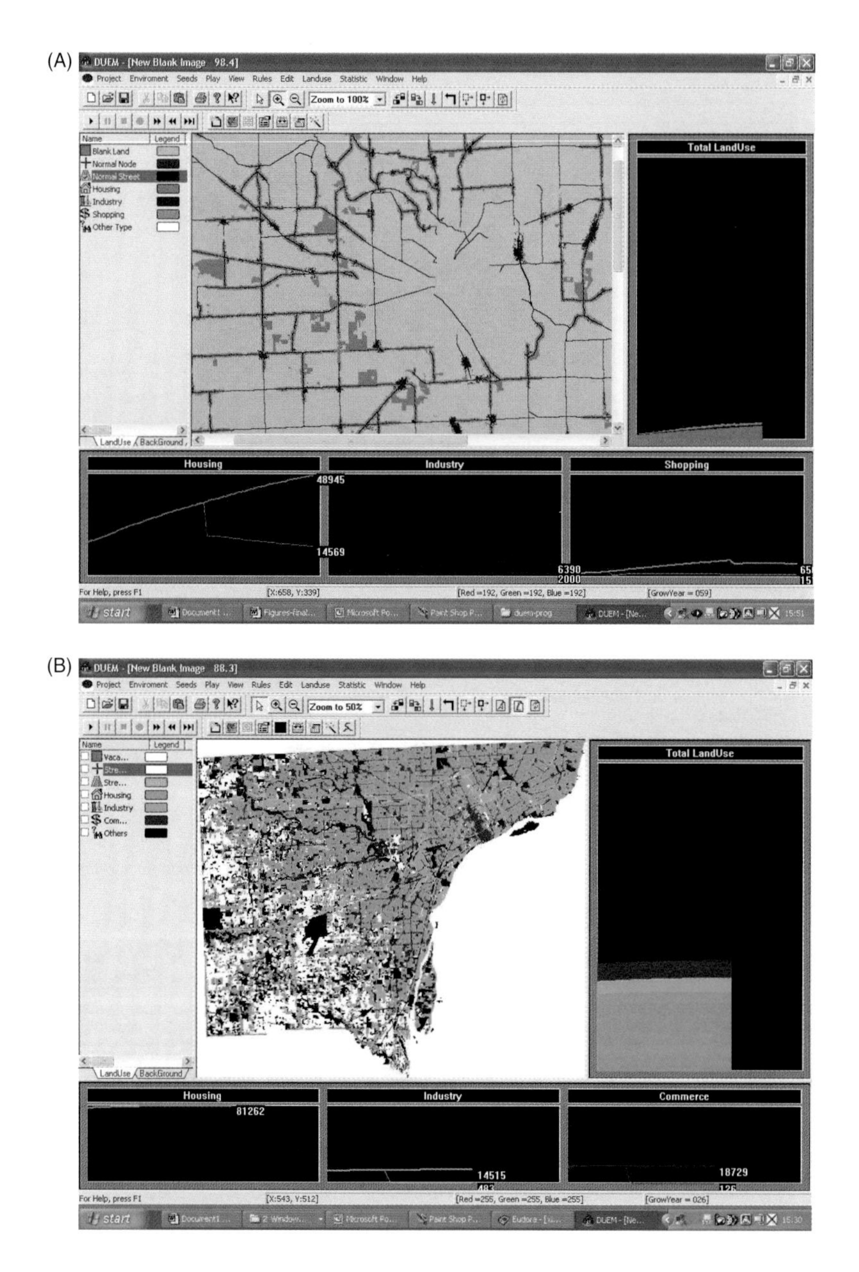

Figure 19.2 *DUEM* simulations in Ann Arbor and Detroit, Michigan: (A) Simulating sprawl in Ann Arbor. Land parcels are hard to assemble into relevant clusters of real development and the simulation predicts growth along roads. The clusters are the existing changes between 1985 and 1990, which spawn the new growth along roads. (B) Simulating long-term housing growth in Detroit. Housing grows in this scenario but in fact Detroit is characterized by decline and abandonment, and it is simply an artifact of the closed space that growth takes place in this fashion.

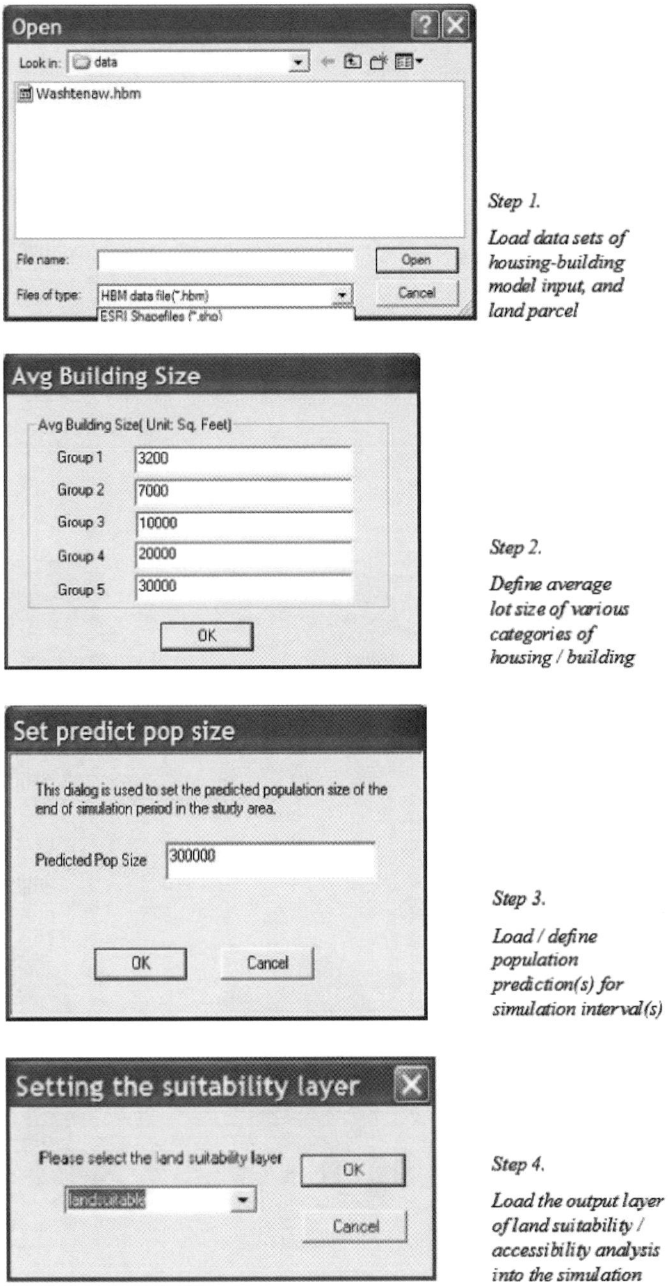

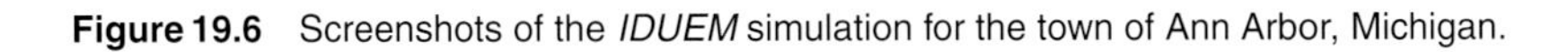

Figure 19.6 Screenshots of the *IDUEM* simulation for the town of Ann Arbor, Michigan.

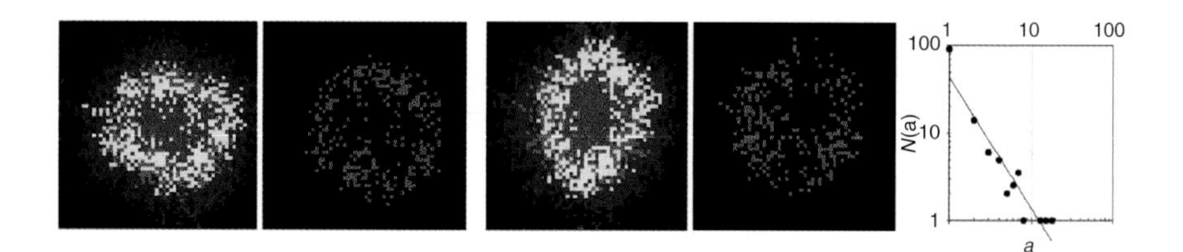

Figure 21.10 Distribution of settlements in the city of slums model with fragmentation graph.

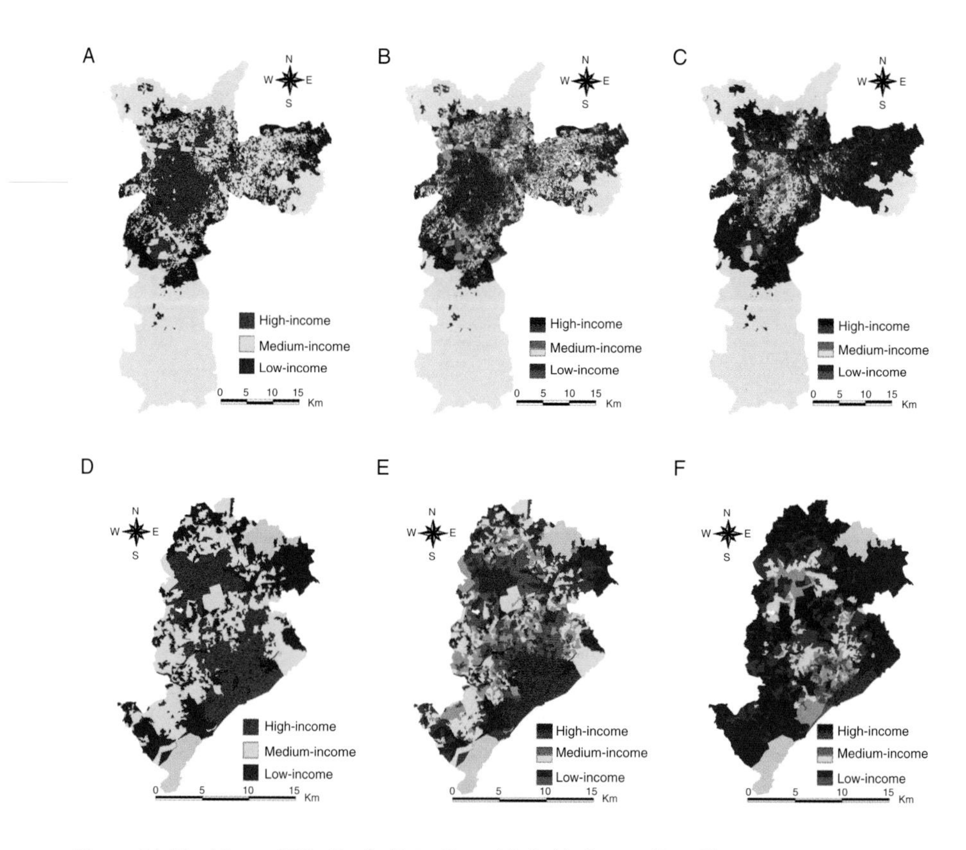

Figure 21.11 Maps of São Paulo (A to C) and Belo Horizonte (D to F) showing distributions of income in the urban area. Maps A, B, D, and E use quantile breaks and maps C and F use natural breaks.

CHAPTER 13

A Three-Dimensional, GIS-Based Model of Historical Groundwater and Slope Stability

Jochen Schmidt

CONTENTS

13.1 INTRODUCTION

Recent climate impact research has been directed towards the assessment of the effects of climatic variability and climate change on geomorphic processes and the related hazards, such as landslides (Collison et al., 2000; Dehn et al., 2000). General circulation models (GCMs) and downscaling techniques have been applied to predict changing climate parameters and their consequences for small-scale hillslope processes (Dehn et al., 2000). These models, however, include high uncertainties because of the unknown boundary conditions and the system complexity, which increases with scale. Moreover, the model output cannot be validated. It is therefore necessary to study also past effects of climatic change on geomorphic processes by

0-8493-2837-3/05/$0.00+$1.50

field evidence and reconstruction (Corominas and Moya, 1999). This gives the chance to validate and to develop sound models reflecting the physical process behavior. Climate is related to landslides via the system of hillslope hydrology. Physically based models of rainfall-induced landslides have been used to study these complex interactions (Brooks, 1997). These studies showed that hydrologic triggering systems for landslides show complicated behavior in relation to geotechnical, hydrological, and climatological parameters (van Asch, 1997). Therefore, more research is needed to understand those relationships, and to relate them to past and future climates (Crozier, 1997).

Landslides heavily affect hillslopes in central and southern Germany (see Wichmann and Becht, Chapter 12, this volume, for an analysis of hillslope processes in southern Germany). A series of recent and historical landslides have been recorded in the Bonn area. Previous studies have shown that precipitation-induced groundwater rises are an important contributor to slope instability in that region (Grunert and Hardenbicker, 1997; Hardenbicker and Grunert, 2001; Schmidt and Dikau, 2004b). However, little is known about the history of landslides and about the link between climate change, precipitation, hillslope hydrology, and landslide processes in that area. It is generally assumed that climatic changes have considerable effects on landslide activity (Grunert and Hardenbicker, 1997), but these effects remain to be verified.

This chapter presents an approach to modeling the effects of climate variability on slope stability for historical time periods. The aim is to assess the stability of hillslopes around Bonn in relation to changing climate by coupling proxy-derived, historic climate series with a physically based model for groundwater-controlled slope stability, that is, climate variables (precipitation, temperature) were used to drive a groundwater/slope stability model. The effects of the different scenarios on model output were compared and used to assess the characteristics of the different hillslopes and the effectiveness of their hydrologic triggering systems with respect to slope stability.

13.2 STUDY AREA

Topography of the area around Bonn is characterized by the plateau of the "Kottenforst" (a horst) west of the Rhein, and the hilly area of the "Siebengebirge" (Figure 13.1) east of the Rhein. As topographic information, contour lines from 1:25,000 Topographic maps and a 10 m grid DTM were available. A Devonian baselayer is overlain with Tertiary sediments, varying from marine clays to sands and fluvial gravels. Under Pleistocene periglacial conditions, terrace material and loess were deposited. West of the Rhein, terrace sediments are found above a series of Tertiary layers (clay, sand, and gravel) and the Devonian baselayer. Pleistocene and Holocene fluvial processes dissected the plateau of the Kottenforst and formed a series of small valleys (e.g., Godesbachtal, Melbtal, Katzenlochbachtal), many of which are incised to the Devonian baselayer. East of the Rhein, layers of volcanic sediments (trachytic tephras) cover large parts of the area and a series of eroded latitic, basaltic, and andesitic intrusions form the hills of the Siebengebirge. The slopes are

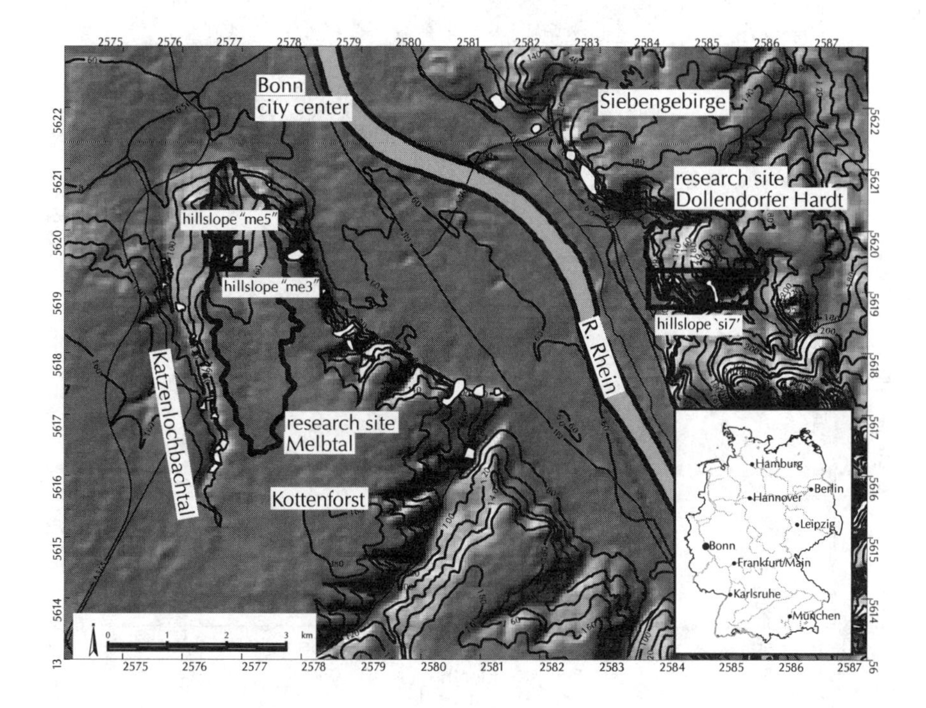

Figure 13.1 Topography and landslides (white polygons) in the Bonn area. Areas of high landslide susceptibility are the hillslopes incised in the horst of the "Kottenforst" and the "Siebengebirge." Three field areas were chosen as detailed study windows.

covered with Pleistocene sediments above volcanic ashes (trachyte tuff). Trachyte tuff is interfingered with Tertiary sediments.

Landslides of varying size and age occur on the hillslopes of the Kottenforst and Siebengebirge (Figure 13.1, compare Grunert and Hardenbicker, 1997). Landslide susceptibility in the Bonn area is influenced by lithology, that is, the sensitive, clay-rich Devonian and Tertiary sediments and layers of volcanic tuffs exposed on the hillslopes. Most of the landslides were interpreted as Holocene mass displacements, with a series of events also occurring in the 20th century (Grunert and Hardenbicker, 1997). Previous studies showed that landslide occurrence is related to rainfall-determined groundwater rises; additionally anthropogenic influences play an important role (Grunert and Hardenbicker, 1997; Hardenbicker and Grunert, 2001; Schmidt and Dikau, 2004b).

Bonn has a moderate maritime climate, dominated by oceanic air masses. Average annual temperature is about 9°C and annual rainfall is about 600 to 750 mm. Temperatures are characterized by mild winters (2°C monthly average) and moderately warm summers (ca. 18°C monthly average) (Figure 13.2). The long-term monthly precipitation totals show a minimum in winter (February, about 40 mm) and a maximum in summer (July, about 70 mm). Holocene climate variability indicates a series of climatic fluctuations, including cool and humid periods, which are of particular relevance for landslide occurrence (Brooks, 1997; Hardenbicker and Grunert, 2001).

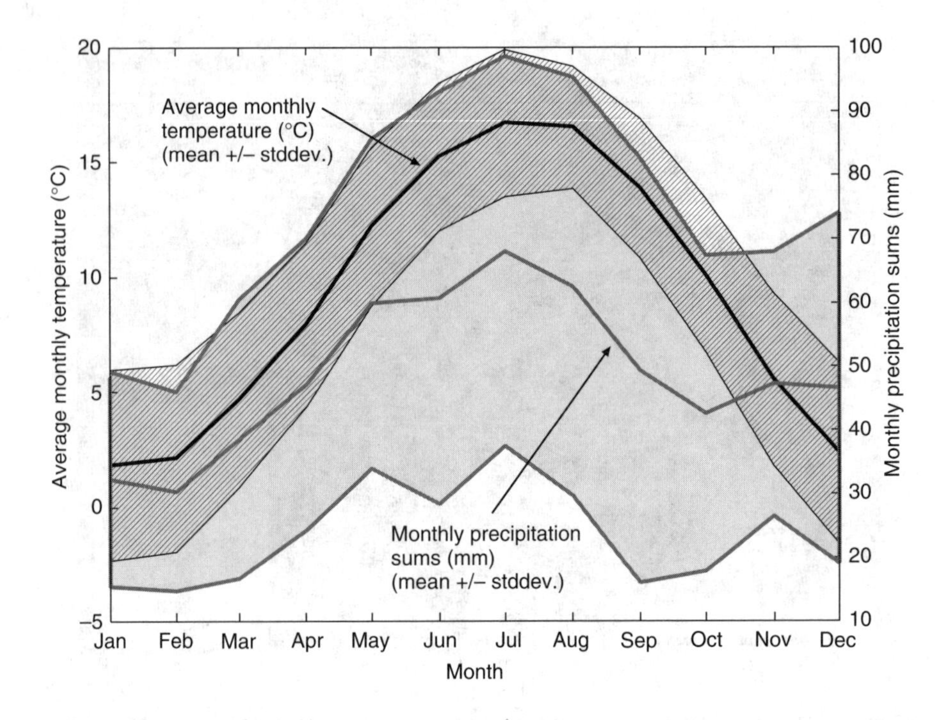

Figure 13.2 Climatic characteristics of the Bonn area. Shown are long-term average temperature and precipitation (and their standard deviations).

Land surface and lithology of the Bonn area show a complex pattern. As a simplification, three study sites were chosen, which serve as representative locations for different geomorphological situations in the Bonn area. Intensive field studies were carried out to explore subsurface stratigraphy and material properties (Schmidt, 2001; Schmidt and Dikau, 2004a, b). The aim was to explore the sensitivity of the different lithological and geomorphometric conditions to changing climatic conditions. The "Melbtal" is a small valley west of the Rhein, cut into the Kottenforst plateau (Figure 13.1). The field investigations (Schmidt, 2001) showed that stratigraphy is a sequence (from top to bottom) of terrace and loess sediments, Tertiary layers (clay, sand, and lignite), and a Devonian baselayer. Loess and terrace sediments associated with Pleistocene periglacial processes can be found on the valley side-slopes. Sensitive Tertiary layers of clay, sand, and lignite introduce considerable slope instability, and several landslides are located on the valley sides (Schmidt and Dikau, 2004b). The valley shows a distinct asymmetry with gentler west-facing valley side-slopes, which also show more extensive loess accumulation. Hillslope "me3" (Figure 13.3) served as a representative location for hillslopes west of the Rhein where Pleistocene and Tertiary sediments are exposed. Land surface and lithological information were taken from a west-facing lower-valley hillslope position of the Melbtal. The average gradient for the site is 7.4°. According to the borehole logs and lab analysis for this site, Tertiary lignite and clays are overlain by Tertiary sand and loess (Schmidt, 2001; Schmidt and Dikau, 2004a, b). The lithology was modeled as a sequence (from top to bottom) of

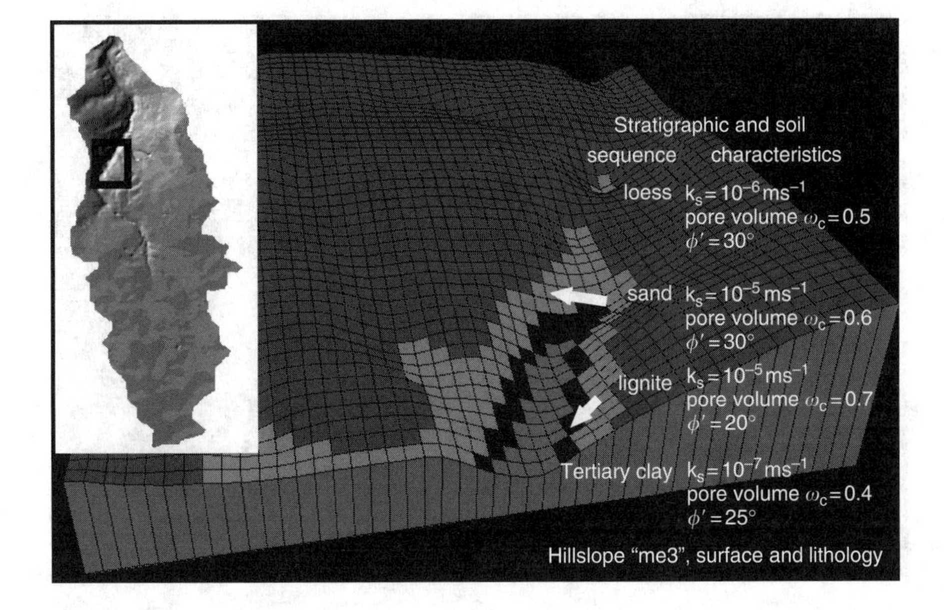

Figure 13.3 Field site hillslope "me3." Shown are landform (DTM) and lithology used in the groundwater model.

loess, Tertiary sand, and lignite and Tertiary clay layers (Figure 13.3). Lignite and sand have significantly higher permeability and pore content than the clay layers.

Hillslope "me5" (Figure 13.4) served as a representative location for hillslopes west of the Rhein where only Tertiary sediments are exposed. Land surface and lithological information were taken from an east-facing, lower-valley hillslope position of the Melbtal, where Tertiary sediments are exposed on the valley sides. The hillslopes were steeper than the first site: 8.1° average gradient. The lithology was modeled as a sequence (from top to bottom) of Tertiary clay, lignite, Tertiary sand, lignite, and Tertiary clay (Figure 13.4).

The "Dollendorfer Hardt" is a volcanic hill in the Siebengebirge (Figure 13.1). Landform and lithology are dominated by a basaltic dome forming the top of the hill, and trachyte tuff, Tertiary sediments (clays, sands) and the Devonian baselayer (from top to bottom) exposed on the southern hillslopes. The south- and west-facing hillslopes show high slope angles (up to 40°), whereas the gentle north- and west-facing hillslopes are contiguous with the northern Siebengebirge. Three landslides are documented for the area of the Dollendorfer Hardt. The largest (landslide "si7," affected area: 30,000 m^2) has been investigated in a range of studies (Hardenbicker and Grunert, 2001; Schmidt, 2001; Schmidt and Dikau, 2004b). The south-facing hillslope of the Dollendorfer Hardt, hillslope "si7" (Figure 13.5), served as a representative location for sensitive hillslopes of the Siebengebirge where Tertiary and volcanic sediments are exposed. The gradient is considerably higher than at the other sites (12.2°). The lithology was modeled according to the drilling results as a sequence (from top to bottom) of basaltic and trachytic layers, Tertiary sediments and the Devonian baselayer (Figure 13.5).

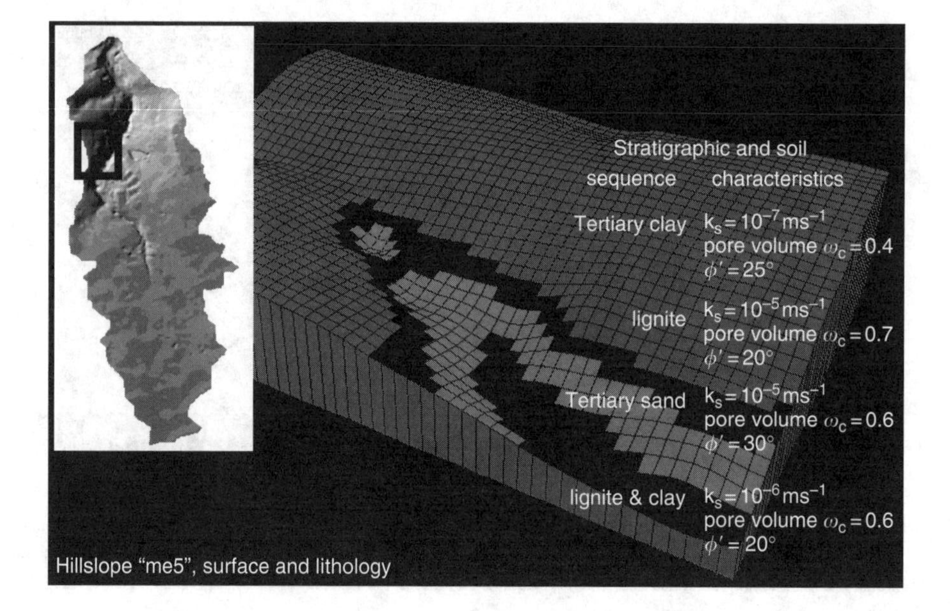

Figure 13.4 Field site hillslope "me5." Shown are landform (DTM) and lithology used in the groundwater model.

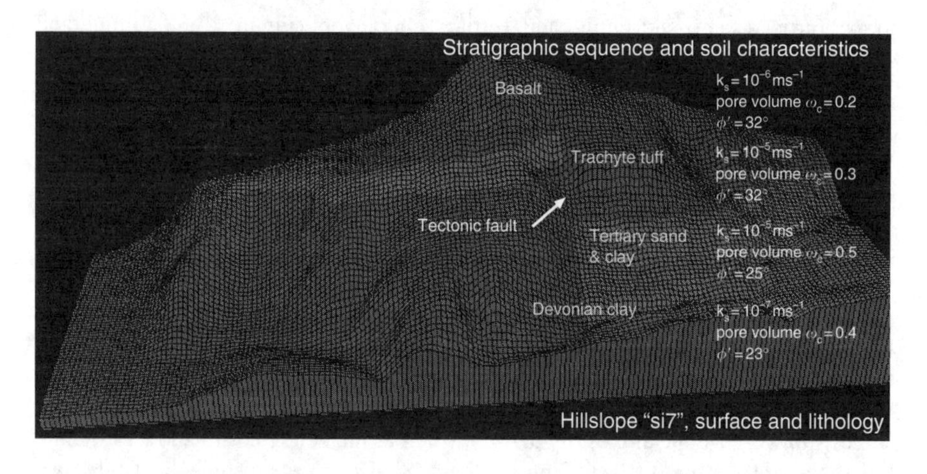

Figure 13.5 Field site hillslope "si7." Shown are landform (DTM) and lithology used in the groundwater model.

13.3 METHODOLOGY AND MODELS

The methodology of this study consists of several parts (Figure 13.6). A model for historical climatic variability delivered scenarios for past climatic conditions. The three hillslopes of the Bonn area as described above were used as spatial

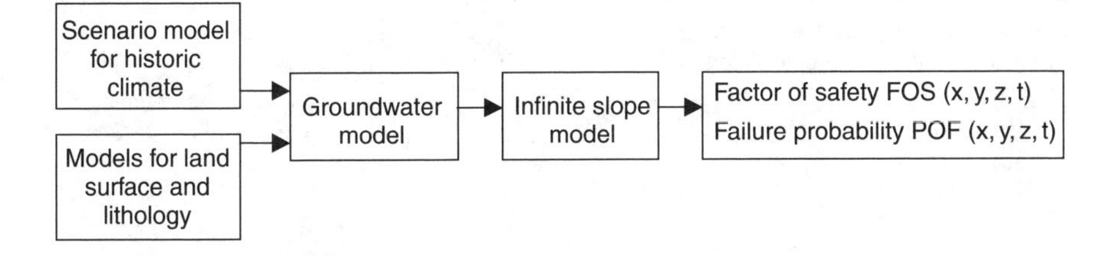

Figure 13.6 General methodology.

scenarios for land surface and lithology. A three-dimensional groundwater model was developed and applied using the spatial and temporal scenarios. The derived groundwater scenarios were used to calculate slope stability using the "infinite slope model."

13.3.1 Modeling Climate Variability

Historical climatic conditions for the study area were modeled as annual climate series of temperature and precipitation, using statistical analysis of available weather records and proxy data. The data used in this study were long-term daily meteorological records (precipitation, temperature) from the Bonn area and inferred paleo data for seasonal temperature and precipitation (from proxies) representing Middle European conditions since AD 1500 (Glaser et al., 1999; Schmidt and Dikau, 2004a). The paleo data were classified into years with similar seasonal patterns, which are related to general climatic trends. The identified clusters were then applied to classify the data sets of weather records in the Bonn area (Figure 13.7). The derived time series of precipitation and temperature were used as model inputs for the groundwater model. Each time series has daily resolution and represents the "average year" for the three scenario classes. Details of the classification procedure are described in Schmidt (2001) and Schmidt and Dikau (2004a).

13.3.2 A GIS-Based Groundwater Model

A groundwater model for small-scale areas (up to several square kilometers) was developed using the modeling environment of the GIS PCRaster (Wesseling et al., 1996; van Beek and van Asch, 1999). PCRaster provides a grid-based, algebraic macro language for development of dynamic models and visualization of environmental processes. Additionally, PCRaster provides graphical user interfaces (GUIs) for visualization and exploration of the model data. The model is a three-dimensional-tank model, simulating lateral and vertical saturated flow in daily timesteps. The model is based on a spatial discretization in equal-sized 10 m cells according to the spatial resolution of the digital terrain model (DTM) available. Additional elevation raster maps provide the information about lithological layers, giving heights of lower-layer boundaries for each cell (Figure 13.8). These three-dimensional subsurface models are produced from the borehole logs of the field investigations and spatial (polynomial) interpolation routines. Each lithological unit is parameterized

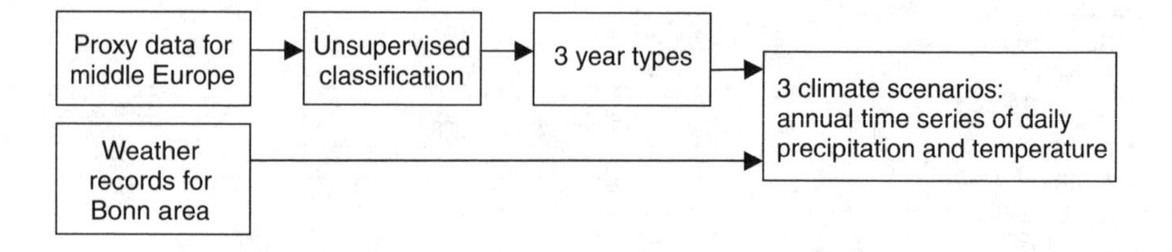

Figure 13.7 Methodology for deriving paleo-climate scenarios.

by material properties: saturated hydraulic conductivity and maximum water content. Precipitation, evaporation, and interception are used as spatially uniform daily time series for the simulated area to calculate net precipitation. Evaporation is calculated using the method of Thornthwaite (1948). Interception was derived from literature values, as the study areas show spatially homogeneous vegetation cover. Infiltration is modeled as the minimum of net precipitation and maximum infiltration capacity. Maximum infiltration capacity is determined by soil water storage of the top layer and a top-layer permeability term. Soil water transport processes are modeled as vertical fluxes between layers, and lateral fluxes between the columns of each cell using Darcy's law for saturated conditions. Vertical fluxes are modeled according to the difference in degree of saturation and hydraulic conductivity between layers (Schmidt, 2001). Vertical flux leaving the lowest layer is modeled as flow in an infinite storage (base flow) and is limited by the hydraulic properties of the lowest layer (conductivity and soil water content) and an additional conductivity term. The groundwater table is derived from the degree of saturation of the uppermost unsaturated layer for each cell. Only one lateral flux is modeled for each soil column (i.e., not layer specific), dependent on the effective height difference in saturated layers (i.e., groundwater table) and the effective conductivity of the adjacent columns.

13.3.3 A GIS-Based Model of Failure Probability

A module calculating local safety factors (FOS), based on the common infinite slope model extended the groundwater model as described in Section 13.3.2. Moisture content of the soil layers and the groundwater table simulated from the groundwater model were used as input for the stability model. Additional material parameters were required for each lithological unit: dry unit weight, effective cohesion, and effective angle of friction. Moreover, an approach for calculating probability of failure (POF) was implemented (Lee et al., 1983). The output of the model were values for the factor of safety $\mathrm{FOS}(x, y, z, t)$ and probability of failure $\mathrm{POF}(x, y, z, t)$ for each node in the four-dimensional mesh, that is, for each cell (x, y) of the used grid, a series of depths z, and for each model day t (Schmidt and Dikau, 2004a). This model output was recalculated to allow comparison of different scenarios: maximum failure probability, $\mathrm{POF}_{\mathrm{m}}$ and minimum factor of safety, $\mathrm{FOS}_{\mathrm{m}}$ for all modeled depths and timesteps describe the highest instability for each cell of a hillslope and a climate scenario, and therefore one map is produced for a model run.

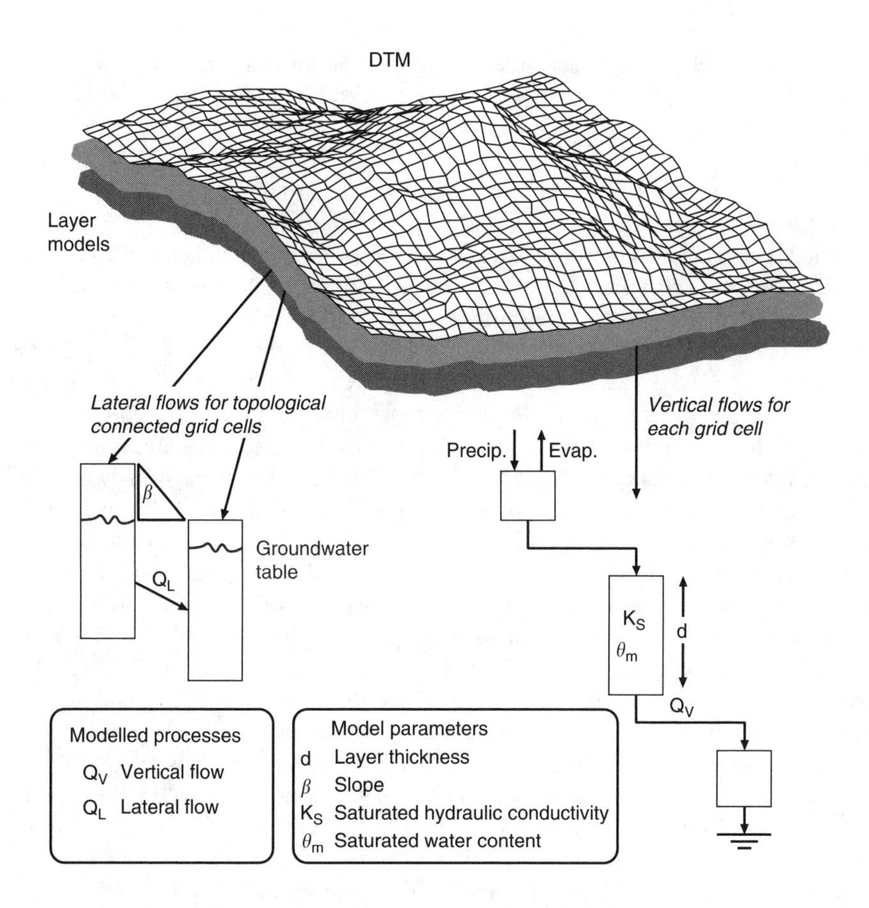

Figure 13.8 Physically based, three-dimensional groundwater model.

13.4 RESULTS

Long-term trends in the paleo data indicate three major stages in Middle European climate (Figure 13.9): (1) a phase of decreasing temperature and precipitation before ca. AD 1740, (2) a transition phase with a relatively distinct temperature rise of more than 1°C from ca. AD 1740 to AD 1850, and (3) the time period since AD 1850, indicating higher temperature level, and lower, but increasing, precipitation. Classification of the paleo data delivered three "year types." The three types ("classes") of years clearly represent the detected climatic trend as described above (Figure 13.9).

- Class 1 is a year type with high precipitation and comparatively low temperatures (especially summer precipitation and temperatures) revealing high frequencies before AD 1750.
- Class 2 shows intermediate temperatures (but low winter temperatures) and lower precipitation (especially in winter) and is dominant in the period between AD 1750 and AD 1850.

- Class 3 indicates an annual pattern of high temperatures and relatively low precipitation (especially in summer); frequencies have risen in the period since AD 1850.

The three average year types were used as input in the groundwater and slope stability models based on the three hillslopes (Figure 13.3 to Figure 13.5). The modeled spatial patterns of maximum failure probability POF_m indicate sensitivity of high slope angles and Tertiary layers. Figure 13.10 to Figure 13.12 show the modeled groundwater pattern for the three hillslopes and climate scenarios. Figures 13.13 displays the percentage of modeled unstable areas (safety factor <1) for the different hillslopes and climate scenarios. The modeled scenarios led to the following results.

Generally, low average values for failure probability and for effective failure depth are due to the high frequency of nodes of lower sensitivity (e.g., flat areas) that were included in the calculations (Schmidt and Dikau, 2004a). However, groundwater and slope stability show distinct differences for the three hillslopes. Hillslope "me3" indicates lower (approximately half) average maximum failure probabilities and less unstable area than hillslope "me5" (Figure 13.13), consistent with the lower slope angle and less sensitive layers (loess). Hillslope "si7" delivered low failure probabilities, because combination of weak substrate and steep slopes (trachyte, Tertiary sediments) cover only small parts of the area (Figure 13.5). The modeled groundwater patterns for the three areas show the groundwater concentration in hollows and valleys. According to the increasing precipitation in climate scenarios 1 and 2, the saturated areas are more extended for those scenarios (Figure 13.10 to Figure 13.12), therefore those scenarios also show increased failure probability (higher slope instability) (Figure 13.13). The results reveal the higher geomorphic effectiveness of climate scenario 2, although climate scenario 1 has higher annual precipitation sums, which can be attributed to the occurrence of a few intensive precipitation events in the winter in scenario 2 (Schmidt and Dikau, 2004a). Climatic years of scenario 2 dominate the "climatic transition phase" from Little Ice Age to recent conditions. This climatic transition phase is prone to fluctuations in annual weather patterns that are likely to induce slope failure. The analyses showed that for all three hillslopes the increasing failure probability for climate scenario 2. Figure 13.13 indicates that the effect of scenario 2 on increasing failure probability is more significant for hillslope "me3" than for hillslopes "me5" and "si7," that is, hillslope "me3" is more sensitive to climatic changes (high groundwater tables). For hillslope "me3," the effect of climate scenario 2 leads to a considerable extension of the saturated areas. These saturated areas extend especially to steep hillslopes, which in turn leads to higher slope instabilities for this spatial scenario. The failure probability for scenario 2 and hillslope "me3" reaches comparatively high values (Figure 13.13).

For hillslope "me5" the saturated areas extend along the valley bottoms, the hillslopes are less affected. Consequently slope instabilities do not increase as much as for hillslope "me3" (Figure 13.13). The reason for that behavior being a more permeable baselayer of Tertiary clay and lignite and less groundwater recharge due to low permeable top clay layers.

For hillslope "si7," sensitive areas are predominantly the Tertiary layers, which occupy only a small part of the steeper hillslopes. Convergent areas on steeper

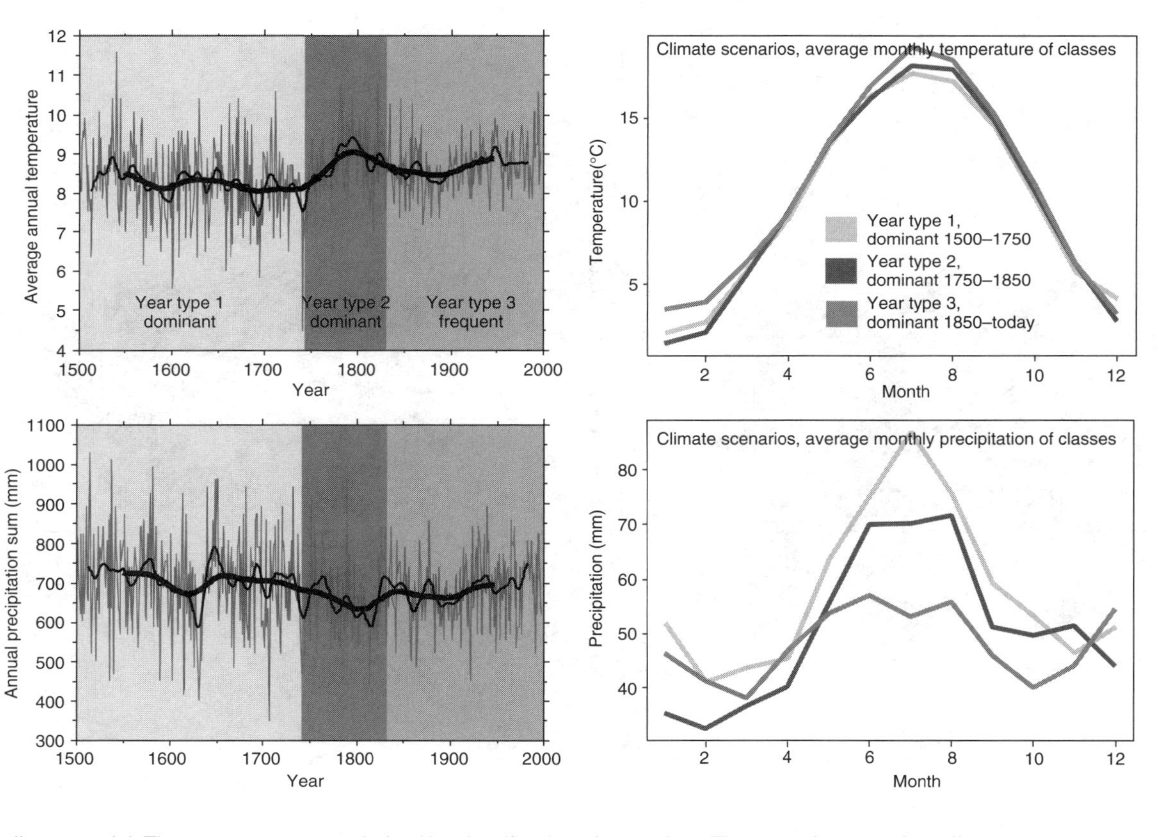

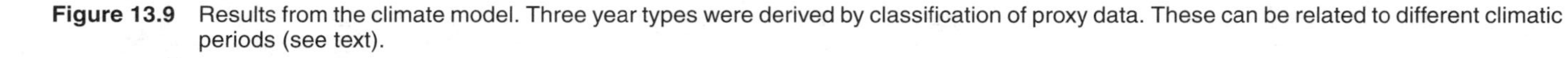

Figure 13.9 Results from the climate model. Three year types were derived by classification of proxy data. These can be related to different climatic periods (see text).

Figure 13.10 Groundwater pattern for hillslope "me3" and the three climate scenarios. Top: climate scenario 1, middle: climate scenario 2, bottom: climate scenario 3 (compare Figure 13.9). Dark: groundwater close to surface.

Figure 13.11 Groundwater pattern for hillslope "me5" and the three climate scenarios. Top: climate scenario 1, middle: climate scenario 2, bottom: climate scenario 3 (compare Figure 13.9). Dark: groundwater close to surface.

hillslopes in Tertiary sediments show therefore the highest instabilities (Schmidt and Dikau, in review). The effect of changing climate leads to only minor extended saturated areas in these hollows (Figure 13.12). Only a low increase in slope instabilities is recognizable (Figure 13.13).

Figure 13.12 Groundwater pattern for hillslope "si7" and the three climate scenarios. Top: climate scenario 1, middle: climate scenario 2, bottom: climate scenario 3 (compare Figure 13.9). Dark: groundwater close to surface.

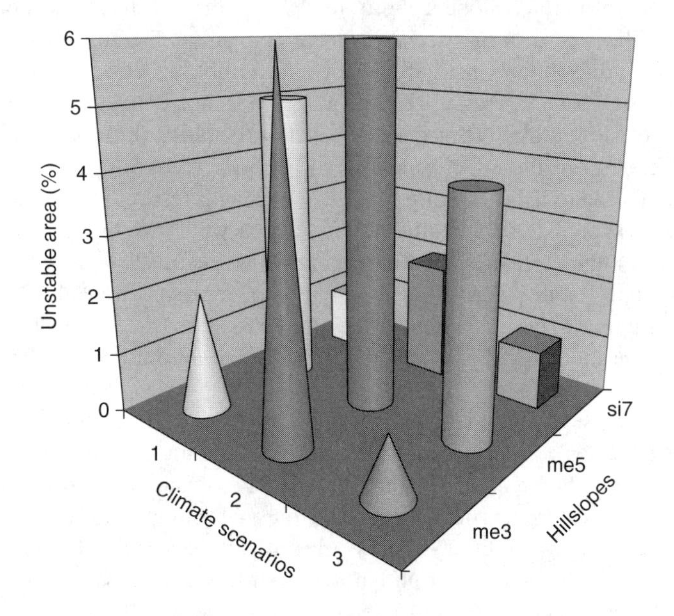

Figure 13.13 Modeled unstable areas (FOS < 1) for the different climate scenarios (Figure 13.9) and hillslopes (Figure 13.3 to Figure 13.5). The results indicate the effectiveness of climate scenario 2 for creating slope instabilities, particularly for hillslope "me3."

These results show that sensitivity to failure shows a clear relationship to general site properties such as gradient and lithology. However, sensitivity to climatic variations is dependent on the spatial configuration of each site. The occurrence of low-permeability clay layers below lignite and sands results in a potential groundwater rise for more sensitive (steeper) areas for hillslope "me3," whereas for hillslope "si7," groundwater rise above low-permeability Devonian layers lowers slope stability only for less sensitive areas. Hence, the location of permeable layers (prone to groundwater rise) in relation to sensitive layers (lower strength) and higher gradient areas (higher stress) determines the sensitivity of a site with respect to climatic changes.

13.5 DISCUSSION AND CONCLUSION

The uncertainty in the assessment of historical environmental conditions is high. The results presented in this study showed the potential and utility of proxy data for assessing past climatic conditions and their effects on geomorphic processes. A method for deriving annual patterns of climate parameters for past periods was developed and applied (Schmidt and Dikau, 2004a). Three "climatic year types" were derived by statistical analysis of paleo climate data. These types are related to past climate periods.

It was shown that this scenario approach is a valuable method to identify different climatic regimes with respect to their effectiveness in driving geomorphic processes (in this study landslides). The identified "representative climatic years" were used to drive groundwater and slope stability models for three different hillslopes in the study area. The nine resulting scenarios were compared. The models showed considerable variations in the sensitivity to slope instabilities of the modeled landscapes in relation to climate variations. The results give indications of (1) differences of the pattern of slope stability for the three spatial scenarios (hillslopes), (2) considerable differences of the effectiveness of the climate scenarios, and (3) variations of the spatial sensitivity of the different hillslopes with respect to climatic changes. The modeled spatial slope stability patterns match well with the spatial pattern of observed existing landslides in the field (Schmidt and Dikau, 2004a, b). The temporal pattern of slope stability derived in this study, however, was not validated because of a lack of appropriate field data for the study area (i.e., dated landslides). It was shown that the modeled climate regime 2 (representing dominantly the transition phase from Little Ice Age to recent climatic conditions) was generally more effective leading to increased slope instability for all tested sites. This can be explained by the high frequency of intensive rainfall events for the unstable transition phase of climate regime 2 (Schmidt and Dikau, 2004a). The sensitivity of slope stability to climatic change, however, is dependent on the internal configuration of the landscape. This means the slope stability system is "filtered" by the system of hillslope hydrology, which is largely dependent on landscape structure, that is, topographical and lithological convergences, and sensitive layers.

The described model is solely based on the effect of climate on changing process behavior. Precipitation and temperature were used as the only variable boundary conditions. In historical time periods, however, certainly landcover conditions alter in response to changing climatic and anthropogenic conditions. Those feedbacks should be included in extending the presented modeling approach. Using historical and recent remote sensing data in combination with integrated evolutionary models (see Svoray and Nathan, Chapter 10, this volume) is a prospective step forward in that direction. These models would be able deliver landcover simulations that could be applied in combined scenario models of climate and land cover variability and their effects on hillslope hydrology and slope stability.

It was shown that for landslide sensitivity, besides the general geomorphological and lithological properties (especially shear strength), landscape configuration is an important determinant. A more sensitive landscape in terms of actual slope stability is not generally more sensitive to climatic changes. Therefore, landscape sensitivity should be viewed as a dynamic feature and has to be specifically connected to processes and climatic changes. Scenario approaches are useful in assessing these sensitivity changes. Despite the simplifications of the presented approach, the method is capable of representing the effect of climate changes in relation to landscape sensitivity for landslide processes on a static level. It can be used to quantify differences in slope stability for different landscapes and their changes with climatic variations. Therefore, simplified two-dimensional modeling approaches as often used in landslide hazard assessments are not suitable to represent a situation as shown in this study. Climatic variations can potentially have complex interactions with the geomorphologic system requiring the utilization of process-based, three-dimensional models for an adequate prediction of the effects of future climate change.

ACKNOWLEDGMENTS

The research was carried out within research project Collaborative Research Center (SFB) 350, which is supported by the German Research Foundation (DFG), Bonn. We are grateful to Prof. J. Grunert, Dr V. Schmanke and Dr U. Hardenbicker for providing some of the digital maps for the Bonn area. Prof. Rüdiger Glaser (Department of Geography, University of Heidelberg) provided the proxy data and paleo data sets for Middle Europe.

REFERENCES

Brooks, S. 1997. Modelling the role of climatic change in landslide initiation for different soils during the Holocene. In: Matthews, J.A., Brunsden, D., Frenzel, B., Gläser, B. and Weiß, M.M. (Eds.): Rapid mass movement as a source of climatic evidence for the Holocene. *Palaeoclimate Research* 19: 207–222. Gustav Fischer. Stuttgart Jena Lübeck Ulm.

Collison, A., Wade, S., Griffiths, J., and Dehn, M. 2000. Modelling the impact of predicted climate change on landslide frequency and magnitude in SE England. *Engineering Geology* 55: 205–218.

Corominas, J. and Moya, J. 1999. Reconstructing recent landslide activity in relation to rainfall in the Llobregat River basin, Eastern Pyrenees, Spain. *Geomorphology* 30(1–2): 79–93.

Crozier, M.J. 1997. The climate-landslide couple: a Southern Hemisphere perspective. In: Matthews, J.A., Brunsden, D., Frenzel, B., Gläser, B., and Weiß, M.M. (Eds.): Rapid mass movement as a source of climatic evidence for the Holocene. *Palaeoclimate Research* 19: 333–354. Gustav Fischer. Stuttgart Jena Lübeck Ulm.

Dehn, M., Burger, G., Buma, J., and Gasparetto, P. 2000. Impact of climate change on slope stability. *Engineering Geology* 55: 193–204.

Glaser, R., Brazdill, R., Pfister, C., Dobrovolny, P., Barriendos Vallve, M., Bokwa, A., Camuo, D., Kotyza, O., Limanowka, D., Racz, L., and Rodrigo, F.S. 1999. Seasonal temperature and precipitation fluctuations in selected parts of Europe during the sixteenth century. *Climatic Change* 43(1): 169–200.

Grunert, J. and Hardenbicker, U. 1997. The frequency of landsliding in the north Rhine area and possible climatic implications. In: Matthews, J.A., Brunsden, D., Frenzel, B., Gläser, B., and Weiß, M.M. (Eds.): Rapid mass movement as a source of climatic evidence for the Holocene. *Palaeoclimate Research* 19: 17–31. Gustav Fischer. Stuttgart Jena Lübeck Ulm.

Hardenbicker, U. and Grunert, J. 2001. Temporal occurrence of mass movements in the Bonn area. *Zeitschrift für Geomorphologie, N.F., Supplement Band* 125: 13–24.

Lee, I., White, W., and Ingles, O. 1983. *Geotechnical Engineering*. Pitman. London.

Schmidt, J. 2001. The role of mass movements for slope evolution — conceptual approaches and model applications in the Bonn area. PhD Thesis. Department of Geography, University of Bonn, Germany. http://hss.ulb.uni-bonn.de:90/ulb_bonn/diss_online/math_nat_fak/2001/schmidt_jochen

Schmidt, J. and Dikau, R. 2004a. Modeling historical climate variability and slope stability. *Geomorphology*. In press.

Schmidt, J. and Dikau, R. 2004b. Preparatory and triggering factors for slope failure: analyses of two landslides in Bonn, Germany. *Zeitschrift für Geomorphologie*. In press.

Svoray, T. and Nathan, R. Chapter 10, this volume.

Thornthwaite, C.W. 1948. An approach toward a rational classification of climate. *Geographical Review* 38: 55–94.

van Asch, T.W.J. 1997. The temporal activity of landslides and its climatological signals. In: Matthews, J.A., Brunsden, D., Frenzel, B., Gläser, B., and Weiß, M.M. (Eds.): Rapid mass movement as a source of climatic evidence for the Holocene. *Palaeoclimate Research* 19: 7–16. Gustav Fischer. Stuttgart Jena Lübeck Ulm.

van Beek, L. and van Asch, T.W.J. 1999. A combined conceptual model for the effects of fissure-induced infiltration on slope stability. In: Hergarten, S. and Neugebauer, H. (Eds.): Process modelling and landform evolution. *Lecture Notes in Earth Sciences* 78: 147–167. Springer-Verlag. Berlin Heidelberg, New York.

Wesseling, C., Karssenberg, D., Burrough, P.A., and Deursen, W. 1996. Integrating dynamic environmental models in GIS: the development of a dynamic modelling language. *Transactions in GIS* 1: 40–48.

Wichmann, V. and Becht, M. Chapter 12, this volume.

CHAPTER 14

Prediction Uncertainty in Floodplain Elevation and its Effect on Flood Inundation Modelling

Matthew D. Wilson and Peter M. Atkinson

CONTENTS

14.1 INTRODUCTION

Flood awareness in the United Kingdom has risen dramatically in the last few years after several major flood events. Flood inundation models allow river discharge upstream to be related directly to flood extent downstream and are, therefore, potentially very useful predictive tools that can be used in a variety of real and "what-if" scenarios. However, all data used (and hence parameters and variables) in flood inundation models have inherent uncertainty. The challenge is to quantify this uncertainty and, perhaps more importantly, assess the effect that uncertainty may have on model predictions.

0-8493-2837-3/05/$0.00+$1.50

Other than the main river channel, floodplain topography is the principal variable that affects the movement of the flood wave and is, therefore, critical to the prediction of inundation extent. Since river floodplains are usually characterised by a low spatial variation in topography, a small degree of uncertainty in elevation may have a relatively large effect on model predictions. Small changes in floodplain elevation may determine the horizontal location of the predicted flood boundary and the timing of inundation. Ideally, a flood inundation model requires elevation data that represent closely the true ground surface. High quality remotely sensed elevation data are often unavailable for the area of interest and may contain features higher than the true land surface (e.g., buildings and vegetation). Any land feature that restricts (but not prevents) the flow of water (e.g., forest) should be accounted for in the friction terms of the model rather than represented by an area of higher elevation.

In this chapter, readily available contour data were supplemented with Differential Global Positioning System (DGPS) measurements of elevation on the floodplain. To assess the effect of elevation prediction uncertainty on inundation extent, multiple plausible digital elevation models (DEMs) were generated. Each DEM was then used to predict flood inundation for the Easter 1998 flood event on the river Nene, Northamptonshire, England, using the grid-based model LISFLOOD-FP (Bates and De Roo, 2000).

14.2 LISFLOOD-FP MODEL OF FLOOD INUNDATION

LISFLOOD-FP is a raster-based flood inundation model (Bates and De Roo, 2000; De Roo et al., 2000; Horritt and Bates, 2001). Channel flow is approximated using the one-dimensional linear kinematic Saint-Venant equations (e.g., Chow et al., 1988). Cells in the domain that are identified as channel are incised by the bankfull depth, and a hydrograph is routed downstream from the domain inflow. When channel depth reaches a cell bankfull level, flood inundation commences.

Flow on the floodplain is based on a simple continuity equation, which states that the change in volume in a cell over time is equal to the fluxes into and out of it (Estrela and Quintas, 1994):

$$\frac{\mathrm{d}h^{i,j}}{\mathrm{d}t} = \frac{Q_x^{i-1,j} - Q_x^{i,j} + Q_y^{i,j-1} - Q_y^{i,j}}{\Delta x \Delta y} \tag{14.1}$$

where $h^{i,j}$ is the water free surface height at the node (i, j) at time, t, and Δx and Δy are the cell dimensions. Floodplain flow is described by Horritt and Bates (2001):

$$Q_x^{i,j} = \frac{h_{\text{flow}}^{5/3}}{n} \left(\frac{h^{i-1,j} - h^{i,j}}{\Delta x} \right)^{1/2} \Delta y \tag{14.2}$$

$$Q_y^{i,j} = \frac{h_{\text{flow}}^{5/3}}{n} \left(\frac{h^{i,j-1} - h^{i,j}}{\Delta y} \right)^{1/2} \Delta x \tag{14.3}$$

where $Q_x^{i,j}$ and $Q_y^{i,j}$ are the flows between cells in the x and y directions, respectively. The depth available for flow, h_{flow}, is defined as the difference between the maximum free and bed surface heights in the two cells. Values for the Manning friction coefficient, n, are published for a wide variety of land covers (see Chow, 1959; Chow et al., 1988).

The design philosophy of LISFLOOD-FP was to produce the simplest physical representation that can accurately simulate dynamic flood spreading when compared with validation data (Horritt and Bates, 2002). Within the limitations of available validation data, LISFLOOD-FP has been found to be capable of similar accuracy to more complex models such as TELEMAC-2D (Horritt and Bates, 2001; Horritt and Bates, 2002). However, due to the simplification of the process representation, the model is more dependent on high quality input data. One advantage of a simple flood inundation scheme (such as that used in LISFLOOD-FP) over finite element models (such as TELEMAC 2D) is computational efficiency, with approximately 40 times fewer floating-point operations per cell, per time step (Bates and De Roo, 2000). Despite high-speed modern computers, this becomes important when multiple simulations are conducted as part of a sensitivity analysis or uncertainty assessment. Further, the use of a raster data structure makes the incorporation of multiple data sets in the model relatively straightforward, particularly from remotely sensed sources.

14.3 ELEVATION DATA

Land-form PROFILE™ contour data were obtained from the Ordnance Survey, and were then supplemented by DGPS measurements. A Trimble ProXRS GPS unit was used to gather elevation measurements along the channel and across the floodplain. The ProXRS is a 12 channel unit with carrier-phase filtered measurements, able to obtain sub-decimetre accuracy in all directions. The location of DGPS measurements in relation to the contour data is shown in Figure 14.1. Measurements were obtained on foot along accessible paths along the channel and across the floodplain, and by vehicle along roads in the area. Some problems were encountered in wooded areas as a clear line of sight is needed between the DGPS unit and the satellite platforms. However, as large parts of the area are open, it was still possible to cover a wide area. The density of DGPS measurements was approximately one point every 5 m when collected on foot, or one point every 80 m when collected by vehicle.

Experimental variograms were predicted for the PROFILE™ contour data, both inclusive and exclusive of the DGPS measurements (Figure 14.2A). In addition, areas above the 15 m contour line were removed and variograms predicted (Figure 14.2B). In both cases, the inclusion of DGPS with the PROFILE™ contour data increased the variance. For the area below the 15 m contour line, the inclusion of DGPS data increased the variance at shorter lags than globally. The increase in spatial variation may indicate that the new data set is more representative of the floodplain, and, therefore, more suitable for use in flood inundation modelling.

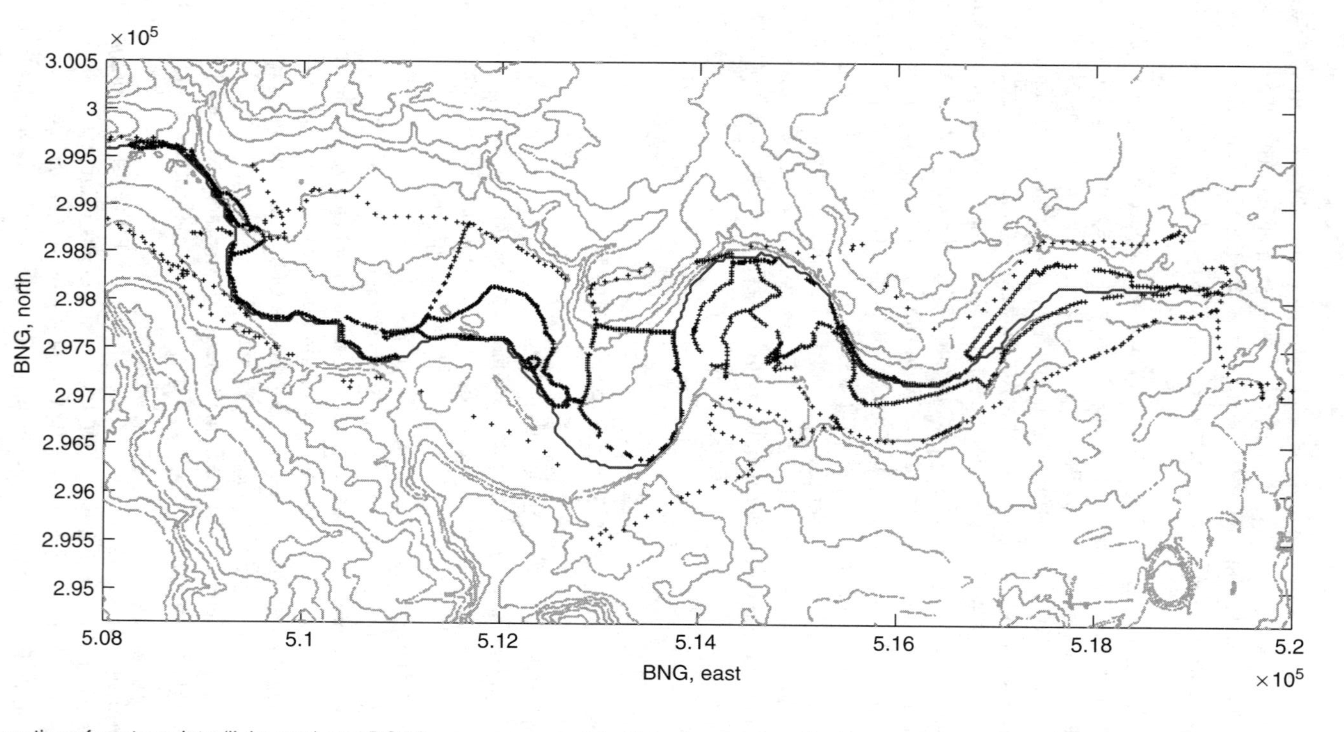

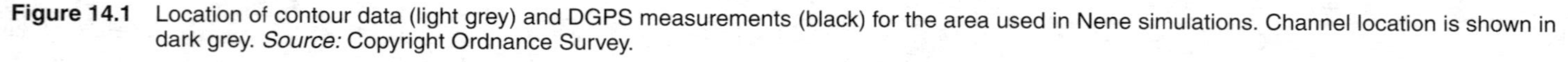
Figure 14.1 Location of contour data (light grey) and DGPS measurements (black) for the area used in Nene simulations. Channel location is shown in dark grey. *Source:* Copyright Ordnance Survey.

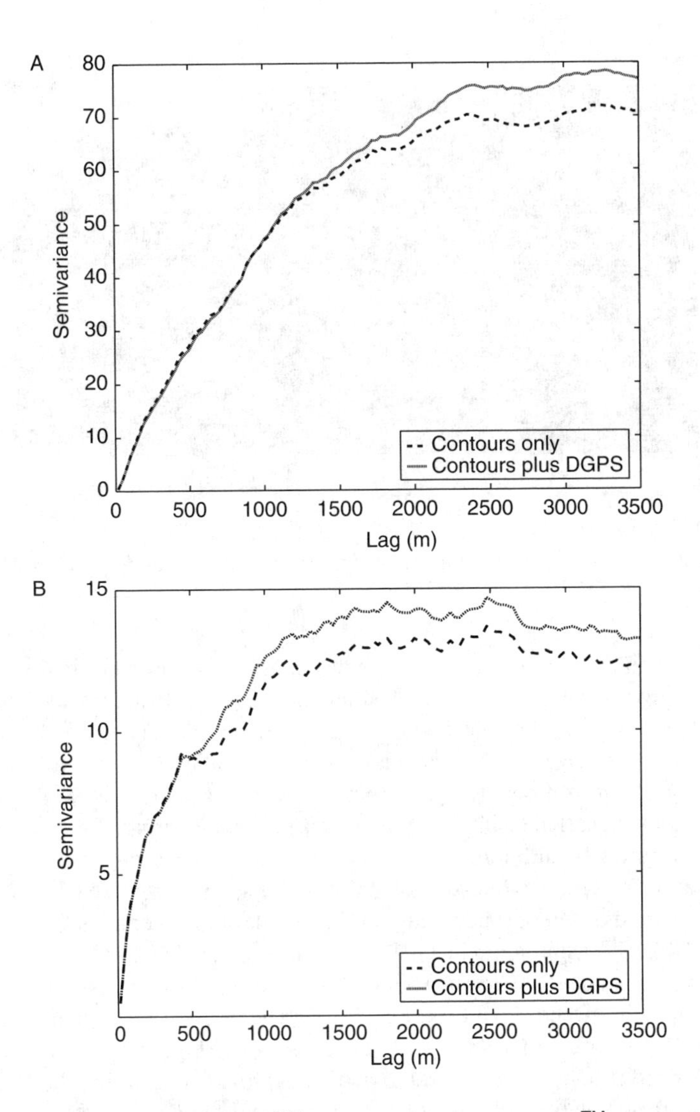

Figure 14.2 Variograms of PROFILE™ contour data, and PROFILE™ contour data after the inclusion of additional DGPS measurements, (A) for the full area, and (B) for the floodplain area below the 15 m contour line.

14.4 GENERATION OF ELEVATION SCENARIOS

Spatial prediction uncertainty in the combined PROFILE™ contour and DGPS data was assessed using the geostatistical method of stochastic imaging or conditional simulation. This enabled the sensitivity of LISFLOOD-FP to small changes in topography to be assessed. Conditional simulation (Deutsch and Journel, 1998) was used to generate elevation scenarios as it honours the values of the data at their original locations (Figure 14.1), and aims to reproduce global features and statistics of the

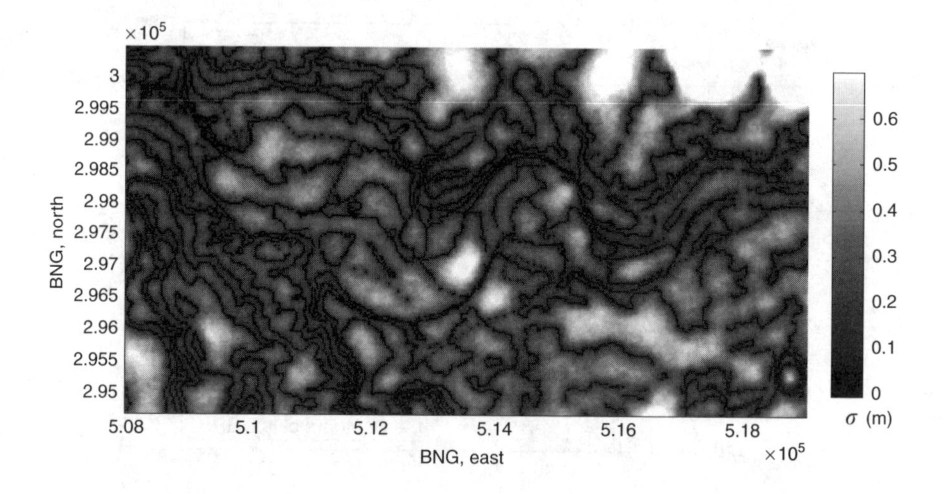

Figure 14.3 Standard deviation of elevation scenarios generated using sequential Gaussian simulation.

data. In particular, each simulation aims to maintain the same original variogram. Therefore, given the original data, each simulation can be said to have an equal probability of representing the true floodplain surface. Importantly, each simulation is superior to a Kriged surface, which is not a possible reality: by predicting optimally (i.e., taking the most probable value from the posterior distribution), Kriging produces a smoothed surface with a variogram very different to that of the original data (Goovaerts, 1997). In addition, conditional simulation is superior to methods such as Monte Carlo simulation, which do not maintain the spatial structure of the data.

One hundred different (but equally probable) elevation scenarios from the combined PROFILE™ contour and DGPS data were generated at a spatial resolution of 30 m using the Sequential Gaussian simulation (SGSim) program, part of the GSLIB software package (Deutsch and Journel, 1998). The original variogram of the combined PROFILE™ contour and DGPS data was used (Figure 14.2A) as scenarios were generated from the full data set to avoid assumptions regarding the location of the floodplain boundary. The standard deviation of all the DEMs generated is shown in Figure 14.3. Standard deviation is low in the vicinity of data points, and progressively rises with increased distance (and, hence, increased uncertainty). Figure 14.4 shows a smaller area of floodplain for clarity. By comparing model results obtained using the multiple elevation scenarios generated, the effect of spatial prediction uncertainty on the prediction of flood inundation was assessed.

14.5 SENSITIVITY OF LISFLOOD-FP TO PREDICTED ELEVATION

All 100 elevation scenarios were used to predict inundation, on a large Beowulf cluster at the University of Southampton, using the hydrograph shown in

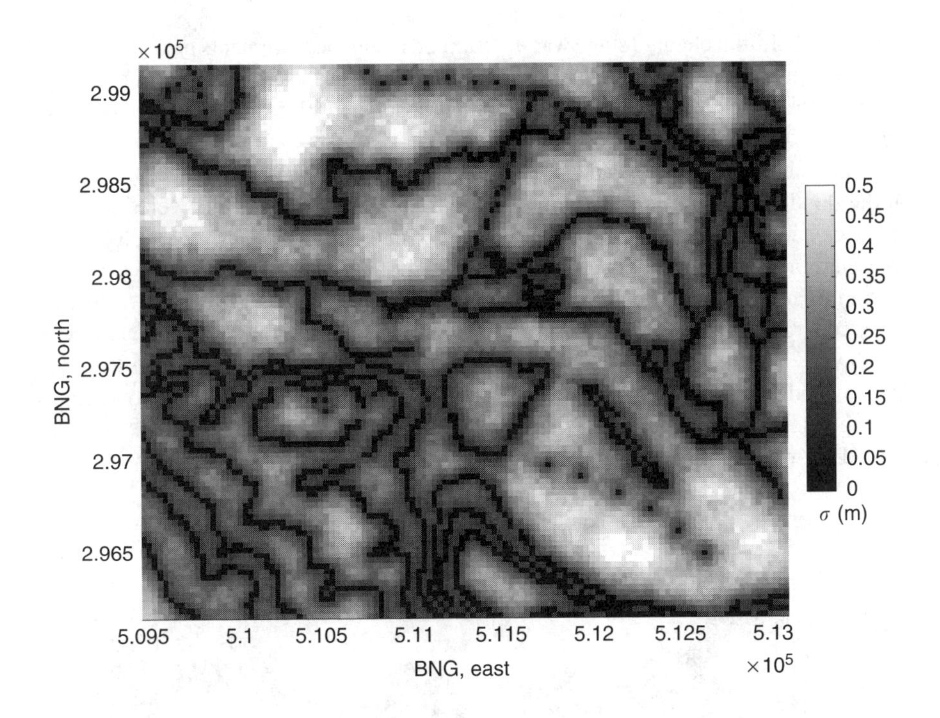

Figure 14.4 Standard deviation of elevation scenarios for a smaller area of floodplain, generated using sequential Gaussian simulation.

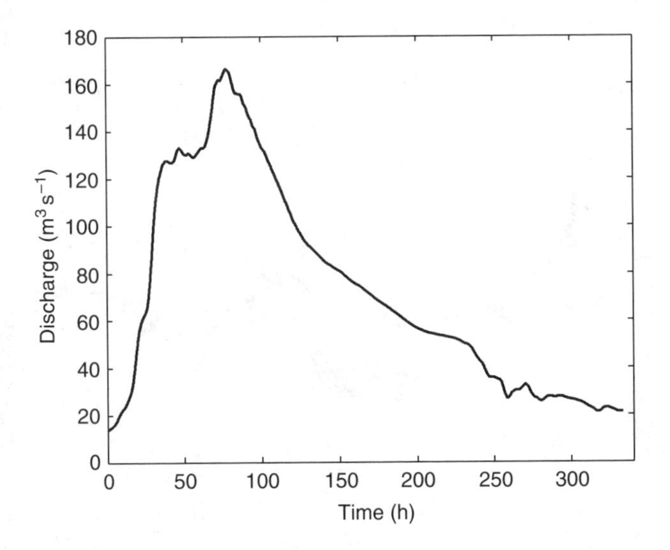

Figure 14.5 Inflow hydrograph for the upstream end of the channel reach, obtained from the Environment Agency, Anglian Region. Authority reference: U32610; Station name: Wansford.

Figure 14.5. Initial channel stage was determined using measurements provided by the Environment Agency, and channel width and bankfull level were determined using channel cross-sections. Channel friction was set at 0.035 and static floodplain friction at 0.04.

The standard deviation and coefficient of variation (CV) of inundation depth at the flood peak (80 h) are shown in Figure 14.6. Only flooded cells were included in the calculations, which prevented non-flooded cells from skewing the results. Cells that were not flooded in some simulations would have decreased both the standard deviation and CV. Variation in predicted inundation depth was generally greatest in areas of large elevation uncertainty. For example, the maximum standard deviation in depth was ~0.8 m at BNG 513,500 E; 297,000 N, which corresponded well with the larger standard deviation in elevation for the same area. In addition, along some lines of small elevation uncertainty, variation in predicted flood depth was correspondingly low. This was not always the case, however, as the pattern was highly complex.

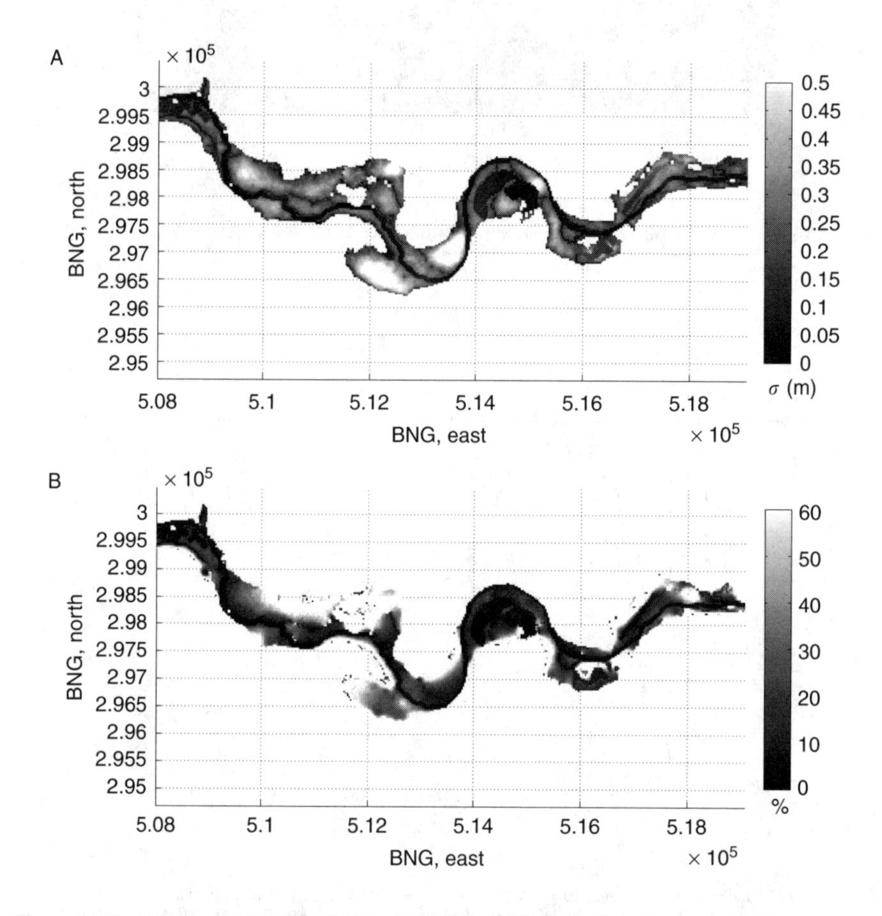

Figure 14.6 Variation in depth at flood peak (80 h): (A) standard deviation, and (B) coefficient of variation. Nonflooded cells in each simulation were discounted to avoid skewing the results.

Points of small elevation uncertainty may still have resulted in a variable prediction of flood depth due to the surrounding topography. When mean depth was taken into account using the CV, it was the shallowest areas (particularly at the edge of the flood envelope) that exhibited the greatest variation in depth.

Uncertainty at the edge of the flood envelope was also observed by calculating the percentage of simulations in which each cell was flooded at the flood peak (Figure 14.7). A shallow gradient at the edge of the floodplain resulted in an uncertain prediction of flood extent, such as in the area at BNG 512,000 E; 298,500 N. Conversely, areas with steep slope at the edge of the floodplain (such as the outside of the meander bend at BNG 514,500 E; 298,750 N) have a high degree of certainty in flood extent.

The standard deviation and CV of the area of inundation is shown in Figure 14.8. Maximum standard deviation occurred shortly after the time of maximum inundation extent, as the flood wave moved downstream. The maximum value of 0.17 km^2 was equivalent to $\sim$189 cells. Although the standard deviation in area decreased as the flood wave receded, the CV continued to rise to a maximum of 2.5%. A similar trend was observed in the mean depth of flooding (Figure 14.9). Although the maximum standard deviation of 0.0225 m was small, the CV continued to rise throughout the simulation to a maximum of 3.2%. Likewise, the CV of volume of flood inundation (Figure 14.10) continued to rise to a maximum of 4.4%. These figures suggest that variation in predictions become compounded through time. Given flood simulations longer than the 340 h duration here, variation (and, hence, uncertainty) may further increase.

A different trend was observed in channel outflow discharge (Figure 14.11). Here, the greatest variation occurred during the rising limb of the hydrograph, before dropping back to a relatively low level. The maximum standard deviation was 3.82 m^3/s, which represented the maximum CV of 4.5%. Large variation in outflow was expected during the rising limb, as this is when flood inundation commences. Variation also

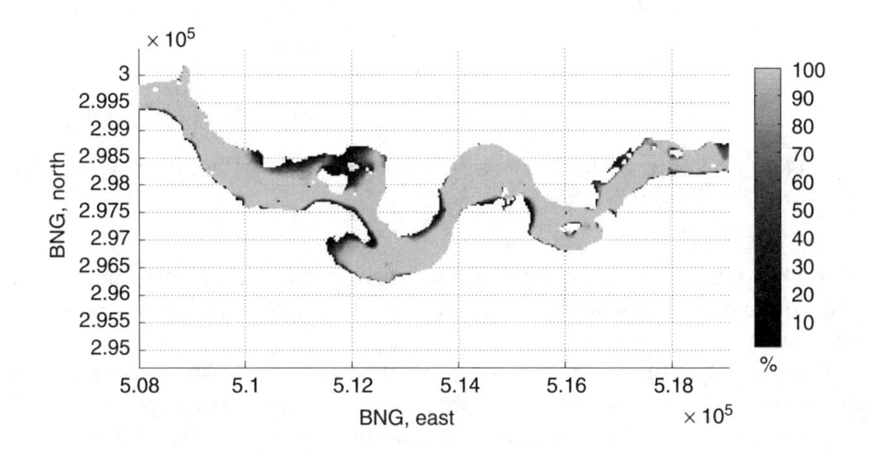

Figure 14.7 Spatial extent of inundation at flood peak: percentage of simulations in which each cell was inundated. Cells which did not flood in any simulations have been removed for clarity.

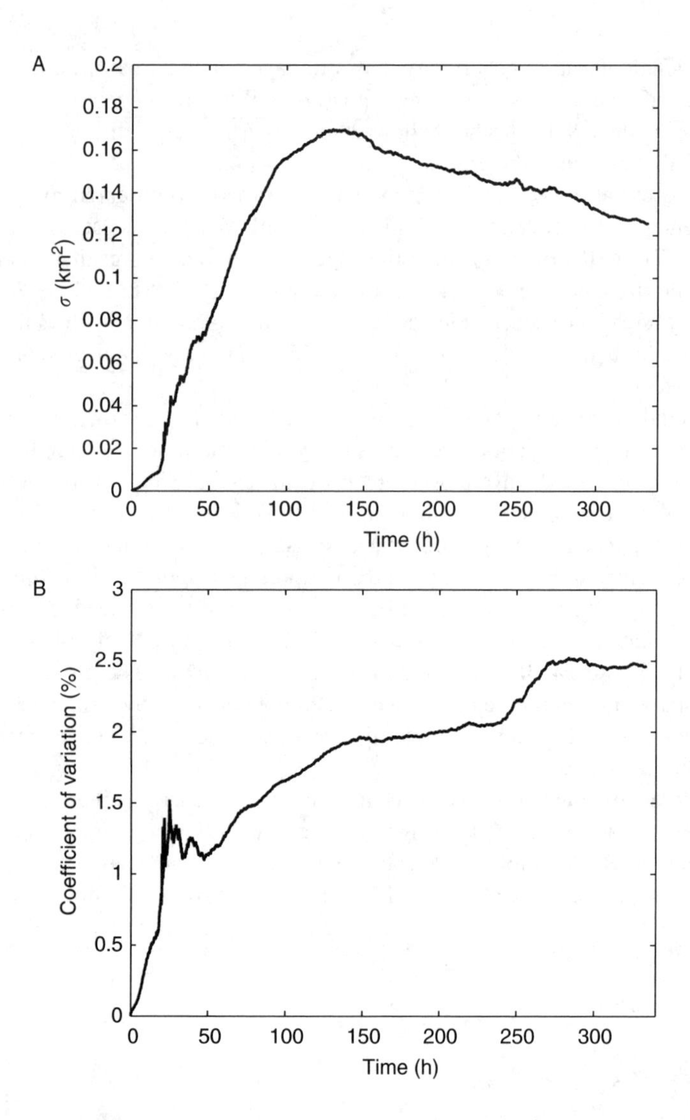

Figure 14.8 Variation in flooded area for all simulations: (A) standard deviation; and (B) coefficient of variation.

occurred during the recession limb of the hydrograph. However, as flood inundation occurred more quickly than flood recession, the variation is restricted to a shorter period of time.

The mean time of maximum flood depth increased progressively downstream, other than in outlying areas with topographic restrictions. As with the depth at the flood peak, only flooded cells were included in the calculations. High variability in time of maximum flood depth occurred in topographically restricted areas, where the flood wave took longer to reach. The greatest standard deviation in time of maximum depth

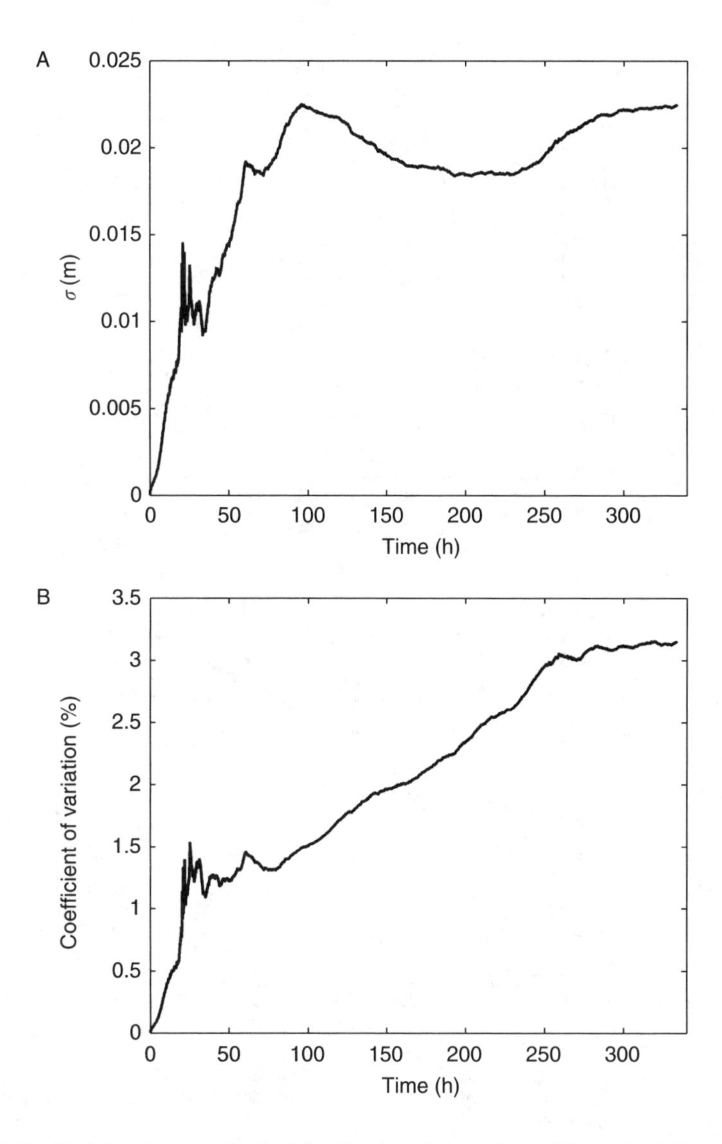

Figure 14.9 Variation in mean depth of flooding for all simulations: (A) standard deviation; and (B) coefficient of variation.

(Figure 14.12A) was >100 h, observed in the area at BNG 512,000 E; 298,500 N. This equated to a CV of ~70% (Figure 14.12B). Variation in the time of maximum flood depth for the near-channel floodplain is more clearly shown in Figure 14.13, and for the channel itself in Figure 14.14. As the flood wave progressed downstream, variation in time of maximum flood depth increased to a maximum standard deviation of ~0.6 h, which was equivalent to a CV of 0.6%. This uncertainty may have increased further given a longer river reach.

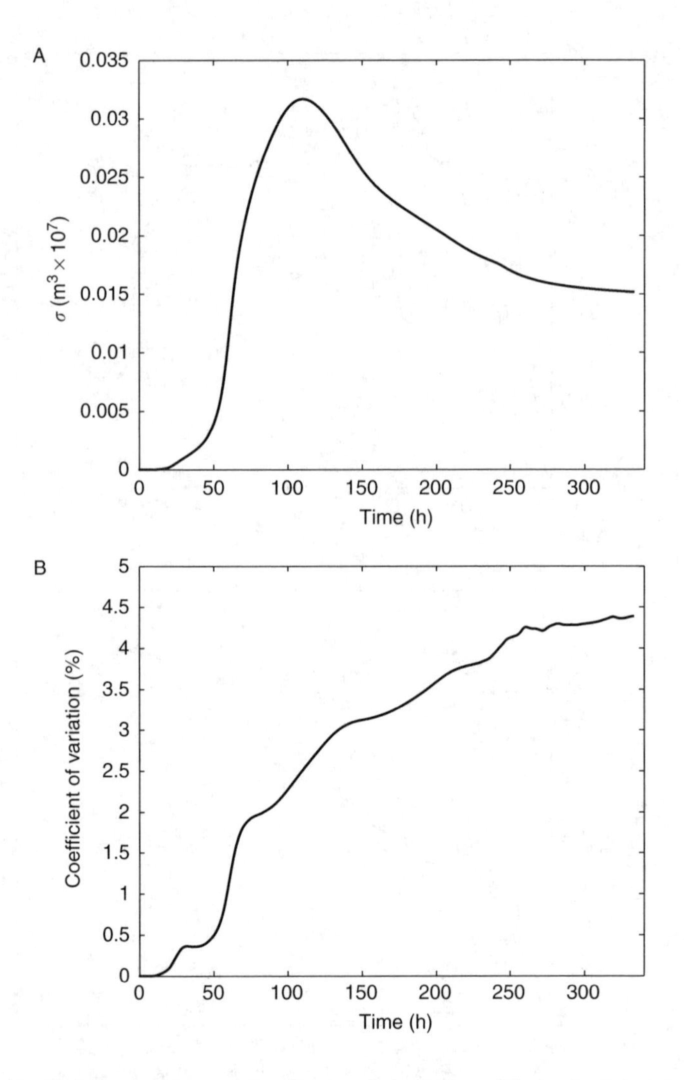

Figure 14.10 Variation in volume of flooding for all simulations: (A) standard deviation; and (B) coefficient of variation.

Variations in the time of initial flooding are shown in Figure 14.15. Areas furthest from the channel that tend to flood last, generally have the largest standard deviation. By using the CV, however, it is clear that there are many areas on the near-channel floodplain in which the time of flooding onset has a high degree of variability of 10 to 20% (Figure 14.15B). Local (high-frequency) topographic variations were the principal control over where flood water initially flowed. Therefore, the time of initial flooding was particularly influenced by the small changes in local topography across the elevation scenarios. In addition, during the early stages of the flood event, near-channel topography (especially bank topography) controlled the timing of inundation.

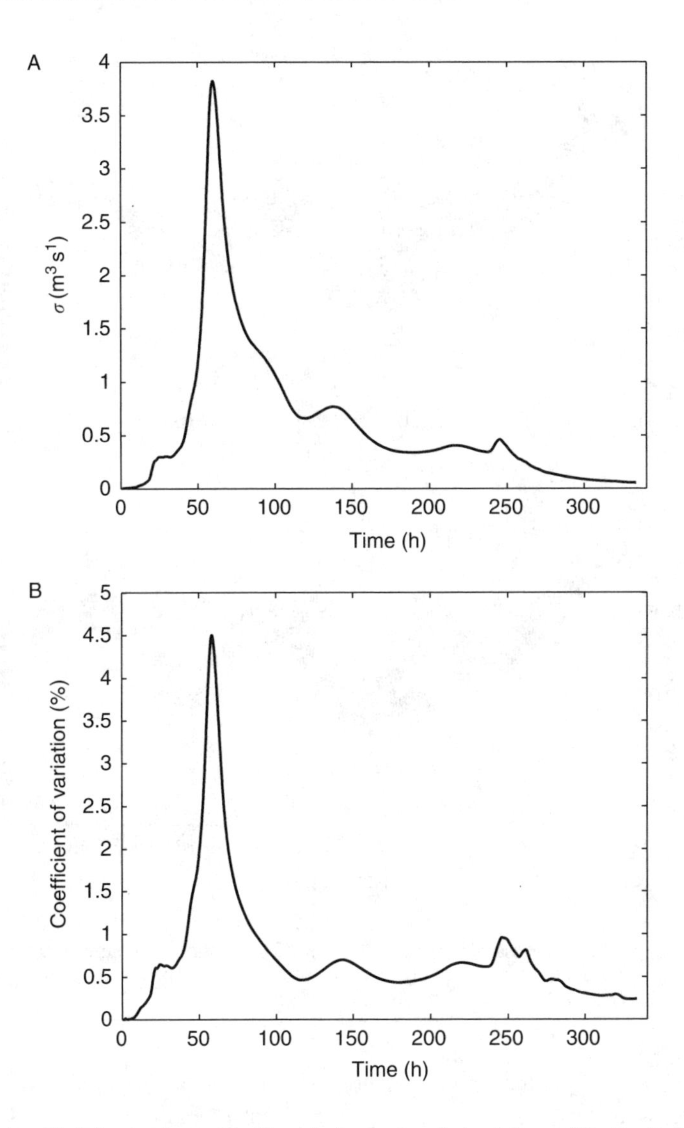

Figure 14.11 Variation in channel outflow discharge for all simulations: (A) standard deviation; and (B) coefficient of variation.

14.6 DISCUSSION

There are large areas where elevation is still uncertain despite additional DGPS data, as indicated by the standard deviation of elevation realisations (Figure 14.3). Some areas of high standard deviation had access problems, and it was not possible to make DGPS measurements within them. Although a denser network of DGPS may reduce the amount of prediction uncertainty in elevation, this may not be feasible for all areas.

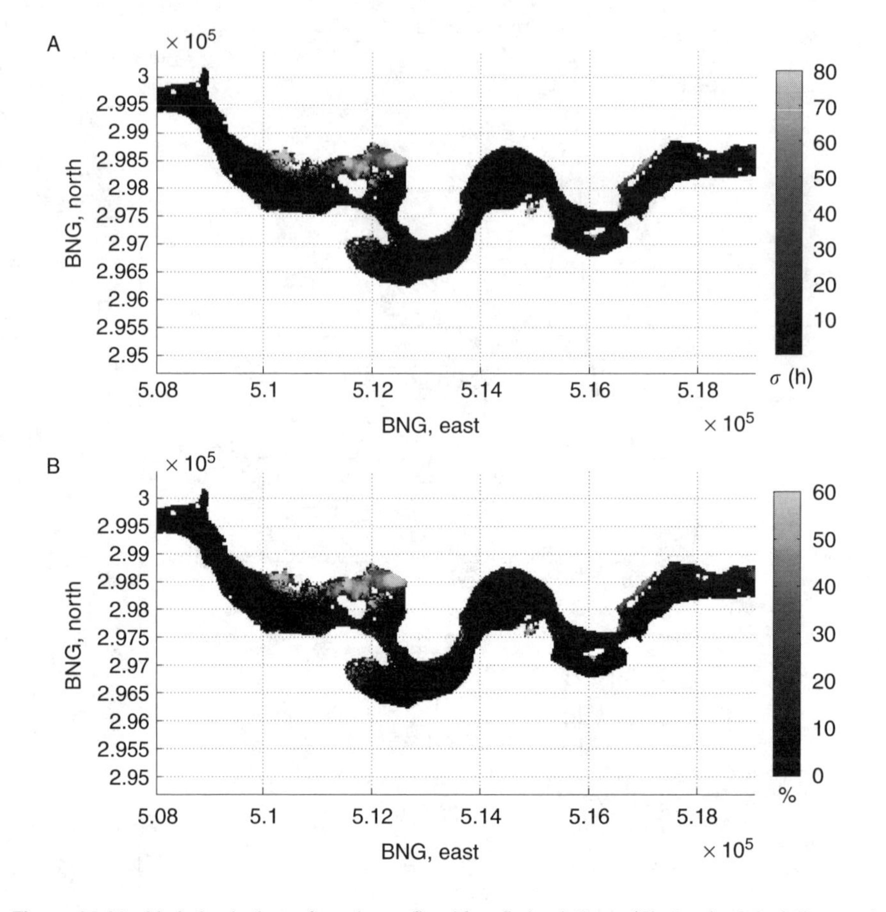

Figure 14.12 Variation in time of maximum flood for all simulations: (A) standard deviation; and (B) coefficient of variation. Non-flooded cells in each simulation were discounted to avoid skewing the results.

The effects of prediction uncertainty in elevation on the prediction of flood inundation were observed both locally and globally, and increased both through time and with distance downstream. For example, during the simulations the uncertainty in the predicted area of inundation increased to a maximum of 2.5%. For a large flood event, this degree of uncertainty may represent a substantial area. Locally, uncertainty in predicted inundation extent was greatest where elevation gradients were smallest. The location of the flood shoreline is critical for the flood insurance and reinsurance industries, and for floodplain management. It is in these areas that uncertainty must be reduced. In particular, topographically important features such as ditches or embankments (natural or anthropogenic) should be captured by the elevation data since they may control the volume of water entering such areas.

At the downstream end of the reach, uncertainty in the time of maximum flood depth was 0.6 h. This was despite the large influence the channel had on the flood

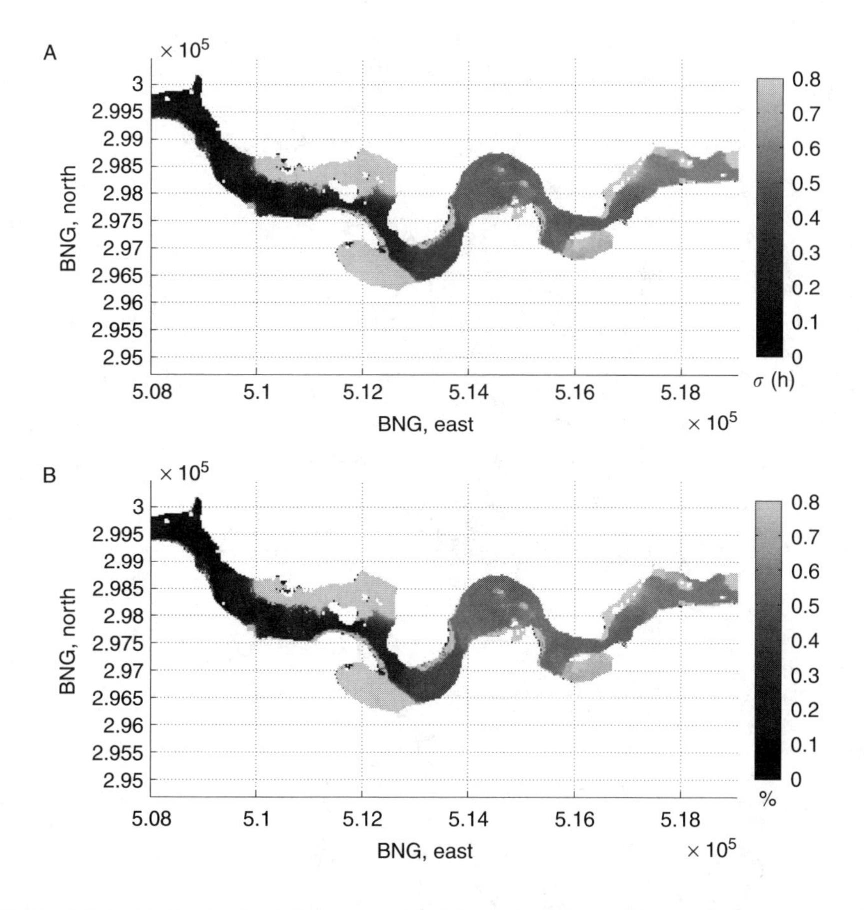

Figure 14.13 Variation in time of maximum flood for all simulations: (A) standard deviation; and (B) coefficient of variation, scaled to highlight the near-channel floodplain.

wave. For larger reaches, and for flood events where the rising limb lasts longer, this uncertainty may be greater. Away from the channel in areas which were topographically restricted, uncertainty in flood wave timing was greater. This has implications for the real-time modelling of flood events when the forecasting of the arrival of the flood wave is essential for emergency management.

14.7 CONCLUSIONS

In this chapter, the PROFILE™ contour data were obtained from the Ordnance Survey and supplemented by measurements of elevation on the floodplain and along the channel using Differential GPS. The aim was to assess the effect of uncertainty in predictions of elevation on the prediction of inundation extent. The addition of the DGPS measurements to the contour data increased the amount of spatial variation

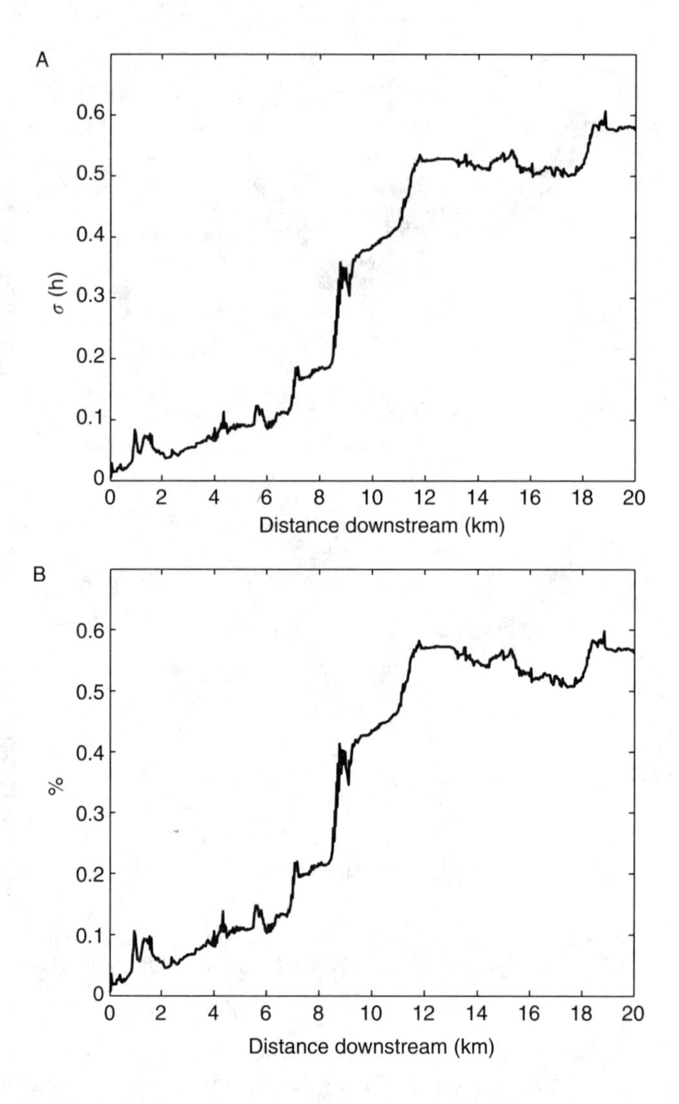

Figure 14.14 Variation in time of maximum flood for all simulations along the channel: (A) standard deviation; and (B) coefficient of variation.

in elevation in the domain, suggesting that the data were more representative of the floodplain.

Sequential Gaussian simulation was used as a novel method of generating multiple realisations of elevation based on combined contour and DGPS data, whilst maintaining their spatial character. Importantly, each realisation was equally representative of floodplain topography, and model predictions were, therefore, equally valid. The effect of uncertainty in the DEM on the prediction of flood inundation increased both with distance downstream and throughout the simulation. This is important as uncertainty is likely to increase for predictions of flood events at larger scales or of longer duration.

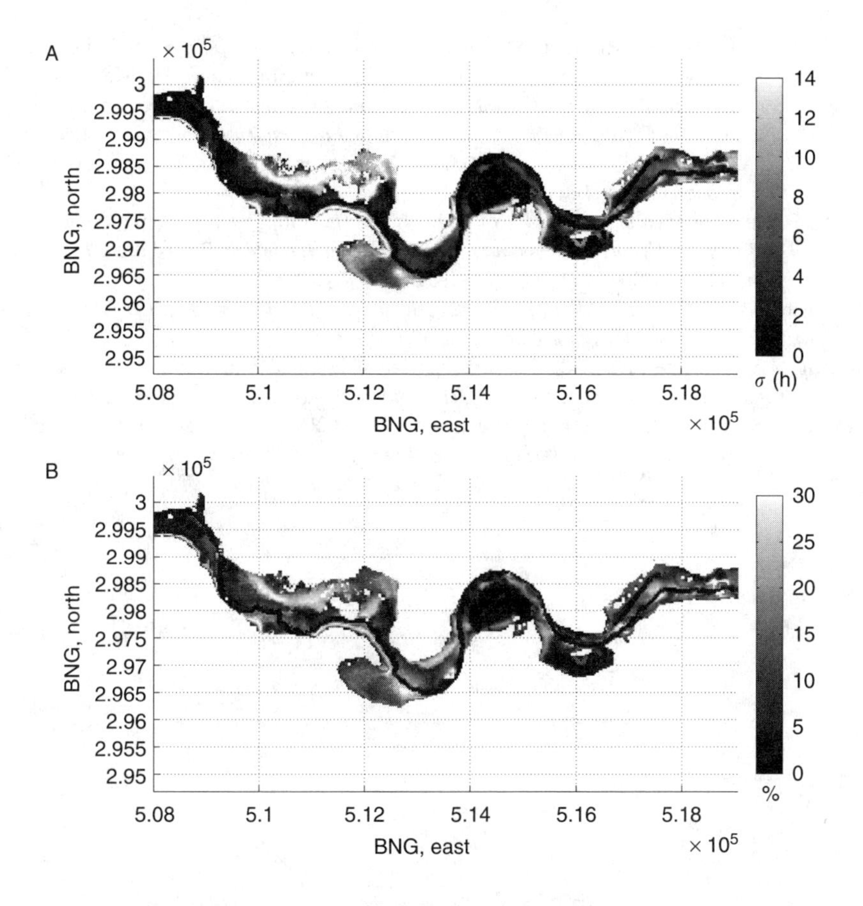

Figure 14.15 Variation in time of initial flood for all simulations: (A) standard deviation; and (B) coefficient of variation. Non-flooded cells in each simulation were discounted to avoid skewing the results.

ACKNOWLEDGMENTS

The authors thank the Ordnance Survey for providing contour data, and Phil Allen of Hampshire County Council for the loan of the Trimble ProXRS unit. Research was funded by a High Performance Computing bursary from the University of Southampton, where simulations were conducted on the Iridis Beowulf cluster.

REFERENCES

Bates, P.D. and De Roo, A.P.J., 2000, A simple raster-based model for flood inundation simulation, *Journal of Hydrology*, **236,** 54–77.

Chow, V.T., 1959, *Open-Channel Hydraulics* (McGraw-Hill Inc, New York).

Chow, V.T., Maidment, D.R., and Mays, L.W., 1988, *Applied Hydrology* (McGraw-Hill Inc).

De Roo, A.P.J., van der Knijff, J., Schmuck, G., and Bates, P., 2000, A simple floodplain inundation model to assist in floodplain management, In Maione, U., Majone Lehto, B., and Monti, R., Eds, *New Trends in Water and Environmental Engineering for Safety and Life: Eco-Compatible Solutions for Aquatic Environments* (Balkema, Rotterdam).

Deutsch, C.V. and Journel, A.G., 1998, *GSLIB Geostatistical Software Library and User's Guide*, 2nd edition (Oxford University Press, New York).

Estrela, T. and Quintas, L., 1994, Use of a GIS in the modelling of flows on floodplains, In White, W. and Watts, J., Eds, *2nd International Conference on River Flood Hydraulics* (John Wiley & Sons Ltd., York, England).

Goovaerts, P., 1997, *Geostatistics for Natural Resources Evaluation* (*Applied Geostatistics Series*), (Oxford University Press, New York).

Horritt, M.S. and Bates, P.D., 2001, Predicting floodplain inundation: raster-based modelling versus the finite element approach, *Hydrological Processes*, **15**, 825–842.

Horritt, M.S. and Bates, P.D., 2002, Evaluation of 1D and 2D numerical models for predicting river flood inundation, *Journal of Hydrology*, **268**, 87–99.

Part III

Human Processes

CHAPTER 15

Introduction — Urban Simulation

Fulong Wu, Peter M. Atkinson, Giles M. Foody, and Stephen E. Darby

CONTENTS

15.1 PROGRESS IN CA DYNAMIC MODELLING

It has been over 10 years now since we saw initial diffusion of interests in using cellular automata (CA) to model human and environmental changes (in particular urban land development). The special sessions of the Annual Meeting of the Association of American Geographers in 1994 marked the development of the paradigm of highly disaggregated CA modelling, which is very different from "traditional" land use and transport modelling. The sessions subsequently resulted in a special issue in *Environment and Planning B* (Batty et al., 1997). Advances in GIS visualization and data processing capacities have further added momentum to CA applications. Since then, there have been widespread applications of various sorts of CA.

Moreover, there have been great improvements in model calibration (Clarke et al., 1997; Clarke and Gaydos, 1998; Li and Yeh, 2001; Silva and Clarke, 2002; Wu, 2002;

0-8493-2837-3/05/$0.00+$1.50

Straatman et al., 2004), irregular presentation of cellular space (O'Sullivan, 2001), vector (Shi and Pang, 2000), and multidimensional representation (Semboloni et al., 2004), together with a series of attempts to expand CA into theoretical modelling tools (Webster and Wu, 1999a, b, 2001; Wu and Webster, 2000; Wu, 2003), as well as a diversity of definitions of transition rules (White et al. 1997; Wu and Webster, 1998; Batty et al., 1999; White and Engelen, 2000; Li and Yeh, 2001; Yeh and Li, 2002). Of course, the above reference list is only partial and does not include many recent publications in CA (e.g., Wu and Martin, 2002; Arentze and Timmermans, 2003; Barredo et al., 2003; Liu and Phinn, 2003; and papers in the special issue of *Computers, Environment and Urban Systems*, **28**, 2004), which are truly proliferating.

Emerging from a CA workshop in CASA (Centre for Advanced Spatial Analysis), an editorial in *Environment and Planning B* (Torrens and O'Sullivan, 2001) asks "Where do we go from here?" in cellular automata and urban simulation. The authors identify a few areas for enhancement such as exploring spatial complexity, infusing urban CA with theory, outreaching education roles, developing hybrid CA-agent operational models, and validating CA models. Recently, there has been tremendous progress in CA research, or broadly speaking, in geodynamic modelling. These issues have been addressed to various extents in the *GeoComputation 2003* conference held in Southampton, from which this volume has been developed.

The strength of CA lies in its heuristic, transparent, and flexible rules, which have a potential link to the decision making process, plus its visualization of future spatial forms. The weaknesses of CA are its rigid notion of cell states, geographical localness, and validation difficulty. While there is always room for making adjustments such as changing the size of neighbourhood and fine-tuning the transition rules, CA models are surprisingly robust in different sorts of model construction. While the performance might not be up to the aspiration of real-life decision support, the output often presents a "signature" of pattern, which reveals some interesting underlying mechanism. Without fine toning, the complex system built upon the recursive and locally defined rules often leads to "self-organized criticality" (Bak, 1996). This is intuitively appealing, as our city is robust against major catastrophe brought by individual events and large changes are actually triggered by "ordinary" and "insignificant" actions. With regard to CA research itself, however, it seems that it has reached an evolutionary plateau where there are many meaningful improvements but without an evolutionary breakthrough.

Perhaps the notion of GeoDynamics can inject some new energy — CA modelling listed here is after all a dynamic modelling approach to changing human and natural environments. More significant development in the modelling methodology would come from the practical task of managing these environments rather than intelligent and innovative twists by modellers. Although CA is based on the notion of grid or cells, it is not necessarily confined to cell structure. The model can be developed based on any configuration of spatial units. The proximity (local) relationship between these units can be mapped by, for example, the list of "neighbouring" units. This notion in fact brings us very close to the "traditional" form of urban modelling and thus opens a door to reuse of many useful model structures in future hybrid CA models. Xie and Batty (Chapter 19, this volume) as well as other agent-based modelling advocators (e.g., Parker et al., 2003; Beneson and Torrens, 2004) have developed new forms or approaches beyond the constraint of CA. However, Xie and Batty's approach differs from agent-based models in that its basic block is more like an aggregated unit (a cell containing households and jobs) than the individual household and thus is closer to the operational urban models. This structure has its strength because dynamic modelling ties very closely to the spatial analyses that are routinely carried out by GIS in a static way.

Considering the strengths and weaknesses of CA, it can be seen that its weaknesses have not been overcome by introducing more flexibility. Dynamic modelling through a multi-agent system approach has become more flexible, after introducing the concept of "agents," cognitive actions and non-grid space. Its ability to attach the behaviour to the real object originating such behaviour (i.e., a household presenting the behaviour of migration, which is different from a census tract having different attributes) is appealing. But the challenge of validation becomes even greater. Although CA and other spatially explicit dynamic models have gained popularity, there is still a long way to go before they can satisfy the aspiration for supporting policy making.

15.2 GEODYNAMIC MODELLING IN HUMAN ENVIRONMENTS

In the Section III, the chapters are concentrated on the development of dynamic modelling in human settlements and the interaction between human and natural environments.

Clarke raises a serious issue of developing/applying the model of "simplicity" and its limit in dealing with complex and rich human and natural environments. His extensive research in CA and GIS in a policy context (Clarke et al., 1997; Silva and Clarke, 2002) helps him to observe that the general public prefers an incomprehensible model that gives useful results to an understandable model. The challenge is, therefore, for the modellers to make the underlying assumptions (required by model simplicity) more explicit. Through comparing three widely available models (SLEUTH, SCOPE, and What if?), he found significant variations in the model outputs. Moreover, each model uses "scenarios" to communicate their results but there is a question to what extent these scenarios are just hypothetical assumptions rather than being rooted in the "real" dynamics of environments. In fact, the use of scenarios is a tool of model reduction. He concludes that rather than rushing towards model reduction, it is important to emphasize the ethics of modelling and in turn to make models more open to scrutiny.

Since the SLUETH model has gained such a popularity recently (see application such as Silva and Clarke, 2002; Yang and Lo, 2003), the Chapters 16 and 17 address the spatial and temporal variation of rule definition and new calibration methods. Dietzel uses the example of San Joaquin Valley, California, to show that the coefficients estimated from different geographical areas vary significantly. The research further raises the question of "scale" at which derived models are applicable. There might be different kinds of model (model parameters) in predicting coarse or finer GeoDynamics of natural and human environments. This research is a further step from Silva and Clarke (2002) to "fine tune" the parameter space according to the suitable geographic region. This spatial–temporal limitation of model coefficients is somewhat similar to the modifiable areal unit problem (MUAP), where the relationship between two variables depends upon the spatial unit partition as well as the scale. Thus, the rules derived/calibrated from one region might not be appropriate to another region. The same is true of time — rules valid in one period may not be appropriate in another.

Calibration continues to be a substantial issue in dynamic modelling. Goldstein introduces a generic algorithm (GA) into calibration of the SLUETH model. The calibration method developed by Clarke et al. (1997, see also Clarke and Gaydos, 1998) is based on historical urban spatial data. There are five coefficients (integers ranging from 0 to 100) controlling the morphology. They are the coefficients of diffusion, bread, spread, slope resistance, and road gravity. The results of the coefficients are usually evaluated by a "product metric." The search for the best values is computationally intensive. The method, called "Brute Force," basically steps through coefficient space by a coarse step (e.g., 25 units) through the entire coefficient space, then a finer step (e.g., 5 to 6 units), followed by even finer step such as 1 unit. The GA thus has its advantage because it searches through the coefficient space in an adaptive manner and thus reduces the computational burden. Furthermore, the use of GA in model calibration opens the door to using artificial intelligence methods to calibrate dynamic models. There have recently been various reports of using artificial neural network (Li and Yeh, 2001) and the difference in cells (Straatman et al., 2004). More broadly speaking, GA can be used in the search of coefficient space without necessarily being confined within the existing SLUETH model structure and range of parameters (i.e., from 0 to 100). For many parameters from real life, such as population density and the distance to the transport network, their values are not within the defined range of SLEUTH. GA as well as other artificial intelligent approaches can thus still search for the best range of parameters more flexibly.

The CA model traditionally is strongly associated with urban morphologic studies. To a lesser extent the model uses factors dealt with in traditional urban models, such as land price, travel cost and behaviour, and land development. In contrast, Strauch et al. report a large-scale modelling project, named Integrated Land-Use Modelling and Transportation System Simulation (ILUMASS), which is reminiscent of large-scale urban modelling but uses microsimulation as the core method to simulate the interaction between urban land use development, transport demand, dynamic traffic assignment, and environment impacts. The microsimulation approach allows the modelling at the highest disaggregated units, for example, synthetic households and jobs. From these synthetic households and jobs are estimated travel demand and goods transportation, which in turn are manifested in the trips and assigned into the network dynamically forming network flows. Emissions are estimated from traffic flows, and the impacts of air quality on population and firm locations can be assessed. The model reported by Strauch et al. here shares some features of microsimulation-based models such as UrbanSim (Waddell, 2000) and California Urban Futures (CUF) (Landis and Zhang, 1998). As seen in large-scale urban modelling, integrating many model components might prove to be a challenging task.

As one of the pioneers who introduced CA into the domain of urban studies, Batty (1997), together with his colleagues, has developed a series of simulation models in the GIS environment (Batty, 1998; Batty et al., 1999). Xie and Batty systematically evaluate the progress of CA modelling and in particular review the prototype and applications of Dynamic Urban Evolutionary Model (DUEM), one of the earliest urban CA models (Xie, 1996). They discuss five key design issues in their new model, Integrated Dynamic Urban Evolutionary Modelling (IDUEM), which keeps many attractive features of DUEM but is enhanced with a significant breakthrough in terms of design philosophy and implementation. Recognizing the limitation of current CA models in their inability to handle socio-economic attributes and spatial interaction, they urge a closer linkage of CA models to traditional cross-sectional land use and transport models such as population and employment forecasting models, which are still widely used in local governments. They build demographic and economic attributes into the cells, and through microsimulation and an agent-based modelling approach, relate the attributes to land use and transport models. Through iterative proportional fitting the attributes are partitioned into the grid space. The relocation and flow of households and jobs are then modelled through space-filling or new subdivision, which produces new urban structures. The overall programming approach benefits from object-based simulation. In short, the model will increase the use of dynamic modelling in operational management and urban policy making by integrating the CA fundamentals with the zonal system in which policy making processes usually take place.

The agent-based modelling approach appears to have some appealing features, as it can model directly the behaviour of agents. In the Chapter 21 by Barros, she uses StarLogo to simulate urban expansion in Latin American cities. Similar to other agent-based simulation, the research is aimed to find some "generic" features of urban development, which has been modelled before as in "polycentric urban structure" (Wu, 1998) (in this case, the "peripheralization" phenomenon in the Third World cities). Although the model itself is implemented in an abstract computer space, the output shares similarity with the fragmentation patterns in many cities in developing countries. The question is, however, whether such a fragmented spatial structure is unique to this simulation. In other words, does the "signature" reflect the unique simulation rules (and, in this case, the behaviour of land development in the developing countries)? The chapter, again, shows that validation continues to be a challenge.

Despite simulation as an increasingly popular method, there are few studies specifically interrogating the error and uncertainty. In contrast, the issue of error propagation has become an established topic in GIS. Yeh and Li build upon their extensive research in CA (Li and Yeh, 2000, 2001; Yeh and Li, 2001, 2002) to examine the effect of error in CA modelling. They differentiate two types of error: errors from data sources and errors from transformations. Furthermore, they examine model uncertainties in terms of discrete space and time, neighbourhood configuration, model structure and transition rules, parameter values, and stochastic variables. Yeh and Li find that errors are propagated in CA simulation and that uncertainties are further increased by model transformation. However, the error mainly exists in the fringe area of urban clusters, which shows the robustness of CA model. That is, different stochastic disturbance might change the urban form at the edge but not in its overall pattern.

Given the popularity of agent-based modelling, Benenson and Torrens (2004) propose a modelling framework of a "geographic automata system" (GAS) to combine CA with multi-agent systems. As a useful formalism, their framework allows a "tight" coupling of CA with GIS (in this sense, it is similar to the Geo-Algebra, see Takeyama and Couclelis, 1997). Such formalism is not only useful to communicate model structures between model designer and model programmer but also helps communication among different model builders. Their modelling platform, Object-Based Environment for Urban Simulation (OBEUS), is a further development from agent-based simulation (Portugali and Benenson, 1995; Portugali, 2000; Benenson et al., 2002; Benenson and Torrens, 2004) but thoroughly redesigned.

Chapter 24 by Agarwal considers the ontological stance for agent-based simulation. This potentially forces us to rethink about knowledge representation and qualitative reasoning and in the end to develop reactive agents into cognitive models of human behaviour. It seems that much useful discussion is currently going on under the framework of "agents" (Parker et al., 2003) but there is still a long way to go before fully implementing these agent-based systems for real policy decision making.

15.3 NEXT STAGE CA MODELLING

15.3.1 Geodynamic Modelling as a "Procedural Framework for Temporal Fitting"

Many chapters in this volume share the view that there should be greater transparency and verification of model building and better validation of the output. There is a need to develop some standard testing data, against which different routines can be benchmarked. Such an initiative can only be realized by coordinated research activities to develop modelling standards, similar to that of image compression in computer science where some standard "portrait" is used to test the effectiveness of different compression methods.

There is also a need for a more procedural framework of geodynamic modelling, perhaps not in the sense of "prototypes" as they often embed the designers' conceptualization of how the city works. Such a conceptualization, no matter how flexible, might still be a constraint for building models in a different context. The call for a "procedure" of modelling is not to generate greater "simplicity" but, rather, to remove unnecessary constraints imposed by prototype designer. On the other hand, the use of some generic modelling language without a procedure is felt to be too "flexible" and too difficult for most practical users. What is needed is a procedure or "routine," like $y = \mathrm{f}(x)$ where x could be a series of development factors and y is the dependent variable describing the change in human environments. Similar to regression, there need not to be just one type of modelling procedure. Each would suit different situations with the underlying assumption provided explicitly (just as different regressions suit different probability distributions). Currently, CA research has seen too many model-driven activities rather than data-driven procedures in dynamic modelling. The former often involves encoding a more plausible and attractive conceptualization of urban dynamics and implementation in diverse programming languages.

Just like the spatial interpolation method, dynamic modelling in the simplest sense can be seen as a "temporal interpolation method." As long as the errors associated with these methods are explicitly treated, these methods would be useful for decision making. The procedure of modelling would enhance the development of calibration, because the model itself provides some certainty for improving calibration. More importantly, there is a need for dynamic calibration, that is, calibrating the rules in multiple time periods and adapting them in the new space–time context. Deriving rules from a series of temporal observations would be an important step towards model refinement.

15.3.2 Geodynamic Modelling as an "Analogical Tool for Theory Building"

The call for some procedure to be developed in dynamic modelling is not to reduce it into boring time–space fitting routines. Another important direction of dynamic modelling is to develop an "analogical tool for theory building," in the sense of "artificial worlds." As with the rapidly growing agent-based models (e.g., Epstein and Axtell, 1996), realism in the terms of geographical representation in these artificial worlds is not an issue. But rather, it is the fundamental behavioural characterization of the dynamics and replicable and valid theories building from the simulation that are attractive features.

If two-dimensional spectra in the trade-offs between global versus local and theoretical versus empirical characterization are used to define a typology of urban simulation (Wu and Martin, 2002), then there are four extreme cases: global and theoretical models (e.g., urban land economics such as Alonso model), global and empirical models (e.g., most GIS cartographic modelling, for example through overlay, to derive land suitability and urban operational models such as the Lowry model), local and theoretical models (e.g., physical metaphoric model such as diffusion-limited aggregation), and finally local and empirical model (most hybrid CA models are in this area but often go beyond the local constraint). In contrast to the proliferation of the last type of models, there are not enough "behaviourally rich" theoretical models in the bottom–up paradigm. The power of theoretical model lies in their parsimonious rules and analogical output — the fundamental similarity between the morphology (or signature) developed in artificial worlds and in real cities might provide a plausible explanation (or at least a "discourse") of urban growth. It is rather unfortunate that artificial world simulation does not attract enough attention from the modellers (Webster and Wu, 1999b, 2001; Wu and Webster, 2000). It is not surprising, perhaps, because most model builders are more familiar with GIS and quantitative geography.

As a consequence, quantitative/dynamic modelling has been deprived of the rich language and theoretical thrusts, which are developing rapidly in "mainstream" geography. The latter is now moving away from its scientific root towards critical theories and experiencing the "postmodern cultural turn." There is a need to open up the imagination of dynamic modelling to embrace wider terms such as "regimes of property rights," public–private partnership, urban compaction, social segregation and exclusion, and "clusters" (defined in "new economic geography"). Not all these terms are spatial but many are often played out in a spatial context. Only by moving away from being a satisfied mayor in the SimCity and taking seriously these rich behavioural characterizations and learning from the "mainstream" theoretical perspectives and policy concerns (e.g., exclusion and compaction) can quantitative modelling regain its deserved recognition. Theoretical models such as Schelling (1978) can potentially generate insights to inform policy making. The ability to be involved in policy agenda does not necessarily need to be "empirical" — as the discourse of neoclassical economics through ideal demand and supply curves proves.

15.3.3 Geodynamic Modelling as an "Envision Device for Alternative Futures"

So-called "predication" is merely temporal interpolation; and so-called "envision" is "prescription" — often through visualizing different scenarios, which are probably generated under different conditions. Envision is often in a qualitative rather than quantitative sense. Therefore, it is a matter of "discourse," where the relationship is lucidly explained (such as market equilibrium and the relationship between supply and demand). This differs from the procedural modelling exercise where the purpose is to fit the relationship, a task carried out through essentially data-driven "procedure" in econometrics. Geodynamic modelling as an envision device is to provide an alternative future. In this sense, it also differs from analogical theoretical building because the process of envisioning is grounded in policy contexts and should involve public participation. The envision type of dynamic modelling could not and should not be made parsimonious; but rather there is a need for "GeoComputational honesty" (Clarke, Chapter 16, this volume) to make these assumptions known. Rules, rather than being hidden in the model, should be made explicit and involve users' input. So, calibration through data is important but only serves as a starting point. Parameters can be changed/modified, not to find the better fit but rather to generate impact analysis. The outputs of envision modelling cannot be assessed in terms of "accuracy" or goodness-of-fit in its traditional sense but, rather, should be evaluated by the ability to inform policy making and reveal the hypotheses for future development. This kind of exercise can be built with widespread web-based GIS and public participation in envisioning the future of communities. Geodynamic modelling of this kind can be built upon various media taking advantage of virtual reality and sketch planning. Such a design principle is behind planning support systems such as What if? (Klosterman, 2001) but the emphasis on the "dynamic" side would strengthen the nature of local (inter- and intra-cities) interactions and the recursive process between conscious actions and the consequential environmental changes.

REFERENCES

Arentze, T. and Timmermans, H., 2003, A multiagent model of negotiation process between multiple actors in urban developments: a framework for and results of numerical experiments. *Environment and Planning B* **30**, 391–410.

Bak, P., 1996, *How Nature Works: Algorithms, Calibrations, Predictions* (Cambridge: Cambridge University Press).

Barredo, J.I., Kasanko, M., McCormick, N., and Lavalle, C., 2003, Modelling dynamic spatial processes: simulation of urban future scenarios through cellular automata. *Landscape and Urban Planning* **64**, 145–160.

Batty, M., 1997, Cellular automata and urban form: a primer. *Journal of the American Planning Association* **63**, 266–274.

Batty, M., 1998, Urban evolution on the desktop: simulation with the use of extended cellular automata. *Environment and Planning A* **30**, 1943–1967.

Batty, M., Couclelis, H., and Eichen, M., 1997, Urban systems as cellular automata. *Environment and Planning B* **24**, 159–164.

Batty, M., Xie, Y., and Sun, Z., 1999, Modeling urban dynamics through GIS-based cellular automata. *Computers, Environment and Urban Systems* **23**, 205–233.

Benenson, I. and Torrens, P.M., 2004, Geosimulation: object-based modeling of urban phenomena. *Computers, Environment and Urban Systems* **28**, 1–8.

Benenson, I., Omer, I., and Hatna, E., 2002, Entity-based modeling of urban residential dynamics: the case of Yaffo, Tel Aviv. *Environment and Planning B* **29**, 491–512.

Clarke, K.C., Chapter 16, this volume.

Clarke, K.C. and Gaydos, L.J., 1998, Loose-coupling a cellular automaton model and GIS: long-term urban growth prediction for San Francisco and Washington/Baltimore. *International Journal of Geographical Information Science* **12**, 699–714.

Clarke, K.C., Hoppen, S., and Gaydos, L., 1997, A self-modifying cellular automaton model of historical urbanization in the San Francisco Bay area. *Environment and Planning B* **24**, 247–261.

Epstein, J.M. and Axtell, R., 1996, *Growing Artificial Societies from the Bottom Up* (Washington D.C.: Brookings Institution).

Klosterman, R.E., 2001, The What if? Planning support system. In *Planning Support Systems: Integrating Geographic Information Systems, Models, and Visualization Tools*, edited by R.K. Brail and R.E. Klosterman (Redlands, CA: Environmental Systems Research Institute).

Landis, J. and Zhang, M., 1998, The second generation of the California urban futures model. Part 1. Model logic and theory. *Environment and Planning B* **25**, 657–666.

Li, X. and Yeh, A.G.O., 2000, Modelling sustainable urban development by the integration of constrained cellular automata and GIS. *International Journal of Geographical Information Science* **14**, 131–152.

Li, X. and Yeh, A.G.O., 2001, Calibration of cellular automata by using neural networks for the simulation of complex urban systems. *Environment and Planning A* **33**, 1445–1462.

Liu, Y. and Phinn, S.R., 2003, Modelling urban development with cellular automata incorporating fuzzy-set approaches. *Computers, Environment and Urban Systems* **27**, 637–658.

O'Sullivan, D., 2001, Exploring spatial process dynamics using irregular cellular automaton models. *Geographical Analysis* **33**, 1–18.

Parker, D.C., Manson, S.M., Janssen, M.A., Hoffmann, M.J., and Deadman, P., 2003, Multi-agent systems for the simulation of land-use and land-cover change: a review. *Annals of the Association of American Geographers* **93**, 314–337.

Portugali, J., 2000, *Self-Organization and the City* (Berlin: Springer-Verlag).

Portugali, J. and Benenson, I., 1995, Artificial planning experience by means of a heuristic cell-space model: simulating international migration in the urban process. *Environment and Planning A* **27**, 1647–1665.

Schelling, T., 1978, *Micromotives and Macrobehavior* (New York: Norton).

Semboloni, F., Assfalg, J., Armeni, S., Gianassi, R., and Marsoni, F., 2004, CityDev, an interactive multi-agents urban model on the web. *Computers, Environment and Urban Systems* **28**, 45–64.

Shi, W. and Pang, M.Y.C., 2000, Development of Voronio-based cellular automata — an integrated dynamic model for geographical information systems. *International Journal of Geographical Information Science* **14**, 455–474.

Silva, E.A. and Clarke, K.C., 2002, Calibration of the SLEUTH urban growth model for Lisbon and Porto, Portugal. *Computers, Environment and Urban Systems* **26**, 525–552.

Straatman, B., White, R., and Engelen, G., 2004, Towards an automatic calibration procedure for constrained cellular automata. *Computers, Environment and Urban Systems* **28**, 149–170.

Takeyama, M. and Couclelis, H., 1997, Map dynamics: integrating cellular automata and GIS through geo-algebra. *International Journal of Geographical Information Science* **11**, 73–91.

Torrens, P.M. and O'Sullivan, D., 2001, Cellular automata and urban simulation: where do we go from here? *Environment and Planning B* **28**, 163–168.

Waddell, P. 2000, A behavioral simulation model for metropolitan policy analysis and planning: residential location and housing market components of UrbanSim. *Environment and Planning B* **27**, 247–263.

Webster, C.J. and Wu, F., 1999a, Regulation, land-use mix, and urban performance. Part 1. Theory. *Environment and Planning A* **31**, 1433–1442.

Webster, C.J. and Wu, F., 1999b, Regulation, land-use mix, and urban performance. Part 2. Simulation. *Environment and Planning A* **31**, 1529–1545.

Webster, C.J. and Wu, F., 2001, Coase, spatial pricing and self-organising cities. *Urban Studies* **38**, 2037–2054.

White, R. and Engelen, G., 2000, High-resolution integrated modelling of the spatial dynamics of urban and regional systems. *Computers, Environment and Urban Systems* **24**, 383–400.

White, R., Engelen, G., and Uljee, I., 1997, The use of constrained cellular automata for high-resolution modelling of urban land-use dynamics. *Environment and Planning B* **24**, 323–343.

Wu, F., 1998, An experiment on generic polycentricity of urban growth in a cellular automatic city. *Environment and Planning B* **25**, 731–752.

Wu, F., 2002, Calibration of stochastic cellular automata: the application to rural–urban land conversions. *International Journal of Geographical Information Science* **16**, 795–818.

Wu, F., 2003, Simulating temporal fluctuations of real estate development in a cellular automata city. *Transaction in GIS* **7**, 193–210.

Wu, F. and Martin, D., 2002, Urban expansion simulation of Southeast England using population surface modelling and cellular automata. *Environment and Planning A* **34**, 1855–1876.

Wu, F. and Webster, C.J., 1998, Simulation of land development through the integration of cellular automata and multi-criteria evaluation. *Environment and Planning B* **25**, 103–126.

Wu, F. and Webster, C.J., 2000, Simulating artificial cities in a GIS environment: urban growth under alternative regimes. *International Journal of Geographical Information Science* **14**, 625–648.

Xie, Y., 1996, A generalized model for cellular urban dynamics. *Geographical Analysis* **28**, 350–373.

Xie, Y. and Batty, Y., Chapter 19, this volume.

Yang, X. and Lo, C.P., 2003, Modeling urban growth and landscape changes in the Atlanta metropolitan area. *International Journal of Geographical Information Science* **17**, 463–488.

Yeh, A.G.O. and Li, X., 2001, A constrained CA model for the simulation and planning of sustainable urban forms using GIS. *Environment and Planning B* **28**, 733–753.

Yeh, A.G.O. and Li, X., 2002, A cellular automata model to simulate development density for urban planning. *Environment and Planning B* **29**, 431–450.

CHAPTER 16

The Limits of Simplicity: Toward GeoComputational Honesty in Urban Modeling

Keith C. Clarke

CONTENTS

16.1 INTRODUCTION

A model is an abstraction of an object, system, or process that permits knowledge to be gained about reality by conducting experiments on the model. Constraining modeling to that of interest to the geographer, a model should abstract geographic space and the dynamic processes that take place within it, and be able to simulate spatial processes

0-8493-2837-3/05/$0.00+$1.50

usefully. At the absurd extreme, a model could be more complicated than the data chosen to represent the spatio-temporal dynamics in question. Occam's razor tells us, of course, that when two models are of equal explanatory power, the simpler one is preferred. Einstein stated that "a model should be as simple as possible and yet no simpler." Some earth systems are so predictable that they are indeed amenable to modeling with the simplest of models. For example, the drop in atmospheric pressure as a function of elevation is so predictable as to be suitable for building a measuring instrument for the latter using the former as a proxy. Of course few human systems, or coupled human/environmental systems are this simple. Geography's past is laced with attempts to use overly simple predictive or explanatory models for human phenomena, the gravity model being a good example. In some cases, the simplicity these models offer serves to produce insight, or at least fill textbooks, in the absence of geographic laws. In this chapter, I raise the issue of model simplicity, and discuss Einstein's question of how much simplicity is enough.

16.2 SIMPLICITY IN MODELS

The problem of model simplicity is that models have multiple components and are often complicated. Complex systems theory has shown that even simple models can produce complexity and chaos when given the right initial conditions or rules, though the theory gives little guidance on how to invert this sequence. Formally, models are sets of inputs, outputs, and algorithms that duplicate processes or forms. The more complicated the process or form, the less simple a model can be.

Models have been thought of as consisting of four components. These are (1) input, both of data and parameters, often forming initial conditions; (2) algorithms, usually formulas, heuristics, or programs that operate on the data, apply rules, enforce limits and conditions, etc.; (3) assumptions, representing constraints placed on the data and algorithms or simplifications of the conditions under which the algorithms operate; and (4) outputs, both of data (the results or forecasts) and of model performance such as goodness of fit. Considering modeling itself requires the addition of a fifth component, that of the modeler, and including their knowledge, specific purpose, level of use, sophistication, and ethics. By level of use, we distinguish between the user who seeks only results (but wants credible models and modelers), the user who themselves run the model, the user who seeks to change the model, and the user who seeks to design a model. We term the former two secondary users, since the model is taken as given. The latter two are primary users, and include modelers. Communication between primary and secondary users usually consists of model documentation and published or unpublished papers, manuals, or websites.

Spatial inputs and outputs are usually maps, and the algorithms in geography and geographic information science usually simulate a human or physical phenomenon, changes in the phenomenon, and changes in its cartographic representation. Modeling itself, however, is a process. A modeler builds data or samples that describe the inputs, selects or writes algorithms that simulate the form or process, applies the model, and interprets the output. Among those outputs are descriptions of model performance, forecasts of unknown outputs, and data sets that may be inputs to further models.

Others have sought answers to the simplicity/complexity paradox, although not always from the context of urban modeling. Smith (1997), for example, examined the treatment of simplicity versus complexity evident in a series of papers on population forecasting. Smith noted that "It is not necessary that simpler models be more accurate than complex models to be useful" (Smith, 1997, p. 559). Long (1995) defined three types of complexity (model specification, degree of disaggregation, and assumptions/alternatives), and Smith argued that these could be used to classify models by their complexity in each area. He concluded that "There is a substantial body of evidence, then, supporting the conclusion that more complex models generally do not lead to more accurate forecasts of total population than can be achieved with simpler models" (Smith, 1997, pp. 560–561). He pointed out other criteria than accuracy as being important, and noted that complex models can make the modeler appear sophisticated, "whereas the use of simple models and techniques may make him/her appear to be stupid, lazy, or poorly trained" (Smith, 1997, p. 563). Occam's razor seems to have many dimensions!

Why might it be necessary to seek the limits of simplicity? I illustrate why using a practical example of immense environmental significance in the United States, that of the Yucca Mountain high-level radioactive waste disposal site in Nevada. The problem requiring modeling is that of finding a permanent, safe, and environmentally suitable storage location for the thousands of tons of highly radioactive waste from decades of civilian and military work in nuclear energy. The specifications of this site are that it should remain "safe" for at least 10,000 years. Safety is defined by the level of radionuclides at stream gaging stations a few kilometers downstream from the location. Failure in this case could be at least nationally catastrophic, and safety depends on the waste remaining dry. The recommendation to bury the waste deep underground dates back to 1957. The selection of the Yucca Mountain site came after a very lengthy process of elimination, debate over safe storage and studies by the regulatory agencies and the National Academy of Science. Scientists at the Los Alamos National Laboratory in the Earth and Environmental Sciences division took on the task of using computer simulation to approach the problem of long-term stability at the site.

Simple risk assessment methodology uses probability distributions on the random variables and perhaps Monte Carlo methods to gain knowledge of the overall probability of failure. As the models become more complex, so also does the probability density function. In 1993, a study reporting the detection of radionuclides from Cold War era surface nuclear testing in the deep subsurface at Yucca Mountain introduced the immense complexity of modeling subsurface faults and so-called "fast hydrographic pathways" (Fabryka-Martin et al., 1993). With the earliest, more simple, models the travel times of radionuclides to the measurement point was about 350,000 years. Adding Dual Permeability, allowing rock fracture, generated a new and far more complex model that showed travel times to only tens to hundreds of years, thereby failing the acceptance criterion. Although more complex, these models are more credible because they can duplicate the data on the fast hydrographic pathways.

Nevertheless, highly complex models become cumbersome when dealing with the public decision-making process. Robinson (2003) notes that for a numerical model to be useful in decision making, it must be correct and appropriate to the questions beings asked from the point of view of the scientists and model developers, and understandable and believable from the perspective of decision makers. A report noted that " ... the linkage between many of the components and parameters is so complex that often the only way to judge the effect on calculated results of changing a parameter or a submodel is through computational sensitivity tests." (NWTRB, 2002). The latest model, called TSPA-LA, has 9 major subcomponent models with 9 interacting processes, and a total of 70 separate models, each with multiple input and complex interconnectivity. Some components are apparently similar (e.g., Climate inputs and assumptions), but interact in complex ways (in this case through unsaturated zone flow and saturated zonal flow and transport). Some submodels produce outputs that enter into macro-level models directly, while in other cases the same values are inputs for more models, such as those related to disruptive events.

In such a case, does any one person comprehend the entire model, even at a superficial level? How can it be proven that one or more variables actually contributes to the explanatory power of the model? More important, how can scientists establish sufficient credibility in such a complex model that sensible (or even any) political decisions are possible, especially in a highly charged decision-making environment? Los Alamos scientists have undertaken a process of model reduction as a consequence. Model reduction is the simplification of complex models through judicious reduction of the number of variables being simulated. Methods for model reduction do not yet exist, and deserve research attention. For example, when does a complex model pass the layman understandability threshold, and when it does, how much of the original model's behavior is still represented in the system? What variables can be made constant, and which eliminate or assumed away altogether? How much uncertainty is introduced into a model system by model reduction? In statistics, reducing the sample size usually increases the uncertainty. Is the same true at the model level? What are the particularly irreducible elements, perhaps the primitives, of models in which geographic space is an inherent component? Such questions are non-trivial, but beg attention if we are to work toward models that hold the credibility of their primary and secondary users.

16.3 HONESTY IN MODELS AND MODELING

Sensitivity testing is obligatory to establish credibility in complex models. Sensitivity tests are usually the last stage of model calibration. The modeler is also usually responsible for calibrating the model, by tinkering with control parameters and model behavior, and at least attempting to validate the model. While validation may actually be impossible in some cases (Oreskes et al., 1994), calibration is obviously among the most essential obligations of modeling. Model calibration is the process by which the controlling parameters of a model are adjusted to optimize the model's performance, that is, the degree to which the model's output resembles the reality that the model is designed to simulate. This involves necessarily quantitative assessments of the degree of fit between the modeled world and the real world, and a measurement of the model's resilience, that is, how sensitive it is to its input, outputs, algorithms, and calibration. That the obligation to make these assessments is that of the modeler is part of what the authors of the Banff Statement termed "Honesty in modeling" (Clarke et al., 2002, p. 296). Honest and transparency are coincident, as Myers and Kitsuse stated "The heart of the problem lies in the secrecy with which the forecaster enters assumptions and prepares his models" (Myers and Kitsuse, 2000).

Such an approach could be considered an essential component of modeling ethics generally, and even more so in Geography where the real world is our target. Uncalibrated models should have the status of untested hypotheses. Such is not often the case, however. Even reputable peer reviewed journals occasionally publish papers on models that are only conceptual, are untested, or tested only on synthetic or toy data sets. These papers are often characterized by statements such as "a visual comparison of the model results," or "the model behavior appears realistic in nature." Assumptions go unlisted, limitations are dismissed, data are used without an explanation of the lineage or dimensions, and fudge factors sometimes outnumber model variables and parameters. A common practice is to use values for parameters that are "common in the literature," or looked up from an obscure source. The Earth's coefficient of refraction for the atmosphere, for example, has long been assumed to be 0.13, a value derived from the history of surveying. This number is a gross average for a value known to vary from 0.0 to 1.0 and even beyond these limits under unusual circumstances, such as in a mirage.

While simplicity in model form is desirable, indeed mandatory, what about simplicity in the other four components of modeling? Simplicity is inherent in the model assumptions, as these are often designed to constrain and make abstract the model's algorithms and formulas. Simplicity is not desirable in inputs, other than in parameter specification, since the best data available to the scientist, in their judgment, should always be that used. Nevertheless, what data are used as inputs is often determined by the processing capabilities and the project's budget. Data output is often also complex, and requires interpretation, although our own work has shown that simple visualizations of model output can be highly effective, for example, as animations. Part of data output is performance data. Should there be simplicity in calibration and sensitivity testing?

Reviews of the literature in urban growth modeling (e.g., Wegener, 1994; Agarwal et al., 2000; EPA, 2000) show that precious few models are even calibrated at all, let alone validated or sensitivity tested. Most common is to present the results of a model and invite the viewer to note how "similar" it is to the real world. Such a lack of attention to calibration has not, apparently, prevented the model's widespread use. Doubts are often assumed away because the data limitations or tractability issues exceed these as modeling concerns.

If one uses the argument that we model not to predict the future, but to change it, then the above concerns are perhaps insignificant. The development of new models obviously requires some degree of trial and error in model design. Models are also testbeds, experimental environments, or laboratories, where conceptual or philosophical issues can be explored (e.g., McMillan, 1996). This is often the case with games, in which some models are intended for education or creative exploration. Such uses are similar to the link between cartography and visualization; or science and belief. These uses are not, strictly, of concern in science. Science-based models have to work in and of themselves. Methods for model calibration include bootstrapping and hind casting. If models are to be used for decisions, to determine human safety in design, or to validate costly programs at public expense, then clearly we need to hold models to a more rigorous, or at least more valid, standard. Such issues are of no small importance in geographical models, and in geocomputation at large. Even minor issues, such as random number generation (Van Niel and Leffan, 2003), can have profound consequences for modeling. Whether or not model algorithms and code are transparent is at the very core of science: without repeatability there can be no proof of spurious results, falsification, or progress toward theory.

16.4 A CASE STUDY

The simplicity question can be partially restated as: *What is the minimum scientifically acceptable level of urban model calibration*? Whether this question can be answered in general for all models is debatable. A complete ethics code for modeling is obviously too broad for a single chapter. The domain of concern will be urban models and issues of their calibration and sensitivity testing. This discussion examines the problem in the context of a set of three models as applied to Santa Barbara, California. The three models are the SLEUTH model (Clarke et al., 1997; Clarke and Gaydos, 1998), the What if? Planning model by Richard Klosterman (2001), and the SCOPE model (Onsted, 2002).

Using the case study approach, the three models are analyzed in terms of their components, their assumptions, and their calibration processes. As a means of further exploring the consequences of modeling variants, their forecasts are also compared and the spatial dissimilarities explored in the light of the data for Santa Barbara.

16.4.1 The SLEUTH Model

SLEUTH is a cellular automaton (CA) model that simulates land use change as a consequence of urban growth. The CA is characterized by working on a grid space of pixels, with a neighborhood of eight cells of two cell states (urban/nonurban), and five transition rules that act in sequential time steps. The states are acted upon by behavior rules, and these rules can self modify to adapt to a place and simulate change according to what have been historically the most important characteristics. More details about the SLEUTH model can be found in the Chapters 17 and 18 by Dietzel and by Goldstein.

SLEUTH requires five GIS-based inputs: urbanization, land use, transportation, areas excluded from urbanization, slopes, and hillshading for visualization. The input layers must have the same number of rows and columns, and be geo-referenced accurately. For statistical calibration of the model, urban extent must be available for at least four temporal snapshots, that for four dates, terminating with the most recently available map. Urbanization results from a "seed" urban file with the oldest urban year, and at least two road maps that interact with a slope layer to allow the generation of new nuclei for outward growth. Besides the topographic slope, a constraint map

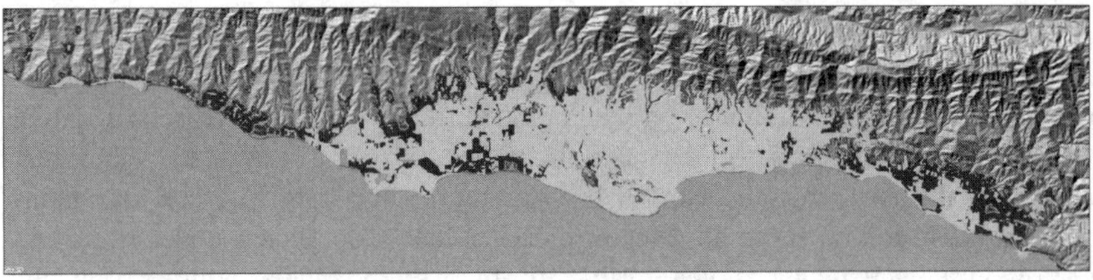

Figure 16.1 Forecast urban growth for Santa Barbara in 2030 using SLEUTH. Yellow was urban in 2000. Red is urban with 100% probability in 2030. Shades of green are increasing urban, starting at 50%.

represents water bodies, natural and agricultural reserves. After reading the input layers, initializing random numbers and controlling parameters, a predefined number of interactions take place that correspond to the passage of time. A model outer loop executes each growth history and retains statistical data, while an inner loop executes the growth rules for a single year.

The change rules that will determine the state of each individual cell are taken according to the neighborhood of each cell, those rules are: (1) Diffusion, (2) Breed, (3) Spread, (4) Slope Resistance, and (5) Road Gravity. A set of self-modification rules also control the parameters, allowing the model to modify its own behavior over time, controlled by constants and the measured rate of growth. This allows the model to parallel known historical conditions by calibration, and also to aid in understanding the importance and intensity of the different scores in a probabilistic way. The land use model embedded within SLEUTH is a second phase after the growth model has iterated for a single pass, and uses the amount of urban growth as its driver. This model, termed the deltatron (Candau and Clarke, 2000), is tightly coupled with the urbanization model. The dynamics of the land cover change are defined through a four-step process in which pixels are selected at random as candidate locations and changed by a series of locally determined rules controlled largely by proximity and the land cover class transition matrix. Forecasts can be for either urban growth alone (Figure 16.1), or for land cover forecasts and their associated uncertainty (Candau, 2000a).

SLEUTH is public domain C-language source code, available for download online. It is not suitable for systems other than UNIX or its variants. A sample data set (demo_city) is provided with a complete set of calibration results. A full set of web-based documentation is on line at http://www.ncgia.ucsb.edu/projects/gig.

16.4.2 SLEUTH Calibration

SLEUTH calibration is described in detail in Clarke et al. (1996) and Silva and Clarke (2002). This "brute force" process has been automated, so that the model code tries many of the combinations and permutations of the control parameters and performs multiple runs from the seed year to the present (last) data set, each time measuring the goodness of fit between the modeled and the real distributions. Results are sorted, and parameters of the highest scoring model runs are used to begin the next, more refined sequences of permutations over the parameter space. Initial exploration of the parameter space uses a condensed, resampled and smaller version of the data sets, and as the calibration closes in on the "best" run, the data are increased in spatial resolution. Between phases in the calibration, the user tries to extract the values that best match the five factors that control the behavior of the system. Coefficient combinations result in combinations of 13 metrics: each either the coefficient of determination of fit between actual and predicted values for the pattern (such as number of pixels, number of edges, number of clusters), for spatial metrics such as shape measures, or for specific targets, such as the correspondence of land use and closeness to the final pixel count (Clarke and Gaydos, 1998). The highest scoring numeric results from each factor that controls the behavior of the system from each phase of calibration feed the subsequent phase, with user-determined weights assigned to the different metrics. Calibration relies on maximizing spatial and other statistics between the model behavior and the known data at specific calibration data years. Monte Carlo simulation is used, and averages are computed across multiple runs to ensure robustness of the solutions.

Recent work on calibration has included testing of the process (Silva and Clarke, 2002), and testing of the temporal sensitivity of the data inputs (Candau, 2000b). Candau's work showed that short-term forecasts are most accurate when calibrated with short-term data. Silva and Clarke showed that the selection process for the "best" model results is itself an interpretation of the forces shaping urbanization in a region. Recent work on Atlanta (Yang and Lo, 2003) has shown that the calibration process steps that take advantage of assumptions of scale independence are invalid, meaning that ideally the calibration should take place exclusively at the full spatial resolution of the data. Discussion of these findings has sometimes taken place on the model's on-line discussion forum. In this book, Chapter 17 by Dietzel examines the impact of changing the geographic extent of the spatial units on the results of the SLEUTH calibration process. Similarly, Goldstein's Chapter 18 examines the substitution of a genetic algorithm for the brute force method that SLEUTH uses. Both are contributions very much in accordance with furthering honesty in modeling by critically examining how the model is applied, and by testing variations in model results that are the consequence of how the input data are aggregated and applied, rather than by parameter control.

16.4.3 The SCOPE Model

The South Coast Outlook and Participation Experience (SCOPE) model was originally based on the UGROW model (EPA, 2000, p. 128), adapted and rebuilt in PowerSim and eventually rebuilt again at UCSB by graduate student Jeffrey Onsted in the STELLA modeling language (HPS, 1995; Onsted, 2002). SCOPE is a systems dynamics model in the Forrester tradition (Forrester, 1969). This type of model posits simple numerical relations between variables that can be simulated with basic equations that link values. STELLA allows these empirical relations to be represented in a complex but logically consistent graphic system, that uses icons and a basic flow diagram to build the model (Figure 16.2). Variables can be stocks and flows, with

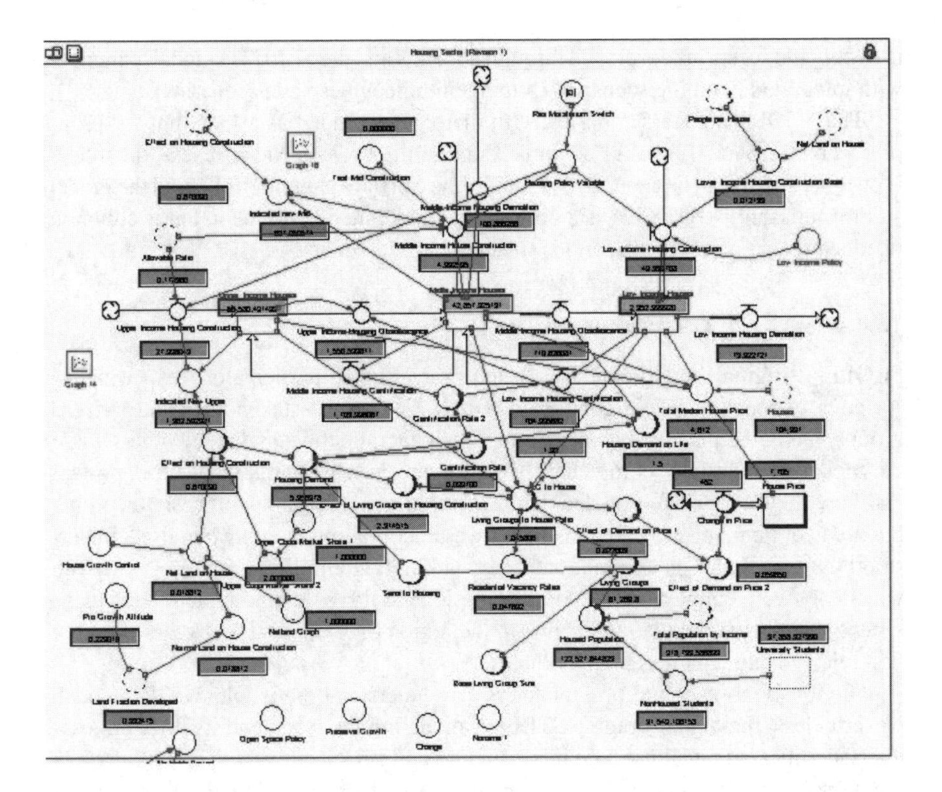

Figure 16.2 Sample STELLA code from the SCOPE model.

measurable interactions with finite units, or factors influencing rates and flows but not contributing to them directly.

The SCOPE model consists of hundreds of variable-to-variable links, expressed by equations. Each of these involves its own set of assumptions. These assumptions can be documented in the model code itself, and in SCOPE's case are also contained in the model's documentation. The complete set of individual assumptions is immense, and many assumptions are simply minor variants on relations and rates separated for one of the five economic sectors in the model (population, quality of life, housing, businesses and jobs, and land use). The housing sector is further divided into low, middle, and upper income, students and senior citizens. In essence, a separate model is maintained for each, with some assumptions enforced to keep aggregates the sum of their parts. The model, when complete, can be used at several levels. At the modeler level, actual parameters and links can be modified and the program used to generate statistical output in the form of charts and tables. At the user level, the model has a STELLA interface that allows users to turn on and off aggregate impacts, such as the application of particular policies, enforcement of rules and zoning, and sets of assumptions leading to the particular scenarios being developed by the model's funders. This is done using graphical icons representing switches. At a third and higher level, this interaction can take place across the world wide web (http://zenith.geog.ucsb.edu). A last level uses only the scenarios developed through both this and the SLEUTH model to interact with integrated planning scenarios (http://zenith.geog.ucsb.edu/scenarios).

The SCOPE model is proprietary in that it was funded as a local not-for-profit project by the Santa Barbara Economic Community Project. Nevertheless, the model is on-line, and so can be examined in totality, as can the assumptions behind the model and formulas used. STELLA is a proprietary modeling environment, but is common in educational environments, and is subsidized for students.

16.4.4 SCOPE Calibration

SCOPE's original and secondary design were public participatory experiments. A series of meetings were conducted with various local stakeholder and citizens groups, including planners, local activists, the general public, and many others while the model was partially complete to solicit input on what variables were important and how they were best modeled. This was a long and arduous process, involving hundreds of hours of work, but the result was encouraging in that when users understood the model and its assumptions, they had high credibility in its forecasts, even when they were counterintuitive. For example, SCOPE's demographic module consistently projected population decline in the region over the next few years, a fact at first rejected and later accepted by the users.

Obviously no user and few modelers can understand the whole SCOPE model. Nevertheless, the highly graphic STELLA modeling language and its different user interfaces meant that all users could zoom into and get specific about critical parts of the model from their perspective. Separate workshops were eventually held to work separately on the major sectors of the model. Final integration was largely a guided trial and error undertaking, with the constraint that as one component changed a single calibration metric was used, that of the Symmetrical Mean Absolute Percent Error (SMAPE) (Tayman and Swanson, 1996). Onsted (2002, p. 140–141) stated "We considered an output 'validated,' at least for now, if it had a SMAPE of less than 10%. This type of calibration has been called "statistical conclusion validity," a type of internal validity, described by Cook and Campbell as "inferences about whether it is reasonable to presume covariation given a specified level and the obtained variances" (Cook and Campbell, 1979, p. 41). While such an approach is similar to bootstrapping, in SCOPE's case rather extensive sets of historical data on a large number of the attributes being simulated were available and could be used for SMAPE calculations. Continued refinements to the model have since improved on the original reported calibration results. Onsted (2002) conducted a sensitivity test of sorts, providing a comprehensive bias and error table, and considering the impacts of the errors on forecasts.

16.4.5 What if?

The What if? modeling and planning support system was developed by Richard Klosterman at the University of Akron. He describes the system as "an interactive GIS-based planning support system that supports all aspects of the planning process: conducting a land suitability analysis, projecting future land use demands, and evaluating the likely impacts of alternative policy choices and assumptions" (Klosterman, 1997, p. 1). The suite of software that constitutes the model is built on top of ESRI's ArcGIS using MapObjects and Visual basic. The software is proprietary, and available from Klosterman's consultancy via their web site (http://www.what-if-pss.com).

For this experiment, a software license was obtained and the data for Santa Barbara, equivalent to the above project, compiled into the ArcView shape files necessary for What if? Data preparation in What if? consists of bringing into the software a large number of data files that contribute to the weighted selection by overlay of the vector polygons that meet various criteria. Initial preparation consists of selecting the layers, creating their union, and eliminating errors such as sliver polygons. This creates a set of Uniform Analysis Zones (UAZs) that are used as the granularity of analysis. Land suitability for urban use (high and low density residential, and commercial/industrial) are then computed using standard weighted overlay multicriterion decision-making methods. The land suitability method was Rating and Weighting (RAW), using current land use, distances to roads, streams, the coastline and commercial centers, topographic slope, the 100 year floodplain, and preserving agricultural land use. The user must select the weights, based on prior experience or intuition, and the result is a land suitability map (Figure 16.3). These maps find their way into an allocation routine that determines future demand for each use (based on exogenous estimates), and assigns development based on probabilities calculated from the weights. The result is a land suitability map, and maps showing development under different assumptions and scenarios about allocations. The software comes on a CD, which also contains data for a simulated data set (Edge City).

The approach was applied to Santa Barbara and used for the land suitability stage only. The software distinguishes between drivers (the growth projection and demand) and factors (the land suitability and assessment, and the land supply). A third stage, allocation, brings demand and supply together. Thus our application used only part of the software, which is a full planning support system. Results are shown in Figure 16.3. Considerable effort was placed into getting the same data as used in SCOPE and SLEUTH into the What if? simulation.

16.4.6 What if? Calibration

The What if? documentation contains no information at all about calibration. References are made to models that use the multicriterion weighted overlay method, such as CUF II (Landis and Zhang, 1998) and MEPLAN (Echenique et al., 1990). This tradition goes back to the work of McHarg and Lowry within planning, and is an established approach. The modeling, or rather forecasting, is an allocation of an exogenous demand using spatially determined input criteria such as buffers, thematic layers, exclusion areas, and other factors.

To be fair, the planning approach of What if? is based on assuming an invariant present and an almost absent past. This is because the model application here dealt with only the land suitability component of the multi-step What if? approach, so eliminating the past as a model input and dealing only with overlay of current data. In the application, past 20-year decadal population totals were used to make linear

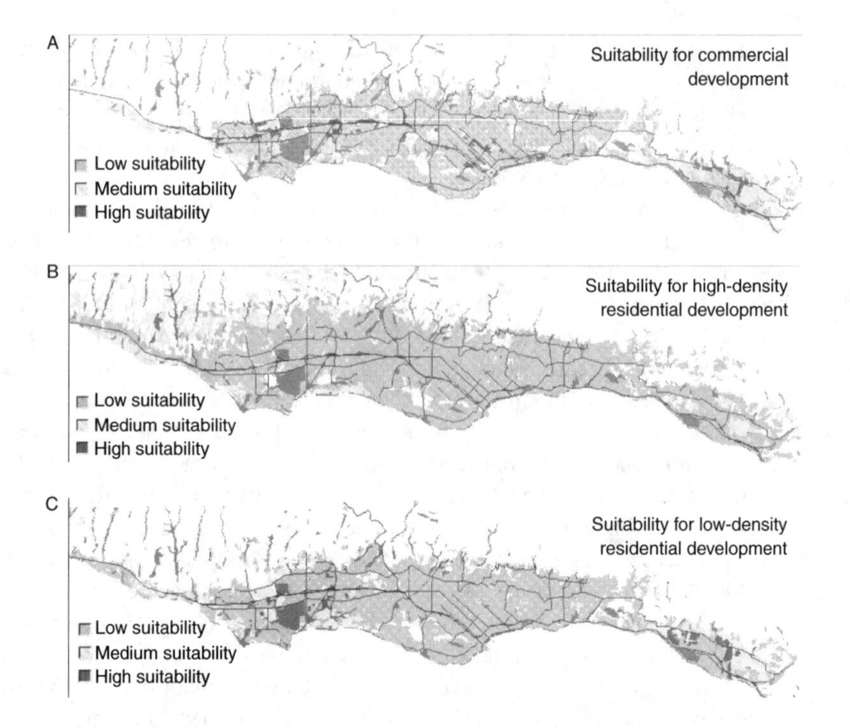

Figure 16.3 (see color insert following page 168) What if? land suitability assessments for Santa Barbara.

projections of future population as part of the housing demand estimation, but these were totals only and not spatially disaggregated. Thus the intelligence passes from the explicit assumptions of the model to the skills and experience of the modeler in compiling accurate input data, in selecting and computing weights, and in combining the requisite layers effectively. Forecasting routines are more deterministic, and even the number of forecast periods is prescribed. The full model bases future projections on past trends and allows the user to modify future conditions. Consequently, every time the model is applied to a new area, its application is completely customized, and while reproducible, is rarely sufficiently documented that the same results would be attained even from the same data. This is because the What if? approach is seen as a way of engaging participants in the scenario-based planning process, a good example of what has been called "urban modeling as story telling" (Guhathakurta, 2002). The model relies entirely on the credibility of the input data and the acceptability of the results. The actual code is compiled, and while the principles are demonstrated clearly in the documentation, the actual executable code is a black-box.

16.5 MODEL COMPARISONS

Each of the models applied to Santa Barbara produced a forecast for the year 2030. It should be pointed out that these models were chosen as being representative of three

Table 16.1 Comparison of Model Results for Additional Developed Area in Santa Barbara (2000–2030)

	What if?	SCOPE	SLEUTH
New urban area (ha)	14,006	4,856	2,047
Assumptions	All commercial and residential needs are filled	Base-line scenario assumed (all existing plans followed)	Base line as SCOPE. All pixels with chance of urbanization >50% assumed to become urban

classes of urban growth modeling strategies: cellular automata, systems dynamics, and rule-based allocation. Whether or not the results are also representative of these classes of models will be left for further work. There is considerable value in comparing models that use different strategies, particularly because agreement between models is often seen as an improvement in the reliability of the forecast, especially in weather and economic forecasting. Comparison between the models can only take the form of a matching of forecasts, since the actual future distribution and amount of housing in Santa Barbara in 2030 remains unknown. Thus we can "simply" compare the models to each other. One of the few values that is an estimate in common among the forecasts is the additional area of built up land. These directly comparable values are listed in Table 16.1.

It should be pointed out that even these forecasts have so many caveats as to be mostly incomparable. The What if? forecasts come from the "What if?" growth report, generated by the software during forecasting. The value is a sum for separate numbers for low-, high- and medium-density and multi-unit housing, plus commercial and industrial land. Of this area, far more is allocated to commercial/industrial than to residential by 8343.5 to 5662.8 ha. It was unclear how applying an exogenous demand estimate would change these values. Spatially, commercial growth was allocated near existing areas of that land use and particularly in Carpinteria (furthest east) and Goleta (west). High-density residential followed the same pattern, while low-density residential was more scattered.

The SCOPE forecasts were generated by flipping the "base-line" policy switch, thus in a black-box fashion incorporating all of the general plan assumptions that go along with this prebuilt scenario. These are listed as follows:

> The Baseline scenario assumes that the current urban limit line and other boundaries shall remain unchanged. Also, the South Coast will continue to build out as outlined in existing General Plans, Local Community Plans and per existing zoning. The market forces and demographics that drove change in the past will continue to influence the construction of new commercial and residential development. Infill will continue as it has along with redevelopment while large parcels will be more limited. There will be some extension of urbanized areas. The Gaviota coast, however (with the exception of Naples), shall be protected as well as the agricultural areas in the Carpinteria Valley.

As such, clearly a greater amount of local knowledge has been built into the forecast, even though we do not know explicitly what values and parameters have been set. Such information is, however, buried deep in the documentation.

For SLEUTH, the assumptions made are summarized as follows: "*This scenario uses the most basic exclusion layer, and uses an additional 'Urban Boundary' GIS layer to constrain growth. The Urban Boundary data was derived through a series of workshops held by the Economic Forecasting Project of Santa Barbara where local stakeholders defined a desired spatial limit to urban expansion. All growth in this scenario is constrained within the urban boundary*." This constraint is the obvious reason why this is the lowest of the three forecasts. Spatially, the growth is allocated as infill in Santa Barbara and Goleta, expansion in Gaviota and Carpinteria, and along a strip following U.S. highway 101 west from Goleta toward Gaviota. Even the enforcement of the urban boundary, however, seems unlikely to stop urban sprawl.

While the model forecasts are at least comparable, they incorporate not only a new set of assumptions about future (rather than assumed algorithm and data) behavior, but also different ways of dealing with uncertainty. Since What if? was not used beyond the land suitability phase, no probabilistic boundaries on the spatial or non-spatial forecasts were given. Nevertheless, the precision of the predictions (the report rounds data to the nearest 1/10th of a hectare) gives an impression of certainty. The predictions were given at decadal increments. SCOPE's forecasts are broad, rounded to the nearest 1000 acres given the error estimates listed in the SMAPE tables. SCOPE does, however, generate annual amounts for the forecast values, allowing the declining rate of addition of new urban land to be quite visible. SLEUTH's forecasts were not forthcoming. Data from the annual forecast images were brought into a GIS for tabulation. SLEUTH had by far the most detailed spatial granularity ($30 \times 30\ \text{m}^2$ pixels), but the Monte Carlo treatment of uncertainty added an averaging quality to the estimates. The new urban areas were ranked by uncertainty. 1606 ha of growth was deemed 95 to 100% certain, with the remainder distributed across the probabilities greater than 50%. Maps of percent likelihood of urbanization for Monte Carlo simulations were computed by the SLEUTH program in its predict mode, and were available for consultation.

So which model's results are the "simplest?" All require further analysis, interpretation, and require deeper understanding of both the input data and the model. From the public's point of view, far more variation in informed decision making has come from variants in graphical presentation of the results than in their statistical validity, level of calibration, or model robustness. It is not even evident that an overlay of the maps, or an averaging of the forecasts for specific targets is any more useful than the same taken independently. Only the passage of time, of course, can actually validate these forecasts. And in the meantime, especially with the SCOPE and SLEUTH models being used in devising a new community plan in the form of the Regional Impacts of Growth Study (SBRECP, 2003) of course the models themselves could — indeed are designed to — influence the result. Thus, Heisenberg's uncertainty principle places the model's use ever further from the scientific goals we set as ethical standards. Where we began with simple ethical directives with data and algorithms, as we move through the primary to the secondary model user, indeed on to the tertiary model "decision consumer," the science of modeling becomes more of an art, and one highly influenced by the methods used for division, simplification, and presentation of the modeling results. Perhaps foremost among urban planning uses in this respect is the use of scenarios as a modeling reduction tool. In each case, the model applications discussed here resorted to scenarios when communicating results. As a method in management science, scenarios indeed have a foundation in experience, yet this can hardly be claimed for Urban Planning and Geography (Xiang and Clarke, 2003).

16.6 DISCUSSION

This chapter posed the question of what is the minimum scientifically acceptable level of urban model calibration? Obviously the minimum (and perhaps the median) current level of calibration is zero, irrespective of validation. Using the truth in labeling approach that has permeated the spatial data and metadata standards arena, an obvious conclusion to answer the question would be that all models should attempt to assess their internal consistency using whatever methods or statistical means that are suitable. The number of digits used to present numerical results should be rounded to levels appropriate to the expected variation, and at least some measure or estimate of variance should be made. At the extreme minimum, the lower base would be that of repetition. The modeler is obliged to say "I ran the experiment twice, and got the same (or different) results each time." Without this, the models are not scientific. At least some kind of report on calibration and sensitivity testing should accompany every model application too, and certainly if the work is to be peer reviewed.

In the examples and discussion, a whole new suite of meanings of "honesty in modeling" have emerged. With the five components of modeling, the primary model user has ethical obligations to the secondary and tertiary users. Thus, data need to be the best available to the scientist, accurate, timely and of the correct precision and spatial resolution; algorithms need to be rigorous, relevant, not-oversimplified and subjected to peer review; assumptions need to be relevant, logical and documented; and output needs to be thorough, sufficient to detect errors and suitable for use, and to include performance measures and success criteria for the model. This discussion introduced a fifth component, the primary modelers themselves. The modeler has the ultimate obligation to ensure that models are used appropriately, not only in the scientific sense of being accurate and correct, but in the sense that their outputs, reductions and simplifications will influence critical decisions, many involving human safety, most involving efficiency or satisfaction with human life. For these results, the modeler is equally to blame as the politician who acts upon them.

In conclusion, simplicity is not always a desirada for models or modeling. Einstein's statement about how simple is profound. In this case, the challenge becomes one of reductionism. How simple can a complex model be made to become understandable? What is lost when such a model reduction is made? Most members of society would rather have an incomprehensible model that gives credible and useful results than an understandable model. Our own work has shown us that the general public seeks to place faith in modelers, and desires credible modelers as much as credible models. This, of course, places great responsibility on the modeler, indeed begs the more comprehensive set of ethical standards that today we instill in our graduate students only casually. We should not only think of ethics when legal problems arise (Wachs, 1982). Furthermore, there are a whole suite of "more simple" reductionist methods that can be applied with minimal effort. Among these are attention to calibration, accuracy, repeatability, documentation, sensitivity testing, model use in decision making, and information communication and dissemination. Modeling with spatial data is hardly in its infancy.

It has been pointed out that the views presented in this chapter assume that decisions regarding urban development are made scientifically, and that policy makers think according to the logic of scientific reasoning, and that this is a dubious assumption. There is little doubt that models, or at least modelers, have influenced urban and regional planning in the past. With increasing use of simulation in all aspects of human life (Casti, 1997), the flood of data now available for mapping of urban phenomena, and the ubiquitous availability of analysis tools such as GIS, it is likely that in the future we will see increasing faith placed by the general public in the models that become integrated into the planning process. At least once in the past, during the 1970s, urban modeling's progress was hampered, and even declared dead, because of a lack of attention to the seven sins (Lee, 1994). Like it or not, urban models have risen from the grave, and been given another chance to work, and work well. Simple models may be understandable to the world at large, but they may not do an adequate job of simulating the complexity of our world. At the same time, urban modelers should not just talk among themselves, regardless of how simple or complex, honest or dishonest their efforts. It is time that GeoComputational models moved out from behind the curtain, and confronted the complex and complicated scientific world of GeoDynamics, in the understanding of which they are a useful part.

ACKNOWLEDGMENTS

This work has received support from the NSF in the United States under the Urban Research Initiative (Project number 9817761). I am grateful to Dr. Bruce Robinson of Los Alamos National Laboratory for information and graphics concerning modeling at the Yucca Mountain Waste Repository. The SLEUTH model was developed with funding from the U.S. Geological Survey, and the SCOPE model with funds from the Irvine Foundation via the Santa Barbara Economic Community Project. The What if! model was developed by Richard Klosterman, and applied to Santa Barbara by UCSB Geography undergraduate student Samantha Ying and graduate student Martin Herold. Thanks to the anonymous reviewers of this chapter for some thought-provoking comments.

REFERENCES

Agarwal, C., Green, G.L., Grove, M., Evans, T., and Schweik, C. 2000, *A Review and Assessment of Land-Use Change Models: Dynamics of Space, Time, and Human Choice* (Bloomington: US Forest Service and the Center for the Study of Institutions, Population, and Environmental Change [CIPEC]).

Candau, J. 2000a. Visualizing modeled land cover change and related uncertainty. *First International Conference on Geographic Information Science. Proceedings,*

GIScience 2000 (Savannah: http://www.giscience.org/ GIScience2000/posters/ 235-Candau.pdf).

Candau, J. 2000b. Calibrating a cellular automaton model of urban growth in a timely manner. *Proceedings (CD-ROM) 4th International Conference on Integrating GIS and Environmental Modeling (GIS/EM4)* (Boulder: GIS/EM Secretariat).

Candau, J. and Clarke, K.C. 2000. Probabilistic land cover modeling using deltatrons. *Proceedings (CD-ROM), URISA 2000 Conference* (Orlando: Urban and Regional Information Systems Association).

Candau, J., Rasmussen, S., and Clarke, K.C. 2000. Structure and dynamics of a coupled cellular automaton for land use/land cover change. *Proceedings (CD-ROM) 4th International Conference on Integrating GIS and Environmental Modeling (GIS/EM4)* (Boulder: GIS/EM Secretariat).

Casti, J.L. 1997. *Would-be Worlds: How Simulation is Changing the Frontiers of Science* (New York: J. Wiley).

Clarke, K.C. and Gaydos, L.J. 1998. Loose-coupling a cellular automation model and GIS: long-term urban growth prediction for San Francisco and Washington/ Baltimore. *International Journal of Geographical Information Science*, **12**, 699–714.

Clarke, K.C., Hoppen, S., and Gaydos, L. 1996. Methods and techniques for rigorous calibration of a cellular automaton model of urban growth. *Proceedings (CD-ROM) 3rd International Conference/Workshop on Integrating Geographic Information Systems and Environmental Modeling* (Boulder: GIS/EM Secretariat).

Clarke K.C., Hoppen, S., and Gaydos, L. 1997. A self-modifying cellular automata model of historical urbanization in the San Francisco Bay area. *Environment and Planning B: Planning and Design*, **24**, 247–261.

Clarke, K.C., Parks, B.O., and Crane, M.P. (Eds.) 2002. *Geographic Information Systems and Environmental Modeling* (Upper Saddle River: Prentice Hall).

Cook, T.D. and Campbell, D.T. 1979. *Quasi-Experimentation* (Boston: Houghton Mifflin)

Echenique, M.H., Flowerdew, A.D., Hunt, J.D., Mayo, T.R., Skidmore, I.J., and Simmonds, D.C. 1990. The MEPLAN models of Bilbao, Leeds and Dortmund. *Transport Reviews*, **10**, 309–322.

Epa, U.S. 2000. *Projecting Land-Use Change: A Summary of Models for Assessing the Effects of Community Growth and Change on Land-Use Patterns* (Cincinnati: U.S. Environmental Protection Agency, Office of Research and Development).

Fabryka-Martin, J.T., Wightman, S.J., Murphy, W.J., Wickham, M.P., Caffee, M.W., Nimz, G.J., Southon, J.R., and Sharma, P. 1993. Distribution of chlorine-36 in the unsaturated zone at Yucca Mountain: an indicator of fast transport paths. *Proceedings, FOCUS '93: Site Characterization and Model Validation*, Las Vegas, Nevada, 26–29 September 1993.

Forrester, J.W. 1969. *Urban Dynamics* (Cambridge: The M.I.T. Press).

Guhathakurta, S. 2002. Urban modeling as storytelling: using simulation models as a narrative. *Environment and Planning B*, **29**, 895–911.

HPS. 1995. STELLA: High Performance Systems. URL: http://www.hps-inc.com/.

Klosterman, R.E. 2001. The What if? Collaborative planning support system. In R.K. Brail and R.E. Klosterman (Eds.), *Integrating Geographic Information Systems, Models, and Visualization Tools* (Redlands: ESRI and New brunswick: Center for Urban Policy Research).

Klosterman, R.E. 1997. The What if? Collaborative planning support system, *Proceedings, ESRI Annual Users Conference*. http://gis.esri.com/library/userconf/ proc97/abstracts/a305.htm

Landis, J.D. and Zhang, M. 1998. The second generation of the california urban futures model: Parts I, II, and III. *Environment and Planning B: Planning and Design*, **25**, 657–666, 795–824.

Lee, D.B. 1994. Retrospective on large-scale urban models. *Journal of the American Planning Association*, **60**, 35–40.

Long, J.F. 1995. Complexity, accuracy, and utility of official population projections. *Mathematical Population Studies*, **5**, 217–234.

McMillan, W. 1996. Fun and games: serious toys for city modelling in a GIS environment. In P. Longley and M. Batty (Eds.), *Spatial Analysis: Modelling in a GIS Environment* (London: Longman Geoinformation), pp. 153–165.

Myers, D. and Kitsuse, A. 2000. Constructing the future in planning: a survey of theories and tools. *Journal of Planning Education and Research*, **29**, 221–231.

NWTRB. 2002. NWTRB Letter Report of Congress and Secretary of Energy, January 24th, 2002. http://www.nwtrb.gov/reports/2002report.pdf

Onsted, J.A. 2002. *SCOPE: A Modification and Application of the Forrester Model to the South Coast of Santa Barbara County*. M.A. Thesis, Department of Geography, University of California, Santa Barbara. http:// zenith.geog.ucsb.edu/title.html

Oreskes, N.K., Shrader-Frechette, K., and Belitz, K., 1994. Verification, validation, and confirmation of numerical models in the earth sciences. *Science*, **263**, 641–646.

Robinson, B. 2003. Use of science-based predictive models to inform environmental management: Yucca Mountain Waste Repository. CARE Workshop, UCSB. Unpublished presentation.

SBRECP. 2003. *South Coast Regional Impacts of Growth Study* (Santa Barbara Region Economic Community Project: http://www.sbecp.org/documents.htm).

Smith, S.K. 1997. Further thoughts on simplicity and complexity in population projection models. *International Journal of Forecasting*, **13**, 557–565.

Silva, E.A. and Clarke, K.C. 2002. Calibration of the SLEUTH urban growth model for Lisbon and Porto, Portugal. *Computers, Environment and Urban Systems*, **26**, 525–552.

Tayman, J. and Swanson, D.A. 1996. On the utility of population forecasts. *Demography*, **33**, 523–528.

van Niel and Leffan, S. 2003. Gambling with randomness: the use of pseudo-random number generators in GIS. *International Journal of Geographical Information Science*, **17**, 49–68.

Wachs, M. 1982. Ethical dilemmas in forecasting for public policy. *Public Administration Review*, **42**, 562–567.

Wegener, M. 1994. Operational urban models: state of the art. *Journal of the American Planning Association*, **60**, 17–30.

Xiang, W.-N. and Clarke, K.C. 2003. The use of scenarios in land use planning. *Environment and Planning B*, **30**, 885–909.

Yang, X. and Lo, C.P. 2003. Modeling urban growth and landscape change in the Atlanta metropolitan area. *International Journal of Geographical Information Science*, **17**, 463–488.

CHAPTER 17

Spatio-Temporal Difference in Model Outputs and Parameter Space as Determined by Calibration Extent

Charles K. Dietzel

CONTENTS

17.1 INTRODUCTION

During recent years, models of land use change and urban growth have drawn considerable interest. Despite past failures in urban modeling (Lee, 1973, 1994), there has been a renaissance of spatial modeling in the last two decades due to increased computing power, increased availability of spatial data, and the need for innovative planning tools for decision support (Brail and Klosterman, 2001; Geertman and Stillwell, 2002). Spatial modeling has become an important tool for city planners, economists, ecologists and resource managers oriented towards sustainable development of regions, and studies have attempted inventories and comparisons of these models (Agarwal et al., 2000; EPA, 2000). These new models have shown potential

0-8493-2837-3/05/$0.00+$1.50

in representing and simulating the complexity of dynamic urban processes and can provide an additional level of knowledge and understanding of spatial and temporal change. Furthermore, the models have been used to anticipate and forecast future changes or trends of development, to describe and assess impacts of future development, and to explore the potential impacts of different policies (Pettit et al., 2002; Verburg et al., 2002). Because many of these models are being used to provide information from which policy and management decisions are made, it is important that modelers have a clear understanding of how the geographic extent at which they are calibrating and modeling influences the forecasts that their models produce. This is directly linked to a larger geographic issue in modeling. Can large-scale (geographic extent) models accurately forecast local growth compared with smaller-scale applications, or should state/nation/global modeling be done at a local level and then aggregated to create a more realistic view?

The concept of geographic extent and how changing it alters a model's parameter space, and subsequently model outputs and forecasts is not something that has been studied extensively. In this work, extent is defined as the geographic boundary of a system. Generally speaking, when the extent of a geographic system is changed, so should the statistics that describe the system, and the interactions that take place within that system. Only in the case of a repeating pattern with small-scale structure, such as a checkerboard or white noise, is this not true. In the case of urban models, the effects of geographic extent on model calibration and outputs may be overlooked or brushed aside due to two constraints that inhibit this type of modeling in general: (1) many times researchers struggle to get the necessary data to run the models, at any spatial extent; and (2) as the spatial extent of data gets larger, the computational time increases, sometimes in more of an exponential manner than a linear one. These two issues have prohibited urban modelers from addressing sufficiently the issue of geographic extent and how it relates to urban model output, but even more importantly, how calibration at different extents can impact model forecasting and final outputs.

With any model, there is an explicit need to calibrate and parameterize the model to fit the dataset. Recently, modelers have increased their focus on the calibration phase of modeling to gain a increased understanding of how models, in particular cellular automata, work (Abraham and Hunt, 2000; Li and Yeh, 2001a; Silva and Clarke, 2002; Wu, 2002; Straatman et al., 2004). The calibration phase of modeling is one, if not the most important, stage in modeling because it allows the fitting of model parameters to the input data, to be further used in forecasting. Failure to calibrate a model to the input data results in an application that is not robust or justifiable. While these efforts have focused on refining the calibration process and the definition of parameters for these models, none of them have focused on how calibration at different spatial resolutions changes the parameter set and the model outputs. For all models, the "best" parameter space is defined as the area or volume of parameters that is searched to find the parameter set. The parameter set is then the "best-fit" set of parameters that describe the behavior of the system within the framework of the model. The parameter space is defined as an area or volume depending on the number of individual parameters. If there are only two parameters, then the parameter space in an area; any more than two then it is a n-dimensional volume. A better understanding of how spatial resolution changes a model's parameter set, and hence its outputs, is an important area of research, especially when many of these models are used in the decision-making process.

Should modelers take into account that an urban area may be a transition zone between two metropolitan areas or is influenced by a larger region in the calibration process? And how does incorporating the influence of these areas change the parameter space, and hence the spatial output of the model? Inclusion of outside influential areas into local urban models is not a new idea (Haake, 1972), but the study of how their inclusion changes the parameter space of current models may be. Advances in computing, especially the advent of parallelization and the cost effective strategy of "clusters" have significantly deflated the GeoComputational cost of modeling larger spatial areas at fine spatial resolutions, so inclusion of possibly influential, but outside, areas is not as much of a taxing task as it once was. Capitalizing on these advances, this research focuses on the relationship between spatial extent and parameter space, and how calibration of an urban cellular automata model at varied spatial extent can allow for forecasts that are more typical of the local–regional interactions taking place.

The SLEUTH urban model is a cellular automaton model that has been widely applied (Esnard and Yang, 2002; Yang and Lo, 2003; also refer Chapter 16 by Clarke, and Chapter 18 by Goldstein) and has shown its robust capabilities for simulation and forecasting of landscape changes (Clarke et al., 1997). The model makes use of several different data layers for parameterization, for example, multi-temporal land use and urban extent data, transportation network routes and digital elevation model data. Application of the model necessitates a complex calibration process to train the model for spatial and temporal urban growth (Silva and Clarke, 2002). This chapter documents work done on the role that geographic extent plays in the calibration of urban models by working with SLEUTH. A large geographic area was calibrated and modeled at three different geographic extents: global (extent of the entire system), regional, and county. The derived parameters were then used to forecast urban growth from 2000 to 2040. The results from the model forecasts were then compared to determine the extent that calibration at different geographic extents had on model forecasts. This analysis was then used to examine some general considerations about the geographic extent over which urban models are calibrated and used, and the implications that this has for using these models to evaluate policies and scenarios.

17.2 CALIBRATING URBAN AUTOMATA

Modeling geographic systems using cellular automata (CA) models is a recent advance relative to the history of the geographic sciences (Silva and Clarke, 2002). Tobler (1979) was the first to describe these models in geography, briefly describing five land use models that were based on an array of regular sized cells, where the land use at location i, j was dependent on the land use at other locations. Applying this method of modeling to urban systems for planning applications was recognized early (Couclelis, 1985), and application of these models has proliferated in the last decade (Ward et al., 2000; Li and Yeh, 2001b; Almeida et al., 2003), including the development of SLEUTH. While the models themselves have proliferated, work on the calibration phase has lagged, and the need for more robust methods for calibrating and validating CA models has been noted (Torrens and O'Sullivan, 2001).

Model calibration has become an increasingly important consideration in the development phase of modeling (Batty and Xie, 1994; Landis and Zhang, 1998). Yet, the high flexibility in rule definition used in CA modeling, and application of these rules to manipulate cell states in a gridded world, makes parameter estimation a more difficult process (Wu, 2002). For CA models where the transition rules consist of equations for calculating future state variables, they generally consist of several linked equations for each land use, and these are complexly linked, so calibrating a model may require the fitting of tens, if not hundreds of parameters (Straatman et al., 2004). The general difficulty in finding the "golden set" of parameter values of CA is due to the complexity of urban development (Batty et al., 1999). Methods for calibration such as the use of off-the-shelf neural network packages have been suggested by some (Li and Yeh, 2001a), but some have argued that these sort of methods produce a "trained" model and not one that has intrinsic meaning in terms of known geographic principles (Straatman et al., 2004). Due to these difficulties and the parameters of CA models being dependent on the transition rules for the model, there has been little research on the parameter space or sets of urban cellular automata and how they are related to the geographic extent of calibration, although the work that has been done on calibration provides a starting point for looking at how the parameter space can be approached. This is in contrast to work on CA in computer science, where rules and parameters impacts on behavior have been studied exhaustively.

17.2.1 SLEUTH Calibration

Calibration of SLEUTH produces a set of five parameters (coefficients) which describe an individual growth characteristic and that when combined with other characteristics, can describe several different growth processes. For this model, the transition rules between time periods are uniform across space, and are applied in a nested set of loops. The outermost of the loops executes each growth period, while an inner loop executes growth rules for a single year. Transition rules and initial conditions of urban areas and land use at the start time are integral to the model because of how the calibration process adapts the model to the local environment. Clarke and Gaydos (1998) describe the initial condition set as the "seed" layer, from which growth and change occur one cell at a time, each cell acting independently of the others, until patterns emerge during growth and the "organism" learns more about its environment. The transition rules that are implemented involve taking a cell at random and investigating the spatial properties of that cell's neighborhood, and then urbanizing the cell, depending on probabilities influenced by other local characteristics (Clarke et al., 1997). Five coefficients (with values 0 to 100) control the behavior of the system, and are predetermined by the user at the onset of every model run (Clarke et al., 1997; Clarke and Gaydos, 1998; Candau, 2000). These parameters are:

1. *Diffusion* — Determines the overall dispersiveness nature of the outward distribution.
2. *Breed Coefficient* — The likelihood that a newly generated detached settlement will start on its own growth cycle.

3. *Spread Coefficient*—Controls how much contagion diffusion radiates from existing settlements.
4. Slope Resistance Factor — Influences the likelihood of development on steep slopes.
5. *Road Gravity Factor* — An attraction factor that draws new settlements towards and along roads.

These parameters drive the four transition rules which simulate spontaneous (of suitable slope and distance from existing centers), diffusive (new growth centers), organic (infill and edge growth), and road influenced (a function of road gravity and density) growth.

By running the model in calibration mode, a set of control parameters is refined in the sequential "brute-force" calibration phases: coarse, fine, and final calibrations (Silva and Clarke, 2002). In the coarse calibration, the input control data are resampled to one quarter of the original size (i.e., 100 m is resampled to 400 m), and then a Monte Carlo simulation of a broad range of parameters are tested for their fit in describing the input data. The results of the calibration run are then analyzed to narrow the range of tested parameters, based on metrics that describe spatial characteristics of the calibration runs against the input control data, specifically using the Lee–Sallee metric because of its "spatial matching" of the control data, although there has been some suggestion that other metrics can be used (Jantz et al., 2002). This metric is a shape index that measures the spatial fit between the model's growth and the known urban extent for the calibration control years. Upon narrowing the range of parameters based on the metrics, the original input data are resampled again, but to one half of the original size (i.e., 100 m is resampled to 200 m), and simulated over the narrowed range of parameters. Again, the results are analyzed, and the range of parameters narrowed. This final set of parameters is simulated with the full spatial resolution original data. The resultant parameters are then used to forecast urban growth. One of the drawbacks of the "brute-force" calibration is the massive amount of computational time required to calibrate the model. Clarke and Gaydos (1998) report using several hundred hours of CPU time to calibrate data for San Francisco, CA. The computational time required to calibrate the SLEUTH model has not decreased significantly with advances in computers, and has led to the search for other methods of model calibration, including the use of genetic algorithms, which have been tested (refer to Chapter 18 by Goldstein).

17.2.2 Study Area and Data

Using the San Joaquin Valley (CA) as a study area (Figure 17.1), input data for modeling urban growth using SLEUTH were compiled at 100 m spatial resolution (Table 17.1). Data sources for historical urban extent are listed in Table 17.1. Urban extent data for San Joaquin Valley for the years 1940, 1954, and 1962 were digitized from historical U.S. Geologic Survey 1:250,000 maps and based on air photo interpretation and supplemental ground survey information. Data from 1974 and later were captured directly from space-based remotely sensed imagery. The urban extent data for 1974 and 1996 were based on Landsat Multispectral

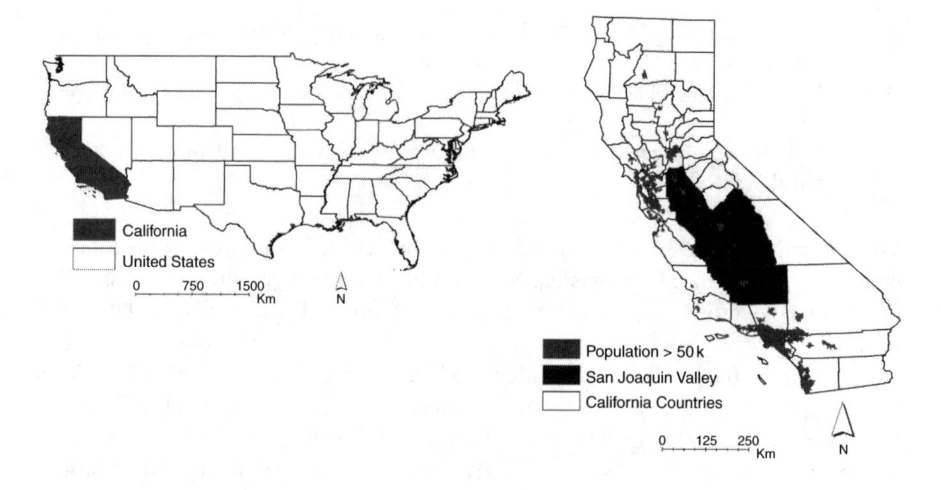

Figure 17.1 Location of California, population centers >50,000 people, and the location of the San Joaquin Valley.

Table 17.1 Sources, Description, and Resolution of Data Used in SLEUTH Modeling of the San Joaquin Valley (CA)

Data Layer	Source	Description	Spatial Resolution
Topography/Slope	USGS	30 m DEM	30 m
Land use	CA-DWR	Not used in this modeling	
Exclusion	CaSIL	Vector coverage of Federal and State owned land	N/A
Urban extent	USGS	Urban extent for 1940, 1954, 1962, 1974, 1996	100 m
	CA-FMMP	Vector coverage of developed land from 1984 to present in 2 year intervals	N/A
Transportation	CalTrans	Vector coverage of functionally classified roads from 1940 in 5 year increments	N/A

Scanner and Landsat Thematic Mapper mosaics compiled by the USGS Moffet Field, California office (http://ceres.ca.gov/calsip/cv/). Additional data for 1984, 1992, 1996, and 2000 were obtained from the California Farmland Mapping and Monitoring Program (CA-FMMP) that utilized aerial photography as a base mapping source (http://www.consrv.ca.gov/DLRP/fmmp/). The 1996 CA-FMMP data were merged with the USGS data to create a composite image of growth. Urban extent through time was treated as a cumulative phenomenon so that each time period built on the previous one, and urban extent was not allowed to disappear once it was established. Use of these data led to control years of 1940, 1954, 1962, 1974, 1984, 1992, 1994, 1996, and 2000. Pre-1974 data were all from cartographic sources, and the remainder were from satellite imagery and high definition aerial photos. All data processing was accomplished within a geographic information system (GIS) environment.

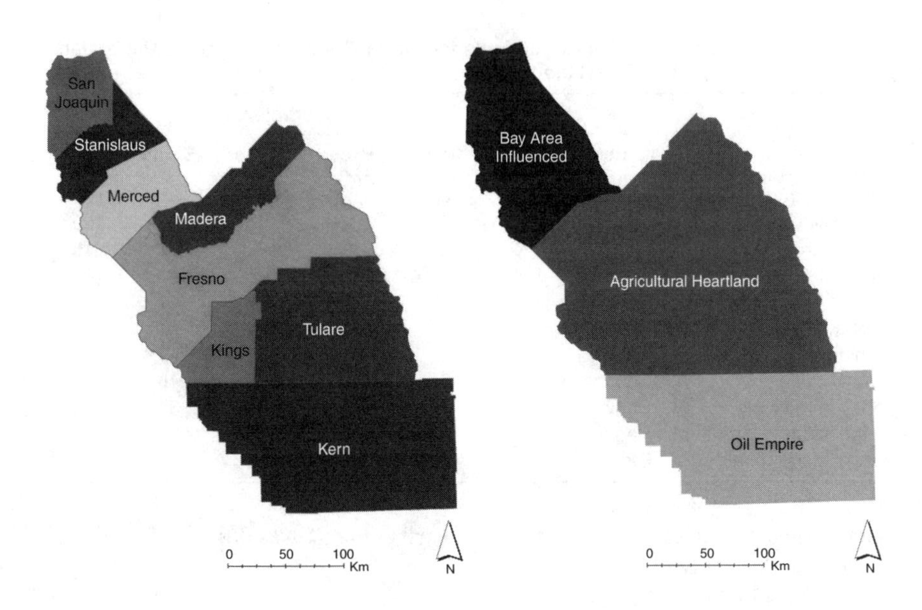

Figure 17.2 The eight independent counties in the San Joaquin Valley (left) and the three economic regions (right).

The San Joaquin Valley, while one large geographic and cultural area, is comprised of eight independent counties (Figure 17.2): San Joaquin, Stanislaus, Merced, Madera, Fresno, Kings, Tulare, and Kern. Additionally, these counties can be grouped into three distinct regions, based primarily on their economy (Figure 17.2). The Bay Area Region (San Joaquin, Stanislaus, Merced) is heavily influenced by the San Francisco Metropolitan Area economy and commuting patterns. Agriculture creates the economic base of Madera, Kern, Kings, and Tulare counties, uniting them as the Agricultural Heartland; and oil dominates Kern County which doubles as a county and region due to its unique natural resource and commuter patterns with the Los Angeles Metropolitan Area.

Input data for SLEUTH were calibrated for the San Joaquin Valley as a whole, each of the three regions, and each of the eight counties. The resulting parameter spaces were then used to forecast urban growth from 2000 to 2040, and using both tabular and graphical outputs, the results of the forecast were compared. Growth for San Joaquin, Madera, and Kern counties was then forecast using the global (San Joaquin Valley), regional, and county parameter sets to determine how that area grew specific to the others. These counties were chosen because they represent the typical historical growth trends in their respective economic region.

17.3 CALIBRATION AND FORECASTING RESULTS

The SLEUTH model was calibrated using the typical routine with self-modifying parameters values as described in Silva and Clarke (2002), and the data layers in Table 17.1. Calibration of the San Joaquin Valley and subsequent datasets resulted

Table 17.2 Resultant Growth Coefficients for the 11 Geographic Extents Calibrated using the SLEUTH Model

Area	Extent	Coefficients				
		Diffusion	Breed	Spread	Slope	Road
San Joaquin	County	2	2	54	1	3
Stanislaus	County	2	7	54	29	100
Merced	County	2	2	41	35	15
Madera	County	2	2	25	83	21
Fresno	County	2	5	58	41	52
Kings	County	2	2	45	1	2
Tulare	County	2	2	32	41	2
Kern	County/Region	2	2	58	46	31
Bay Influenced	Region	2	4	47	30	3
Agricultural Heartland	Region	2	2	45	36	41
San Joaquin Valley	Global	2	2	83	10	4

in a parameter set describing the growth of each area that was different for all areas included in this study (Table 17.2).

San Joaquin, Madera, and Kern counties were forecast using the global parameters along with their respective region and county parameters. Total urban area and new urban growth over time under each of the parameter sets were plotted for these three areas (Figure 17.3).

Total urban area from 2000 to 2040 in San Joaquin County was greater when the local county parameter set was used in forecasting compared with that of the Bay Influenced region and the global San Joaquin Valley (Figure 17.4). This is further supported by the total hectares of new urban growth for the county parameters set than the region and valley parameters (Figure 17.3), yet the difference between the three was slight over the forty year forecast. Total urban area is predicted to cover 125,000 ha under the county parameter set, 127,000 and 135,000 ha under the region and valley parameters.

Growth trends in Madera County under the three parameter set forecasts were opposite of San Joaquin County. Using the global San Joaquin Valley parameters to forecast future growth produced a county that grew three times faster than when the local county parameters were used, and twice as fast when the regional Agricultural Heartland parameters were used in forecasting (Figure 17.5). Total urban area under the county parameters was 36,000 ha, opposed to the 60,000 and 90,000 predicted using the region and valley parameters.

Kern County, a unique area that itself is a region, had growth curves that were lower than those of the global San Joaquin Valley (Figure 17.6). The number of hectares of urban growth is predicted to be 305,000 for the county and region parameters, and 395,000 under the valley parameters. This difference between the total urban area forecast by the different parameter sets was not as great as was found for Madera County, but larger than the minute differences found in San Joaquin County.

Using a GIS to overlay the model outputs, geo-algebra was done to determine the total area that was both unique and common between all parameter set forecasts for the year 2040 for each of the three areas (Figure 17.7). In concurrence with the total

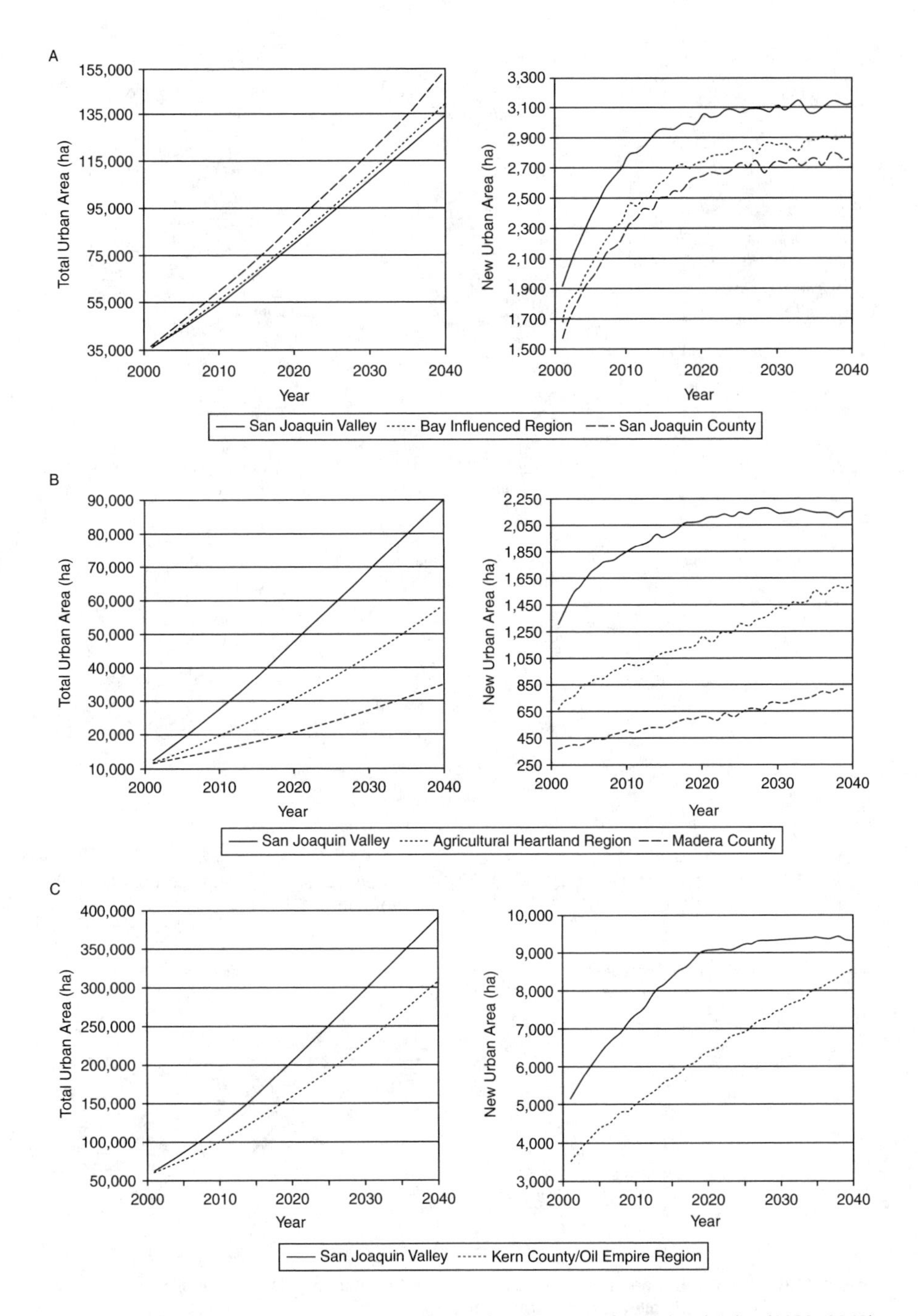

Figure 17.3 Total urban area and new urban area for each year of model projection (2000–2040) for San Joaquin (A), Madera (B), and Kern (C) counties. The model projections were made using the parameters derived from calibration of the entire San Joaquin Valley dataset, as well as those parameters derived from the calibration of the individual county, and the economic region that the county is part of.

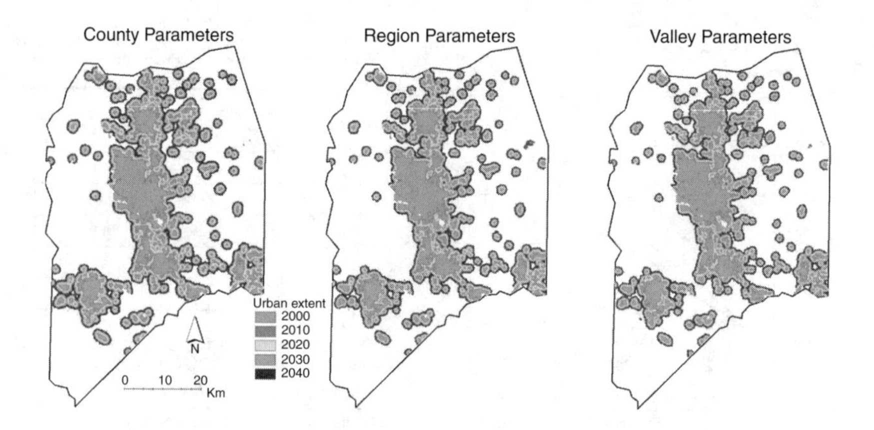

Figure 17.4 Predicted urban growth (2000–2040) for San Joaquin County using the county (left), region (middle), and valley (right) parameter sets.

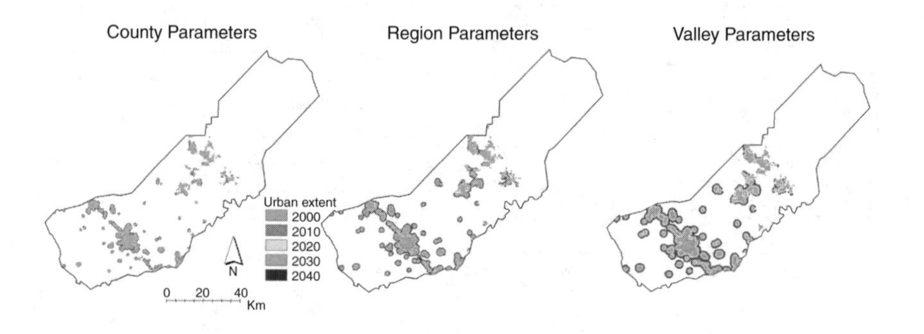

Figure 17.5 Predicted urban growth (2000–2040) for Madera County using the county (left), region (middle), and valley (right) parameter sets.

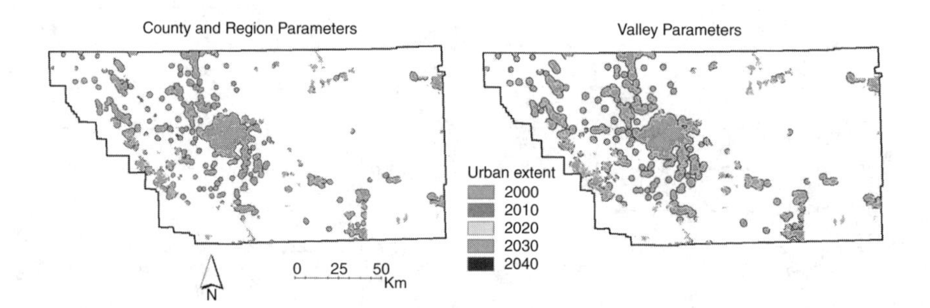

Figure 17.6 Predicted urban growth (2000–2040) for Kern County using the county and region (left), and valley (right) parameter sets.

urban area curves in Figure 17.3, the urban area common between growth forecasts using the county and regional parameters, compared with the valley parameters, were most similar in San Joaquin County. The total urban area common between the county and valley parameters was 89%, while there was 94% similarity between the

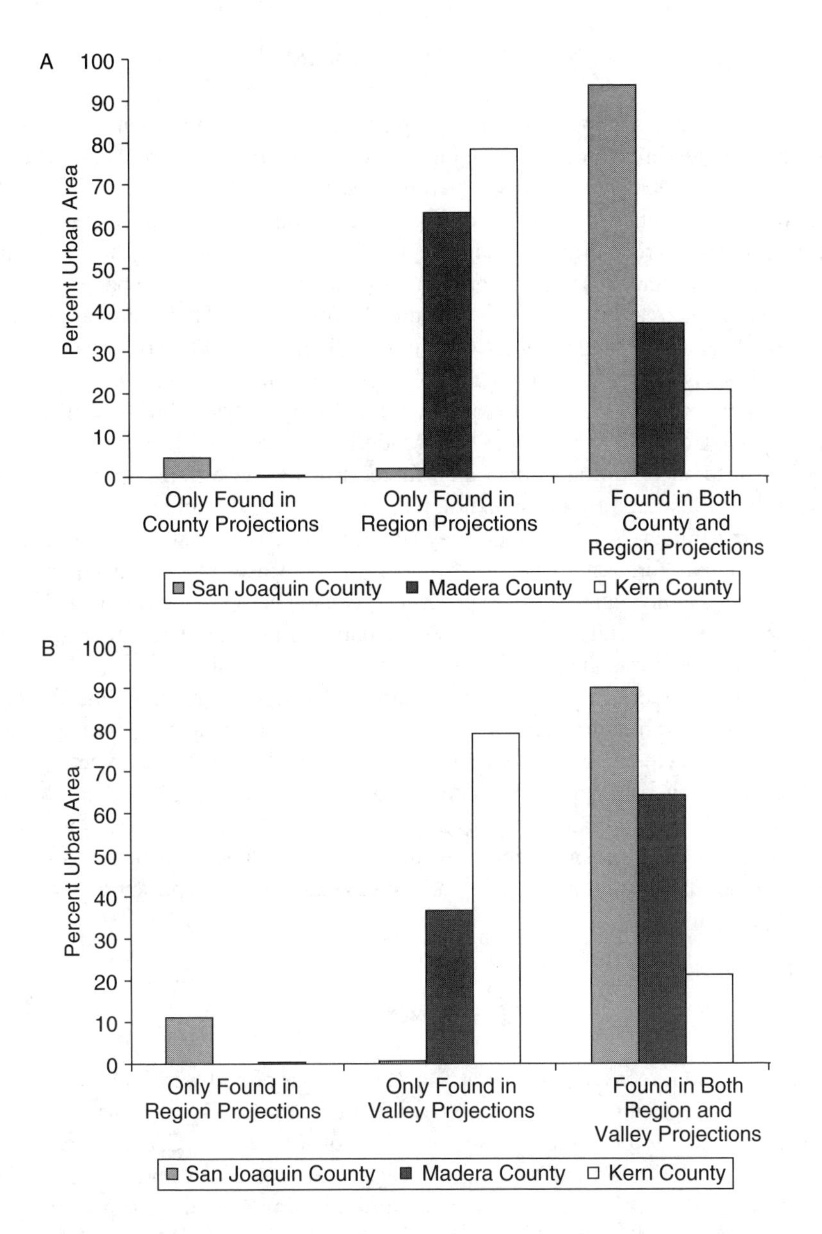

Figure 17.7 Spatial agreement between urban growth forecast (2000–2040) using the county and valley parameters (A) for San Joaquin, Madera, and Kern counties, and between forecast made for those same counties using the region and valley parameters (B). Urban area is measured in hectares.

region and valley parameter forecasts. The contrast between the three forecasts for Madera County was further demonstrated, and their was only a 36% spatial agreement between the county and valley forecasts, and 63% between the region and valley. Urban growth forecast for Kern County had a 79% spatial agreement between the county/region and valley forecasts.

17.4 DISCUSSION

Taking a hierarchical approach to modeling urban growth in three counties, each part of a different region in one large geographic area, resulted in a different parameter set for modeling at each level within the hierarchy. While the differences in the parameter sets were small between the diffusion and breed parameters, the differences in the spread, slope, and road gravity parameters were greatest, and probably had the largest impact on the differences in urban growth forecast under the different parameter sets. The total urban area forecast in San Joaquin County appeared to be similar under all three parameter sets, while Madera County growth differed by two or three times the amount forecast using the local county parameters. County and region parameters for Kern County produced urban forecasts that differed by <20%, which was more than San Joaquin County, but much less than Madera County.

The differences in total urban area forecast using the different parameter sets was further illustrated by the spatial agreement between the forecast of the county and region parameters against the valley (Figure 17.7). Spatial agreement is defined as an area that forecast to be urban under the forecasts of each parameter set. San Joaquin County had the highest percent agreement between the county and valley parameters (89), and the region and valley parameters (94). The large difference in growth under the county and valley parameters for Madera County showed in the low 36% agreement between the county and valley parameters, and the 63% between the region and valley. Kern County, the county that was itself a region, had a 79% agreement. Both Kern and Madera counties, were areas where growth under the valley parameter set produced more urban area in 2040, had a large portion of urban growth that only occurred with the valley parameters. None of the parameter sets used in forecasting growth produced a substantial portion of urban growth under the county and region parameters that was not captured by the valley parameters.

17.5 CONCLUSIONS

Calibration and forecasting of a hierarchical system has provided insight into whether large-scale (geographic extent) models can accurately forecast local growth compared with smaller-scale applications. Using the San Joaquin Valley (CA) as a study area for this investigation provided examples that demonstrated that regional and global calibration and forecasting of models can provide similar outputs (up to 94%) to a local scale application. But there are also cases, such as Madera County, where large-scale (San Joaquin Valley and Agricultural Heartland scale) modeling was grossly different from the local modeling effort. The question is then, how to distinguish when it is appropriate to and when not to use global and regional modeling to look at local spatial phenomena, because it is the local scale at which most policy and land use decisions are made? This question is most likely dependent on the model being used and its parameters, since they are what determine the model output. One noticeable feature that may play a role in most models is topography. In SLEUTH, this is addressed by the "slope resistance" parameter, but it is undoubtedly used differently in many other models. Focusing on Madera County, the local parameterization of the model showed a slope resistance of 83 as opposed to 36 and 10 by the region and valley parameters. These lower slope resistance parameters allowed growth to occur where it would not normally have under conditions characterized by local parameters. This is different from the parameters from San Joaquin and its respective region, where local slope resistance was 1, regional was 30, and the valley parameter was 10; not as strikingly different as in the case of Madera County. Taking this into account, a general rule for determining whether a large-scale application can be useful for looking at a local area might be that if the topography is uniform across the entire geographic area, then the model is more likely to accurately capture local growth patterns at a larger scale. Under this line of thought, it would be possible to model many of the states and geographic regions in the United States, and look at county/metropolitan growth patterns, as well as countries like the Netherlands. But this will most likely not be the case, especially when the modeling is being done at the national, continental, and global scale. For these size applications it will most likely be necessary to model at a smaller state or region level, and then aggregate the results, producing a composite output that is more reflective of local behavior and growth.

The issue of whether large-scale modeling efforts can accurately forecast local growth compared with smaller-scale applications is inevitably also tied to spatial resolution and how changes in spatial resolution change the parameter space and model forecasts. Although this issue has not been addressed directly in this paper, the link between spatial resolution and geographic extent is one that should not be ignored. Coarser spatial resolution modeling will undoubtedly dilute model outputs at the local level, but may capture regional growth better. The converse may also be true, but these are areas of research that need exploration.

Modelers should continue to work to find the optimal geographic extent to model local urban growth, while still allowing the inclusion of influences from surrounding urban areas. While the role of geographic extent issues within the calibration and forecasting routine might appear to be a too finely detailed area to warrant extensive research efforts, this chapter has demonstrated that changes in geographic extent can create different model outputs, and in some cases, gross differences between results using large- and small-scale calibration. Future efforts should continue to research this area using current working models, building on the knowledge gained to further improve them, or create new ones that are more robust in forecasting the future. Only once the models, their behavior, and their proper use are fully understood with honesty (Chapter 16 by Clarke), will they be able to be used in an understandable and believable manner for application to the planning and decision-making process.

ACKNOWLEDGMENTS

The author would like to thank Keith Clarke for comments on the original draft as well as the three anonymous reviewers, and the Public Policy Institute of California for their funding of a larger project, from which data for this work were used.

REFERENCES

Abraham, J. and Hunt J., 2000. Parameter estimation strategies for large-scale urban models. *Transportation Research Record*, **1722**, 9–16.

Agarwal, C., Green, G., Grove, M., Evans, T., and Schweik, C., 2000. *A Review and Assessment of Land-Use Change Models: Dynamics of Space, Time, and Human Choice* (Bloomington: U.S. Forest Service and the Center for the Study of Institutions, Population, and Environmental Change (CIPEC)).

Almeida, C., Batty, M., Monteiro, A., Camara, G., Soares-Filho, B., Cerqueira, G., and Pennachin, C., 2003. Stochastic cellular automata modeling of urban land use dynamics: empirical development and estimation. *Computers, Environment and Urban Systems*, **27**, 481–509.

Batty, M. and Xie, Y., 1994. Modeling inside GIS: part 2. Selecting and calibrating urban models using Arc-Info. *International Journal of Geographic Information Systems*, **8**, 470–541.

Batty, M., Xie, Y., and Sun, Z., 1999. Modeling urban dynamics through GIS-based cellular automata. *Computers, Environment and Urban Systems*, **23**, 205–233.

Brail, R. and Klosterman, R., 2001. *Planning Support Systems: Integrating Geographic Information Systems, Models, and Visualization Tools* (Redlands: ESRI Press).

CALIFORNIA FARMLAND MAPPING AND MONITORING PROGRAM, 2003. WWW document, http://ceres.ca.gov/calsip/cv/.

Candau, J., 2000. Calibrating a cellular automaton model of urban growth in a timely manner. Presented at the *4th International Conference on Integrating GIS and Environmental Modeling* (GIS/EM4). Banff, Alberta, Canada, September 2–8, 2000.

Clarke, K. and Gaydos, L., 1998. Loose-coupling a cellular automaton model and GIS: long-term urban growth prediction for San Francisco and Washington/Baltimore. *International Journal of Geographic Information Science*, **12**, 699–714.

Clarke, K., Hoppen, S., and Gaydos, L., 1997. A self-modifying cellular automata model of historical urbanization in the San Francisco Bay area. *Environment and Planning B: Planning and Design*, **24**, 247–261.

Clarke, K., Parks, B., and Crane, M., 2002. *Geographic Information Systems and Environmental Modeling* (Upper Saddle River: Prentice Hall).

Couclelis, H., 1985. Cellular worlds: a framework for modeling micro–macro dynamics. *International Journal of Urban and Regional Research*, **17**, 585–596.

Epa, U.S., 2000. *Projecting Land-Use Change: A Summary of Models for Assessing the Effects of Community Growth and Change on Land-Use Patterns* (Cincinnati: U.S. Environmental Protection Agency, Office of Research and Development).

Esnard, A. and Yang, Y., 2002. Descriptive and comparative studies of 1990 urban extent data for the New York Metropolitan Region. *URISA Journal*, **14**, 57–62.

Geertman, S. and Stillwell, J., 2002. *Planning Support Systems in Practice* (Heidelburg: Springer Verlag).

Haake, J., 1972. Do cities grow in isolation? Metropolitan expansion and urban corridors. *Journal of Geography*, **71**, 285–293.

Jantz, C., Sheeley, M., Goetz, S., and Smith, A., 2002. *Modeling Future Urban Growth in the Washington, D.C. Area. College Park, MD, U.S.A.* (College Park: University of Maryland Department of Geography, Mid-Atlantic Regional Earth Science Applications Center).

Klostermann, R., 1999. The What if? Collaborative planning support system. *Environment and Planning B: Planning and Design*, **26**, 393–408.

Landis, J. and Zhang, M., 1998. The second generation of the California urban futures model. Part 2: Specification and calibration results of the land use change submodel. *Environment and Planning B: Planning and Design*, **25**, 795–842.

Lee, D., 1973. Requiem for large-scale models. *Journal of the American Institute of Planners*, **39**, 163–178.

Lee, D., 1994. Retrospective on large-scale urban models. *Journal of the American Planning Association*, **60**, 35–40.

Li, X. and Yeh, A., 2001a. Calibration of cellular automata by using neural networks for the simulation of complex urban systems. *Environment and Planning A*, **33**, 1445–1462.

Li, X. and Yeh, A., 2001b. Zoning land for agricultural protection by the integration of remote sensing, GIS, and cellular automata. *Photogrammetric Engineering and Remote Sensing*, **67**, 471–477.

Pettit, C., Shyy, T., and Stimson, R., 2002. An on-line planning support system to evaluate urban and regional planning scenarios, in S. Geertman and J. Stillwell (Eds.), *Planning Support Systems in Practice* (Heidelberg: Springer-Verlag), pp. 331–347.

Silva, E. and Clarke, K., 2002. Calibration of the SLEUTH urban growth model for Lisbon and Porto, Portugal. *Computers, Environment and Urban Systems*, **26**, 525–552.

Straatman, B., White, R., and Engelen, G., 2004. Towards an automatic calibration procedure for constrained cellular automata. *Computers, Environment and Urban Systems*, **28**, 149–170.

Tobler, W., 1979. Cellular geography, in S. Gale (Ed.), *Philosophy in Geography* (Dordrecht: D. Reidel Publishing Group), pp. 379–386.

Torrens, P. and O'Sullivan, D., 2001. Cellular automata and urban simulation: where do we go from here? *Environment and Planning B*, **28**, 163–168.

U.S. GEOLOGICAL SURVEY, 2003. Preliminary assessment of urban growth in California's Central Valley. WWW document, http://ceres.ca.gov/calsip/cv/.

Verburg, P., Soepboer, W., Veldkamp, A., Limpiada, R., Espaldon, V., and Mastura, S., 2002. Modeling the spatial dynamics of regional land use: The CLUE-S model. *Environmental Management*, **30**, 391–405.

Ward, D., Murray, A., and Phinn, S., 2000. A stochastically constrained cellular model of urban growth. *Computers, Environment and Urban Systems*, **24**, 539–558.

Wegener, M., 1994. Operational urban models: state of the art. *Journal of the American Planning Association*, **60**, 17–30.

White, R. and Engelen, G., 1993. Cellular automata and fractal urban form — a cellular modeling approach to the evolution of urban land-use patterns. *Environment and Planning A*, **25**, 1175–1199.

Wu, F., 2002. Calibration of stochastic cellular automata: the application to rural–urban land conversions. *International Journal of Geographical Information Science*, **16**, 795–818.

Yang, X. and Lo, C., 2003. Modeling urban growth and landscape changes in the Atlanta metropolitan area. *International Journal of Geographical Information Science*, **17**, 463–448.

CHAPTER 18

Brains versus Brawn — Comparative Strategies for the Calibration of a Cellular Automata-Based Urban Growth Model

Noah C. Goldstein

CONTENTS

0-8493-2837-3/05/$0.00+$1.50

18.1 INTRODUCTION

18.1.1 Urban Modeling and SLEUTH

It has been recognized that humanity's spread of cities has and will have severe consequences to both human society and to the natural world. From the large shantytowns and slums of the third world, to the thousands killed by natural disasters, cities grow with dire consequences. It is no wonder that we appear to be moving into a period of many approaches to modeling the spatio-temporal phenomenon of urban growth. This has been aided by the increased availability of computers, both in their ubiquity and increased affordability.

One popular urban growth model is the SLEUTH, created by Dr. Keith Clarke at UCSB Geography (Clarke et al., 1996, 1997). SLEUTH is composed of a series of growth rules, which form modified Cellular Automata (CA). When running SLEUTH, the rules of the CAs (the growth rules) are calibrated to historical urban spatial data. SLEUTH can then be used to forecast urban extent under different scenarios. Due to its scale independence, transportability, and transparency, SLEUTH has become a popular tool in modeling the spread of urban extent over time, be it recreating the past or forecasting growth into the future (Yang and Lo, 2003). Yang and Lo also cite the ability of SLEUTH's growth rules to self-modify as it models a region into the future, deviating from monotonically increasing, line-fitting urban growth models. SLEUTH has been used as an alternative to the demographics-driven urban growth models, which are frequently custom built for a particular regions, or too cumbersome (and expensive) to understand or implement. While the lack of demographic and economic output is a drawback to SLEUTH, some efforts have been made to coupling SLEUTH to demographic models of city growth (UCIME, 2001). Lastly, SLEUTH can be used as a powerful planning tool by incorporating different human perceptions or planning options into the model in order to forecast different scenarios of the growing footprint of a city.

SLEUTH has been used to model a growing number of geographical regions. These cities include Porto Alegre City, Brazil, where the model was used to assess the potential demand of refuse disposal (Leao et al., 2001), and Chester County, Pennsylvania, where SLEUTH was loosely coupled to a groundwater runoff model (Arthur, 2001). Other urban regions include Porto and Lisbon, Portugal (Silva and Clarke, 2002), San Francisco (Bell et al., 1995; Clarke et al., 1997), the Mid-Atlantic United States (Bradley and Landy, 2000), the Washington-Baltimore metropolitan region (Clarke and Gaydos, 1998), and California's Central Valley (Dietzel, Chapter 17, this volume). Santa Barbara, California has been the focus of many SLEUTH modeling experiments. These include the loose coupling of SLEUTH to a systems model of population and economics (UCIME, 2001), the comparison of SLEUTH and spatio-temporal interpolation to recreate historical urban extent (Goldstein et al., 2004), and the comparison of SLEUTH to the What If? Urban Growth Model (Clarke, Chapter 16, this volume). Herold et al. (2003) used SLEUTH's back-casting ability to measure how Santa Barbara and the surrounding cities' urban form and shape have changed since 1929.

Like many geographical models accounting for real, data-rich regions, as opposed to simulated regions, SLEUTH is computationally expensive. For SLEUTH it is the calibration stage of the model, when the rules of CAs are parameterized with calibration coefficients, that is the most computationally expensive, taking close to 15 days of CPU time to completion (Yang and Lo, 2003). The long computation time in SLEUTH calibration has been cited as a drawback to using the model (Clarke and Gaydos, 1998). With the increased availability of free remotely sensed products and publicly available GIS data, the geographical extent, and along with it, the array sizes of SLEUTH modeling will increase. Currently Dr. Clarke and his colleagues recommend using what is called the "Brute Force" method of calibration, which entails a slow, yet methodical exploration of the possible values of the coefficients for the growth rules (Silva and Clarke, 2002; Gigalopolis, 2004). What is presented here is an alternative to Brute Force calibration, one that takes advantage of a Genetic Algorithm to be the "Brains" to Brute Force's "Brawn."

18.1.2 An Overview of Genetic Algorithms

Genetic Algorithms (GAs), developed by Holland (Holland, 1975) as a method of mimicking meiosis in cellular reproduction to solve complex problems, were later adapted as a mechanism of optimization (Goldberg, 1989). GAs have been used to calibrate urban growth models as exhibited by Wong et al. (2001), who revisited the primordial Lowry model in an attempt to choose the parameters of household and employment distributions for Hong Kong. A GA was employed as a search tool to find the optimal set of possibilities of land use planning for Provo, Utah (Balling et al., 1999). Colonna et al. (1998) used a GA as a method of generating new rules for a CA in a landuse modeling exercise of Rome, Italy. GA clearly offers a viable alternative in urban model calibration.

The use of GAs as a tool has expanded considerably since their inception. Their strength lies in their ability to "exploit and explore" the search space in a non-random manner, improving after each iteration. This is done by evaluating a suite of potential solutions, then allowing the fittest members of that population of solutions to recombine to form successive populations of solutions. What follows is a generic description of a GA and how it works. For this example, it is assumed that the problem to solve, be it an equation, a model, or a set of solutions, is clearly defined. In addition, it is assumed that the representation schema for that problem is well defined.

First, the Seed Population is initialized, and the initial group of "chromosomes" to be evaluated is chosen (Figure 18.1). A chromosome is a set of characters that encode solutions to the problem. They are the solution's "genotype." Initialization can be done by random, stratified random, or non-random methods. The Initial Population is then evaluated by the problem. The evaluation process tests every chromosome in the population and assigns each potential solution a metric of fit. If the criteria for stopping are met, then the GA ends, presenting the best solution. There are a number of different methods for determining the stopping criteria for a GA, including a set number of model runs, a shrinking variance in the solutions presented, or, ideally, obtaining the sole, optimal solution. However, if the stopping criteria are not met, the

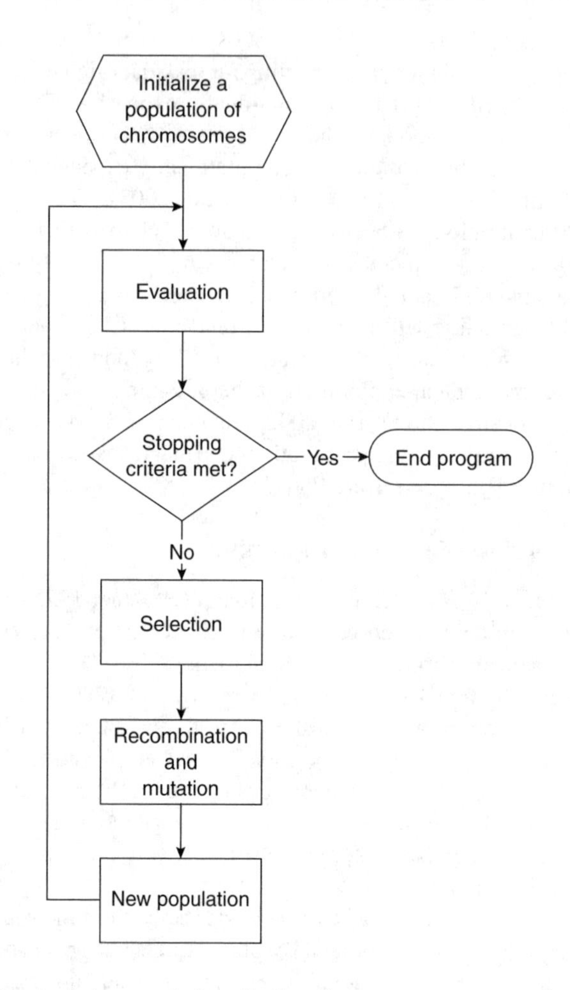

Figure 18.1 Flowchart of a generic Genetic Algorithm.

fittest members of the population, the chromosomes, are selected to "breed" to create the next generation of chromosomes. There are a number of strategies for selection, including elitism, which promotes the best performers, tournament selection, which allows pairs of chromosomes to compete based on their fitness; and roulette-wheel selection, which also judges chromosomes based on their fitness, allowing the better performers better odds of progressing. For review and analysis of selection criteria, see Goldberg and Kalyanmoy (1991).

The chromosomes chosen for selection are then recombined and mutated, in a process aping meiosis in cellular reproduction. Chromosomes undergo crossover and replacement, allowing bits of information, or "genes" from one chromosome to combined with another for a hybrid "offspring." An example of crossover will be demonstrated in Section 18.2.3.3. In mutation, individual genes are perturbed randomly to change their individual and, therefore, chromosomal characteristics. The resulting new population of chromosomes is then evaluated by the "problem" and the cycle begins again, until the stopping criteria are met.

18.2 THE SLEUTH URBAN GROWTH MODEL

18.2.1 General Description of SLEUTH

The urban growth model SLEUTH, uses a modified CA to model the spread of urbanization across a landscape (Clarke et al., 1996, 1997). Its name comes from the GIS data layers that are incorporated into the model; Slope, Landuse, Exclusion layer (where growth cannot occur, like the ocean), Urban, Transportation, and Hillshade. For complete documentation of SLEUTH, see the Gigalopolis website, where the source code (in the C programming language) is available for download (Gigalopolis, 2004).

SLEUTH is an urban growth model that employs CA to represent the change in the system. It deals exclusively with urban shape and form, as opposed to demographic or economic models of urban change. In modeling the change of urban form over time, the other factors that drive urban change are subsumed and therefore not explicitly necessary. SLEUTH results have been compared with the results systems models of economics and demographics and have been shown to yield comparable results of urban change over time (UCIME, 2001). The calibration stage of running SLEUTH is of import in the work presented here. During calibration, the four SLEUTH growth rules are "tuned" by adjusting or exploring the rule parameter values. The growth rules, described below, are separate CAs that operate on one urban landscape. The coefficients of the growth rules are Dispersion, Breed, Spread, Slope, and Road Gravity. The following is a description of the four growth rules which occur in the order presented, as summarized from the Gigalopolis website.

(1) Spontaneous Growth Rule
Spontaneous Growth determines if a random pixel in the urban lattice will be urbanized. The Dispersion and the Slope coefficients are used in this rule.

(2) New Spreading Centers Rule
The New Spreading Centers Rule determines if a newly urbanized pixel (from the Spontaneous Growth Rule) will be a new urban center and if so, will urbanize land. The Breed and Slope coefficients factor into this growth rule.

(3) Edge Growth Rule
The Edge Growth Rule adds new growth adjacent to existing urban pixels. The Spread and Slope coefficients determine the amount of Edge Growth.

(4) Road-Influenced Growth Rule
This final growth rule determines the extent the road (or transportation) network contributes to the urban growth of a city. New urbanized pixels "travel" on the road network and urbanize available pixels. This rule is determined by the Dispersion, Breed, Slope, and Road Gravity coefficients.

The five calibration coefficients are all integers and range from 1 to 100. By running SLEUTH in Calibration Mode, the different combinations of coefficients are used to model the historical urban growth of a geographic locale. Deviation from the city's real extent and shape (as indicated by the historical, or control, years) determines the best and the worst combination of the calibration coefficients. The goal of calibration is to determine which of the 10^{10} (or 100^5) possible combinations of coefficients gives a specific urban region the best fit. The term for all the possible coefficient combinations is called the *coefficient space*. It is unknown, but assumed that the coefficient space is a complex surface, due to the unique properties of individual urban extents, the spatial scale, and resolution chosen for the modeling, but most interestingly, because of the combinatory effects of the calibration coefficients with each other. Current research is underway at UCSB Geography to explore the coefficient space and to examine if some coefficients have systematic, as opposed to chaotic reactions to each other.

SLEUTH calibration makes available 12 metrics, each a different spatial metric of the calibration's fit relative to the control years. In addition a 13th, the product of the 12 metrics is provided. Most SLEUTH modelers use the product metric to determine the fitness of the coefficients (Clarke et al., 1996, 1997; Bradley and Landy, 2000; Arthur, 2001; Silva and Clarke, 2002; Yang and Lo, 2003). Candau (2002) has presented calibration results using only the Lee–Sallee index (Lee and Sallee, 1970). The Lee–Sallee index is a shape index measuring the union of the model estimation of growth, as compared with the known data years' urban growth.

In a typical SLEUTH run (see Figure 18.2), the geospatial data for a region is first organized in the required manner. The model requires that all data be in the GIF data format and that the domain of the data (as well as the pixel count) be identical for all data layers. The geospatial data can include many historical urban extents (four required), one to many historical transportation layers, one hillshade layer, one slope layer, and one excluded layer. Calibration, the focus of the work presented here, is described in more detail in the following sections. However, regardless of the method of calibration, Brute Force or Genetic Algorithm, the final stage is identical. In Parameter Self-Modification, the parameters are calibrated once again, allowing the model to change the value of the coefficient due to over- or under-estimates of simulating an S-curve, a common growth signature of many U.S. cities (Clarke et al., 1996). The S-curve pattern is that of slow growth for long periods, a short period of extremely fast growth, followed by a tapering of new urban growth. Following calibration, the modeler then uses SLEUTH to forecast the future urban growth of the region. Both statistical products are available, as are graphical outputs, which can then be returned to a GIS for further geographical analysis.

18.2.2 Brute Force Calibration

The recommended method of coefficient calibration is the Brute Force method, which steps through the coefficient space in large, and then increasingly smaller steps (Gigalopolis, 2004). The first step, Coarse Calibration, takes steps of 25 units through the entire coefficient space, for all coefficients. The second step, Fine Calibration, takes steps of 5 or 6 units through the coefficient space and the third, Final Calibration, takes steps of 1 (ideally) to 20 units through the coefficient space.

The following is a sample run-through of Brute Force calibration, and the decisions that go along with it. The Coarse Calibration takes steps of 25 units through the coefficient space, with a range of 1 to 100. Table 18.1 shows the top 15 performers,

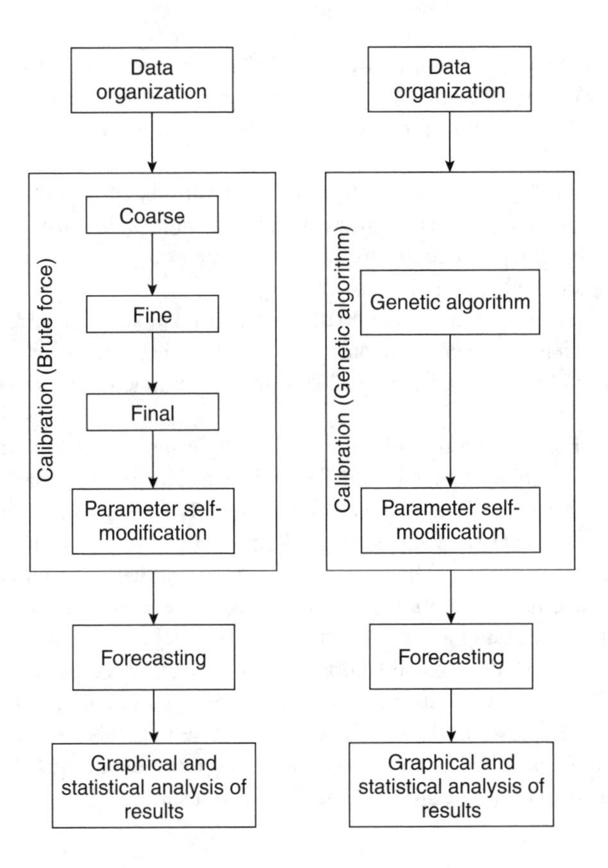

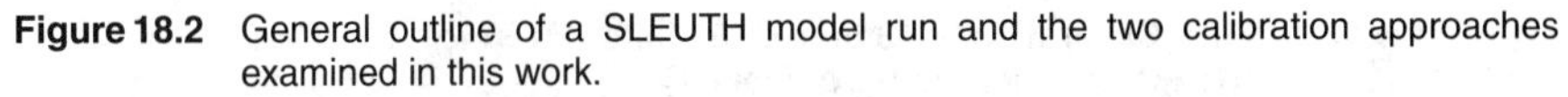

Figure 18.2 General outline of a SLEUTH model run and the two calibration approaches examined in this work.

Table 18.1 Coarse Calibration — Steps of 25 Units Over All the Coefficient Space [0 to 100]

15 Sample Calibration Sets of 3125 (5^5)		
{1–1–1–1–1}	{25–25–25–25–1}	{1–1–75–1–1}
{25–1–1–1–1}	{25–25–25–25–25}[a]	{25–25–75–25–25}
{51–1–1–1–1}	{25–25–25–25–50}	{50–50–75–50–50}
{75–1–1–1–1}	{25–25–25–25–75}[b]	{75–75–75–75–75}[a]
{100–1–1–1–1}	{25–25–25–25–100}	{100–100–75–100–100}

Note: The Sets contain the values of the calibration coefficients, Dispersion, Breed, Spread, Slope, and Road Gravity.

[a] Indicates the top three coefficient sets.

[b] Indicates the top coefficient sets.

according to a specific metric of fit. In this example, the set with the *b* is the best performer, with the sets with the *a* just behind it.

Next, the range is narrowed, to 20 units plus or minus each of the best Coarse Calibration, with a step of 5 units. This process, of Coarse to Fine resolution necessitates some "feel" for the direction and the performance of the calibration coefficients. The range could have been smaller or larger, as could be the step. Table 18.2 shows the results of the Fine Calibration, again showing the best and close performers.

Again, the modeler is forced with a decision. In this case, looking at the first coefficient, Dispersion, the best coefficient set had a value of 30, the next two, a value of 25. When the modeler decides to focus the search on the 30, instead of the 25, potentially better results could be missed.

Table 18.3 shows sample Final coefficients. In this example, 9375 (3125 × 3) different coefficient sets were evaluated. Recent SLEUTH calibration of the Atlanta region (Yang and Lo, 2003) used only 4268 different coefficients due to computational constraints.

The Brute Force method is efficient in searching through the coefficient space in a complete, regular, and reproducible manner. However, it has three major shortcomings. The first is that it is computationally expensive. It takes at least 9375 runs of SLEUTH (as suggested by the model's guidelines) in three calibration modes for each dataset. The second shortcoming is that due to non-linearity in the model for coefficient combinations, the Brute Force method may get trapped in a local maximum, missing a better coefficient set at a global maximum. As seen in the example above in the Coarse Calibration, one of the three best coefficient sets {75–75–75–75–75} is very different from the other two coefficient sets. This set, though potentially closer to the global optima, will be in essence discarded in favor of the more popular {25–25–25–25–25} sets. For more detail on the Brute Force calibration of SLEUTH, refer to the Gigalopolis web site.

Table 18.2 Fine Calibration — Steps of 5 Units "Around" Top Performers [Range of 20 for Each Coefficient]

15 Sample Calibration Sets of 3125 (5^5)		
{15–15–15–15–75}	{25–25–25–25–65}	{25–15–15–25–75}[a]
{20–15–15–15–75}	{25–25–25–25–70}[a]	{25–20–20–25–75}
{25–15–15–15–75}	{25–25–25–25–75}	{25–20–15–25–75}
{30–15–15–15–75}[b]	{25–25–25–25–80}	{25–15–20–25–75}
{35–15–15–15–75}	{25–25–25–25–85}	{25–15–25–25–75}

[a] Indicates the top three coefficient sets.

[b] Indicates the top coefficient sets.

Table 18.3 Final Calibration — Steps of 1 Unit "Around" Top Performers [Range of 5 for Each Coefficient]

15 Sample Calibration Sets of 3125 (5^5)		
{30–13–15–16–75}	{30–13–15–15–75}	{28–15–15–18–73}
{30–14–15–17–75}	{30–14–16–15–75}	{29–15–15–18–74}
{30–15–15–18–75}	{31–15–17–15–75}	{30–15–16–18–75}
{30–16–15–19–75}	{30–16–18–15–75}	{31–15–16–18–76}
{30–17–15–20–75}	{30–17–19–15–75}	{32–15–16–19–77}

18.2.3 The Calibration of SLEUTH Using a Genetic Algorithm

This work presents a new and novel method of calibrating SLEUTH to the historical urban extent data. This method takes advantage of a GAs ability to explore the entire coefficient space in an innovative and systematic way. What follows is an overview of the terminology used for the GA as applied to SLEUTH, and the SLEUTH GA methods.

A string of SLEUTH calibration coefficients is called a *chromosome*, also called a coefficient set as in the Brute Force description. Again, the calibration coefficients are always in the same order: Dispersion, Breed, Spread, Slope, and Road Gravity. Each individual calibration coefficient is referred to as a *gene*. A set of chromosomes, evaluated at one time is a *population*. The first population used at the beginning of the experiment is the *seed population*, described in Section 18.2.3.1. Each successive population is also referred to as a *generation*.

One common issue in constructing a GA is the issue of encoding the process of representing a gene's value in the parlance of the computer program. A common approach is to use binary encoding, whereas all genes' values are translated into their binary value. For example, the number 5 becomes the string 0101 or the number 38 becomes 100110. Since SLEUTH's calibration coefficients (or genes) are already numerical values between 1 and 100, SLEUTH-GA uses natural coding, whereas the numerical values are used as their "raw" value, or 5 is represented as 5 or 38 as 38.

18.2.3.1 Population Initialization

There are three strategies employed in determining the Seed Population. These are Stratified, Partial Random, and Random. For the initial population of 18 chromosomes, 6 were Stratified, 4 were Partial Random, and 8 were Random (Table 18.4). The Stratified chromosomes (1 to 6) were segments of the coefficient space, spread out over all possible solutions. These can be seen in Table 18.4. The Partial Random (7 to 10) chromosomes were alternating low or high values of genes interspersed with randomly chosen genes. The Random chromosomes (11 to 18) are indicated by an X in Table 18.4. Lastly, the Random chromosomes are composed of completely random genes, drawn from a random number generator.

18.2.3.2 Selection Strategies

Two Selection methods were employed in the selection of chromosomes from $\mathbf{P_n}$ to be bred into $\mathbf{P_{n+1}}$. The first was Elitism. Elitism promotes the best-performing chromosome to the next generation. While this has the benefit of ensuring the increase of fit of the populations over time, it can be responsible for the premature arrival of a (false) optimal solution through continual self-promotion (Liepins and Potter, 1991; Goldberg, 2002). In Tournament selection, pairs of chromosomes are randomly chosen and the best performer out of the pair is allowed to breed. This is done twice to obtain pairs of chromosomes for breeding. A sample selection is presented in Table 18.5.

The first pair of tournaments for the population in Table 18.5 randomly chose C5 versus C3, and C1 versus C6 to compete. The "winners" will be C3 and C6, according

Table 18.4 Schema of Seed Population for GA-SLEUTH

Chromosome number	Method of Choice	Dispersion	Breed	Spread	Slope	Road Gravity
1	Stratified	1	1	1	1	1
2	Stratified	25	25	25	25	25
3	Stratified	50	50	50	50	50
4	Stratified	75	75	75	75	75
5	Stratified	100	100	100	100	100
6	Stratified	10	90	10	90	10
7	Partial Random	10	X	10	X	10
8	Partial Random	X	10	X	10	X
9	Partial Random	90	X	90	X	90
10	Partial Random	X	90	X	90	X
11	Random	X	X	X	X	X
12	Random	X	X	X	X	X
13	Random	X	X	X	X	X
14	Random	X	X	X	X	X
15	Random	X	X	X	X	X
16	Random	X	X	X	X	X
17	Random	X	X	X	X	X
18	Random	X	X	X	X	X

Note: An X indicates a random number between 1 and 100.

Table 18.5 Sample Results from SLEUTH Calibration

ID	Coefficients	Metric of Fit
C 1	10–10–14–14–15	0·13
C 2	2–5–1–43–88	0·34
C 3	75–75–75–75–75	0·55
C 4	100–90–100–9–10	0·34
C 5	3–5–35–35–99	0·05
C 6	6–6–6–6–6	0·24

to their higher fit in each tournament, both of which will proceed to Crossover due to their higher metrics of fit. In some cases, the Tournament may tie (as in the case of C2 versus C4). In this case, the first-drawn chromosome will be selected for breeding.

18.2.3.3 Breeding Methods (Crossover)

After the chromosomes are selected, they are then used to determine the successive generation's population through meiosis-like processes called crossover. For the SLEUTH-GA, Uniform crossover was used to breed the winners of the Tournament Selection. In Uniform crossover, each gene in a chromosome has an equally good chance of being chosen to compose each "child." The advantage of this method is that the gene pool can be "shaken, not stirred," exploring the coefficient space in an extreme manner (Syswerda, 1989). A disadvantage of uniform crossover is that chromosomes can be thoroughly disrupted and can produce few beneficial combinations (Davis, 1991).

Take for example the breeding of C3 and C6, from Section 18.2.3.2.

C3: 75–75–75–75–75
C6: 6–6–6–6–6

After random drawings of either a 1 or a 0, the following "template" is created:

"Template of crossover": 1–0–0–0–1

This then results in the following prodigy:

Child 1: 75–6–6–6–75
Child 2: 6–75–75–75–6.

Crossover is an effective method of exchanging generic material between two chromosomes. In the GA used here, another form of crossover was used to explore the coefficient space. Self-crossover is the process of one chromosome "looping back" on itself (Pal et al., 1998). For example, C2, from Table 18.5, will be self-crossed on the second position, between the 5 and the 1:

Original C2: **2–5**–1–43–88
Self-crossed C2: 1–43–88–**2–5**

While self-crossover has few biological homologues, it provides an entirely different mechanism for the genetic material to be exploited, without being lost (as in crossover). Further research on the utility of self-crossover is needed to see if it actually contributes to the path of the optimal solution.

18.2.3.4 Mutation

The final stage of the GA is mutation. In the mutation routine, each gene has the potential to be changed by the addition or subtraction of a random number. The mutation rate used was 10%, though there were many instances of no mutation in a chromosome. After mutation, the entire new population is scanned and all identical chromosomes are subjected to mutation a second time. Since there is a maximum and minimum value for a gene, values >100 and <1 were "looped" back to a valid value. This was by either subtracting 100 (for values >100) or taking the absolute value of the number (for values <0).

The GA used for SLEUTH calibration was comprised of 18 chromosomes changing over 200 generations. The new population, $\mathbf{P}_{n+1}$, can be traced back to $\mathbf{P}_n$ as shown in Table 18.6.

18.2.4 The Sioux Falls Study Area

The Sioux Falls, South Dakota urban region was used to compare the calibration techniques. Sioux Falls is located in the southeast corner of the Mid-Western State (Figure 18.3). Westerners first settled in Sioux Falls in the late 1850s as a part of the great American westward expansion, bringing the railroads and industry with them and displacing the native American inhabitants. The Population of Sioux Falls has grown considerably from 10,300 people in 1900 to over 124,000 people in 2000,

Table 18.6 Summary Table of Breeding of P_n to P_{n+1}

Chromosome number	Method of Selection	Crossover	Mutation
1	Elitism (best)	None	No
2	Elitism (best)	None	Yes
3	Elitism (best)	Self	No
4	Elitism (second best)	Self	No
5	Tournament	Yes	Yes
6	Tournament	Yes	Yes
7	Tournament	Yes	Yes
.	Tournament	Yes	Yes
.	Tournament	Yes	Yes
.	Tournament	Yes	Yes
18	Tournament	Yes	Yes

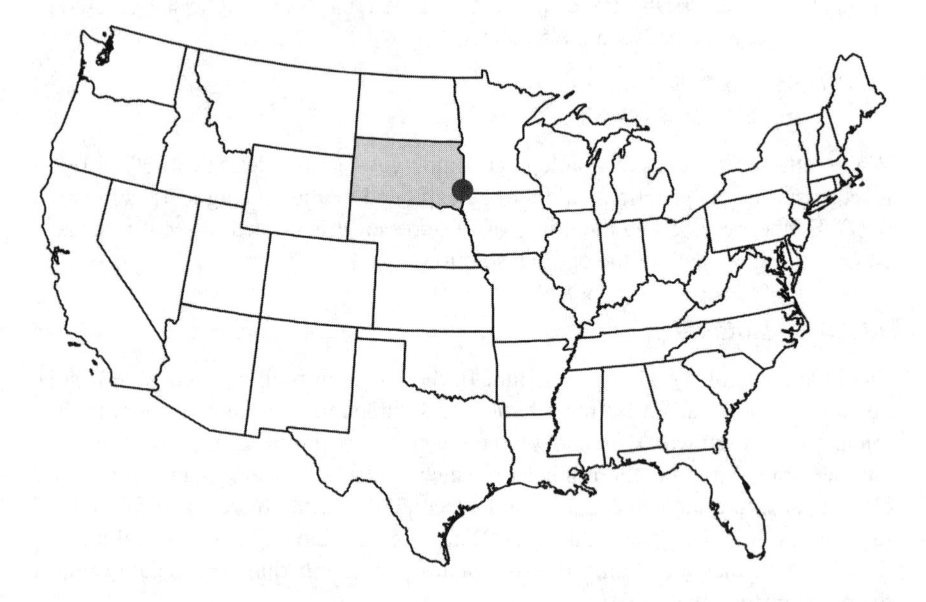

Figure 18.3 Location of Sioux Falls, South Dakota. The circle inside the gray state indicates the Sioux Falls region.

according to the U.S. Census Bureau. The Metropolitan region of Sioux Falls lies mostly within Minnehaha County, yet has expanded into Lincoln County since the 1940s. Since 1979, Sioux Falls has implemented a growth plan called "Sioux Falls 2015" that has been attempting to focus and control growth in an intelligent manner (www.siouxfalls.org/planning). Their efforts have been focusing on infill and development in designated areas, trying to avoid the infamous "urban sprawl" of many American and increasingly European cities.

The terrain of the Sioux Falls region is mostly flat with some moderately sloped areas and some extreme slopes around the waterfalls. Figure 18.4 shows the historical growth of urban extent of Sioux Falls. Over time, the urban extent had expanded into

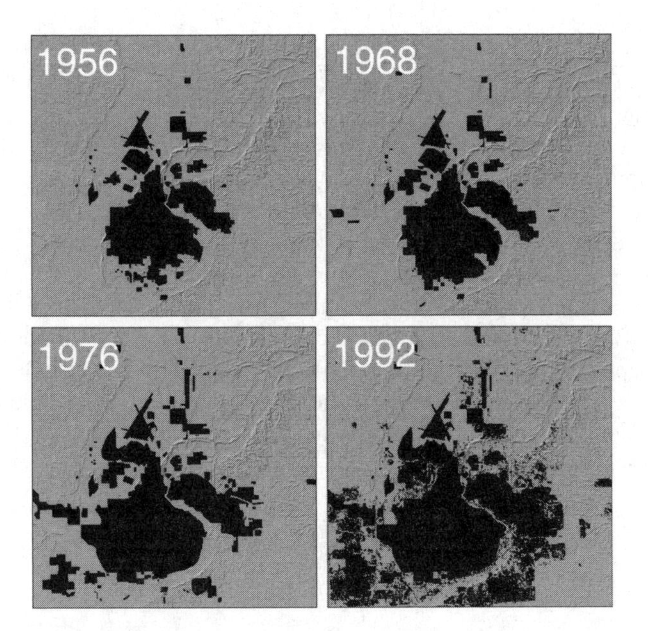

Figure 18.4 Sioux Falls historical urban layers. The width of the study area is 14·58 km along the East–West axis and 13·95 km along the North–South axis. The black regions indicate observed urban extent. The urban extent is overlaid on the hillshade image in grayscale.

the flat regions in the region, avoiding the low, yet steeper sloped to the North. Sioux Falls lies at the intersection of Interstates 90 and 29, which are major transportation routes in the Mid-west, connecting Sioux Falls to other major U.S. cities.

18.2.4.1 Sioux Falls Data

The geospatial data of Sioux Falls were created from USGS GIS polygon coverages of urban extent, transportation networks, parks, and protected lands. All data was rasterized at 30 m spatial resolution, and clipped to a grid of 486 by 465 pixels, or 14·58 km East–West by 13·95 km North–South, comprising 20,339·1 ha.

SLEUTH uses binary urban/not urban data layers to represent the domain of anthropogenic influence on land. For the Sioux Falls urban layers from 1956, 1968, and 1976, urban extent was determined as areas which were indicated in USGS coverage as "urban/built-up lands" or "urban other/open spaces, parks, etc. …." These coverages were created from black and white aerial photography. The spatial resolution of the photography was approximately 91 cm (3 ft).

The 1992 urban extent was obtained from the National Land Cover Database, a product of the 30 m Landsat Thematic Mapper and USGS (www.usgs.gov). For the Sioux Falls region, the polygons that had a value corresponding to "Developed" were converted to urban. This included "Low Intensity Residential," "High Intensity Residential" and "Commercial/Industrial/Transportation" land use classes. Note in Figure 18.4 how the remotely sensed data products led to a very "speckled" texture of the new urban growth.

The historical Transportation Layers were created from the USGS 1:100,000 scale Digital Line Graphs (DLGs). The roads and railroads were classified by use type (major, minor roads, or railroad) and rasterized to a 30 m spatial resolution. Only major roads and railroads were used for this study. More information on DLGs can be found at http://edcwww.cr.usgs.gov/glis/hyper/guide/usgs_dlg. The dataset contains transportation layers for the years 1956, 1968, 1976, and 1984.

The Exclusion Layer is composed of areas in which a city cannot grow. These include parks, natural reserve areas, and bodies of water. The Sioux Falls excluded layer was made from a parks coverage, acquired from the City of Sioux Falls Planning and Engineering Department in Arc/Info coverage format. This was combined with the hydrography coverages taken from the USGS 1:100,000 scale DLGs.

The Slope Layer of Sioux Falls was created from the USGS digital elevation model of South Dakota, compiled at a 30 m spatial resolution. The Arc/Info SLOPE routine was used to calculate the city's slope, in units of percent rise.

18.3 METHOD OF COMPARISON

To compare the different calibration methodologies, a single metric of fit was calculated for each run of calibration coefficient combinations. For this calibration exercise, the metric of fit was comprised of the product of three spatial metrics; number of urban pixels (Pop), number of urban clusters (Clusters), and the Lee–Sallee (LeeSallee) index. SLEUTH provides a diverse number of spatial metrics (13 different metrics) to be used in calibration. However, there has been no consensus on which of these metrics (as well as a combined metric) is the most appropriate to be used in the calibration of SLEUTH (Dietzel, Chapter 17, this volume). The three chosen represent the most challenging components of an urban model (numbers of urban clusters, size of the urban extent, and the spatial overlap of the model versus reality) as well as being simple enough to explain and observe.

The Brute Force and GA methods of SLEUTH calibration were performed on the Sioux Falls dataset in the following ways. The Brute Force method was run through once, at coarse, fine, and final calibration resolutions (resulting in 9375 calibration runs). The GA was run for 200 generations, for 18 chromosomes in each population (resulting in 3600 calibration runs). Both the GA and the Brute Force methods were run for only one Monte Carlo iteration, as opposed to 10–100 Monte Carlo iterations, as suggested by the current Gigalopolis protocol. This was done because of the author's suggestion that in using multiple Monte Carlo runs for calibration, SLEUTH only uses the mean response of the calibration coefficients to the model, and does not account for the variance or the stochasticity in the model for true urban growth behavior. The exploration of the use of multiple Monte Carlo runs will be presented in upcoming research.

The GA was implemented on a cluster of 12 SGI O2s, running Irix release 6.5, running R10000 MIPS processor at 195 MHz with 128 MB of RAM. The Brute Force calibration was carried out on a PC running Redhat Linux release 8.2 on a 700 MHz Pentium III processor with 384 MB RAM.

To compare the different calibration methods, Sioux Falls was calibrated by each method, Brute Force once, the GA method ten times. Calibration included parameter self-modification, for the preparation for forecasting. For parameter self-modification, all coefficient sets were calibrated for 10 Monte Carlo iterations. Following calibration, the Brute Force coefficient set and the top two GA coefficient sets were used to forecast Sioux Falls urban extent to the year 2015. Analysis and comparison of the two calibration techniques include the calibration metric of fit, the number of iterations needed for calibration, and the overall forecasting behavior.

18.4 RESULTS

18.4.1 Results of the Brute Force Calibration

The Coarse Calibration was initialized to explore the entire range of the coefficient space, at steps of 25 units (Table 18.7A). This resulted in 3125 combinations and took close to 8 h of CPU time to evaluate. The results of the Coarse Calibration show that high metric of fit values are associated with low values of the Dispersion and Breed coefficients, and that the Slope coefficient seems to vary greatly (Table 18.7B). In general, the top two performers had a slightly higher metric of fit than the following performers.

The results of the Coarse Calibration show that the range and step of the Dispersion and Breed coefficients should be shrunk, in favor of exploring low coefficient values

Table 18.7 The Parameters of the *Coarse* Calibration as well as the Elapsed Time to Completion and the Ten Best Coefficient Sets from Coarse Calibration and their Related Statistics

A Parameters of Coarse Calibration

	Dispersion	Breed	Spread	Slope	Road Gravity
Start	1	1	1	1	1
Step	25	25	25	25	25
Stop	100	100	100	100	100
Possible units	5	5	5	5	5
Possible combinations			3125		
Elapsed time			7:46:59		

B Coefficient Sets from Coarse Calibration

Rank	Pop	Clusters	LeeSallee	Dispersion	Breed	Spread	Slope	Road Gravity	Metric
1	0·99037	0·99743	0·60125	1	1	50	25	25	0·59393
2	0·98809	0·99915	0·59965	1	1	50	1	50	0·592005
3	0·97925	0·99699	0·60068	1	1	50	75	50	0·586445
4	0·98806	0·97829	0·60447	1	25	1	100	1	0·584286
5	0·98396	0·98886	0·60033	1	1	50	75	1	0·58412
6	0·99124	0·97461	0·60426	25	25	1	100	25	0·583759
7	0·99693	0·96079	0·60774	1	1	25	1	1	0·582118
8	0·98388	0·98238	0·60033	1	1	50	100	25	0·580245
9	0·98415	0·98578	0·5974	1	1	50	75	25	0·579571
10	0·98422	0·99886	0·58709	1	1	75	50	1	0·577167

Table 18.8 The Parameters of the *Fine* Calibration as well as the Elapsed Time to Completion and the Ten Best Coefficient Sets from Coarse Calibration and their Related Statistics

A Parameters of Fine Calibration

	Dispersion	Breed	Spread	Slope	Road Gravity
Start	1	1	35	1	20
Step	5	5	5	20	10
Stop	25	25	60	80	60
Possible units	5	5	6	5	5
Possible combinations			3750		
Elapsed time			7:46:59		

B Coefficient Sets from Coarse Calibration

Rank	Pop	Clusters	LeeSallee	Dispersion	Breed	Spread	Slope	Road Gravity	Metric
1	0·99928	0·99988	0·60418	1	5	35	80	50	0·603673
2	0·99366	0·99339	0·60668	1	1	35	60	30	0·598849
3	0·98701	0·99959	0·60691	1	1	35	1	60	0·598781
4	0·98851	0·99514	0·60727	1	1	35	20	20	0·597375
5	0·98426	0·99842	0·60783	1	1	35	20	30	0·597318
6	0·98431	0·99831	0·60783	1	1	35	20	40	0·597282
7	0·9835	0·99959	0·60671	1	1	40	40	30	0·596455
8	0·98178	1	0·60622	1	5	35	1	20	0·595175
9	0·98713	0·99342	0·60614	1	1	40	1	30	0·594402
10	0·98126	0·99977	0·60573	1	10	35	20	20	0·594242

(Table 18.8A). The Spread and Road Gravity coefficients of the best performers of the Coarse Calibration appear in Table 18.7B to have widespread coefficient values. A path for searching the wide coefficient values was used, as demonstrated in Table 18.8A. The Slope coefficient, still widely variant in the Coarse Calibration did not largely reduce in range and step for the Fine Calibration run.

For the Final Calibration, the Dispersion, Breed, and Spread coefficients' ranges were narrowed, due to the nature of the best performers of the Fine Calibration. The Slope and Road Gravity coefficients still varied considerably across the domain. For the Final Calibration, the ranges of the Dispersion, Breed, and Spread coefficients were narrowed, and their step reduced to one unit (Table 18.9A). For the Slope coefficient, the range was still large (1 to 80) with a large step of 20 units. The wide range in the Slope coefficient may be due to the relatively flat nature of the Sioux Falls region. The exception is, of course, the falls and the river valley. However, the river valley is all included in the Excluded layer, as to not render a significant effect on the growth of the urban extent. The range of the Road Gravity coefficient was kept large, from 30 to 60, with a moderate step of 5 units. This resulted in 5250 runs of the program, taking over 10 h to evaluate.

The Final Calibration honed in on some coefficient sets with high metric of fit values. Most notable is the best performer, while not having the best LeeSalle metric, had a higher combined metric of fit than the other coefficient sets (Table 18.9B), with a metric of 0·60472. It should be noted, however, that the first-ranked coefficient set from the Fine Calibration was ranked third in the Final Calibration run, and only two other sets performed better.

Table 18.9 The Parameters of the *Final* Calibration as well as the Elapsed Time to Completion and the Ten Best Coefficient Sets from Coarse Calibration and their Related Statistics

A Parameters of Final Calibration

	Dispersion	Breed	Spread	Slope	Road Gravity
Start	1	1	32	1	30
Step	1	1	1	20	5
Stop	5	5	37	80	60
Possible units	5	5	6	5	7
Possible combinations			5250		
Elapsed time			10:27:08		
Total number of runs			12,125		
Total elapsed time			27:05:05		

B Coefficient Sets from Coarse Calibration

Rank	Pop	Clusters	LeeSallee	Dispersion	Breed	Spread	Slope	Road Gravity	Metric
1	0·99998	0·99959	0·60498	1	3	33	80	45	0·60472
2	0·99546	0·99988	0·60652	1	2	36	80	50	0·603694
3	0·99928	0·99988	0·60418	1	5	35	80	50	0·603673
4	0·99457	0·99997	0·60651	1	1	35	60	35	0·603199
5	0·99941	0·9993	0·60384	1	4	34	80	50	0·603061
6	0·99888	0·99999	0·60374	1	1	35	80	45	0·603058
7	0·99819	0·99959	0·60399	2	1	35	80	35	0·60265
8	0·99938	0·99787	0·60402	1	2	36	80	35	0·60236
9	0·99661	1	0·60406	1	2	32	60	60	0·602012
10	0·9982	0·99655	0·60493	1	2	35	80	55	0·601758

Results of Parameter self-modification for the Brute Force calibration resulted in the following coefficient set: 1–2–24–83–45.

18.4.2 Results of the Genetic Algorithm Calibration

The GA approach to calibration resulted in seven chromosomes that outperformed the Brute Force calibration, and three that did not (Table 18.9B and Table 18.10). The general trend of all the GA chromosomes after 200 generations was that low values of the Dispersion and Breed coefficient are desirable and the Spread coefficient was relatively stable around values in the high 20s. However, both the Slope and the Road Gravity coefficients are highly variant with relatively large ranges. This could have been due to the complex interactions among the coefficients in the growth rules, or due to the flat nature of Sioux Falls (other than the steep ravines, surround by parklands). In addition, the transportation layer used for Sioux Falls included only the major (interstate) roads and railroads. A more detailed transportation layer including minor roads may have given different results. The average time it took for the GA calibration was 5·364 h, one-fifth the time of the Brute Force calibration, on a much slower processor.

When plotted out over time, the general asymptotic behavior of all the GAs is clear (Figure 18.5). This may be due to the relative high amount of elitism in the GA itself. Most of the GA runs exhibited a large step to a higher metric value, and then slowly stepped up after that.

Table 18.10 Results of Ten Independent GA Calibration Runs

Rank	Dispersion	Breed	Spread	Slope	Road Gravity	Metric	Time (h)
1	1	1	27	16	31	0·608151	5·35
2	1	3	27	17	29	0·60642	5·23
3	2	1	29	24	3	0·606047	5·52
4	2	1	29	24	4	0·606047	5·21
5	1	1	27	24	20	0·605818	5·21
6	1	1	31	70	55	0·605118	5·47
7	1	1	31	77	24	0·603935	5·35
8	1	7	27	31	37	0·603886	5·34
9	1	4	28	29	27	0·599125	5·45
10	3	2	41	25	32	0·598816	5·51
					Average Time		**5·364**

Note: All used the same Seed Populations and ran for 200 generations of 18 chromosomes.

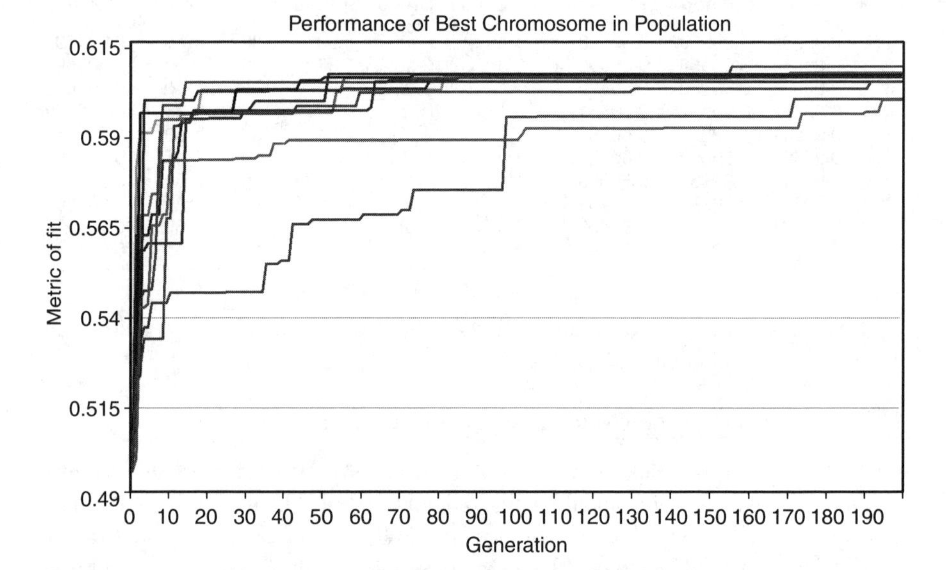

Figure 18.5 Performance of ten individual GA calibrations after 200 generations.

The self-modification of the top two GA calibration coefficients resulted in the following prediction coefficient sets (in order of rank): 1–1–12–34–29 and 1–1–21–26–3.

18.4.3 Comparison of Forecasting Results

According to the forecasted urban extents, Sioux Falls will either grow very little or not at all (Figure 18.6). Most of the growth was focused near the edges of the1992 urban extent, focused on the urban edges, rather than "spotting" and creating new

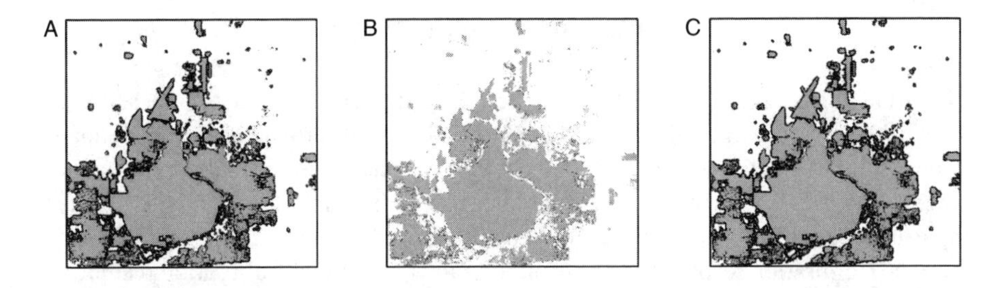

Figure 18.6 Forecasts of Sioux Falls urban extent from 1992 to 2015: Left (A) Using Brute Force coefficients. Right (B) Using best GA coefficients. (C) Using second-best GA coefficients. Gray is urban extent in 1992. Black was forecasted in >50% of the runs. The width of the study area is 14·58 km along the East–West axis and 13·95 km along the North–South axis.

Table 18.11 Statistics from the Three Forecasting Runs

	Organic Growth	Total new Pixels	Number of Pixels as Urban	Clusters	Growth Rate
Brute Force	1000·9	1005·8	97342·2	611·4	1·03
Best GA	91·8	92·8	76691·9	987·5	0·12
2nd best GA	915·3	918·8	97323·8	580·9	0·94

Note: Ten Monte Carlo iterations were used to forecast to 2015.

urban centers. The best GA calibration run, forecasted a minute amount of urban growth in 23 years between 1992 and 2015. This warrants further investigation.

All three forecasting runs exhibited most of their growth from Organic Growth that occurs at the city edges (Table 18.11). This mirrors the Sioux Falls Planning Department's plans to incorporate adjacent lands into the city as residential and commercial centers (www.siouxfalls.org/planning). However, Sioux Falls is itself expecting the urban extent to grow at a rate of 2·5% to 2015 yet SLEUTH does not forecast similar growth.

Other than the Best GA run, the forecasts are very similar in visual and statistical measures. The Best GA run had a much greater number of clusters than the other two, indicating two components. First, the Best GA run was sending out more and smaller urban clusters than the other two. Second, and more likely, the other two runs filled in urban extent where the best GA run could not, merging many clusters together. The Best GA run also had a very low growth rate, relative to the other two forecasts. The low growth rate may have been partially due to a lower value for the Spread parameter (a "12" for the Best GA, relative to a "24" for the Brute Force and a "21" for the second best GA), the low growth rate clearly demonstrates the challenges of the SLEUTH calibration process. The SLEUTH parameters combine in a non-linear and potentially chaotic manner, and the forecasting of an urban area using different coefficient sets is potentially non-linear as well. What would aid clarification of the coefficient space would be a "complete calibration," where all combinations are explored, followed by a validation of all coefficient sets, where SLEUTH is used to forecast an urban region of a known date, with validated remotely sensed data.

18.5 DISCUSSION AND CONCLUSION

The Brute Force method of SLEUTH calibration helps determine coefficient sets with an increasing metric of fit, fitting the model to the historical urban data layers. It hones in on a suite of solutions by parsing the coefficient space up into smaller and smaller pieces. However, the Brute Force method leaves much room for subjectivity. Human input is necessary twice in the coefficient calibration process, from the Coarse to the Fine Calibration resolution, and from the Fine to the Final Calibration resolution searches. It is up to the user to determine what the domain of the successive search will be, and what size the steps will be through that space. For Sioux Falls, it was evident from the Coarse Calibration that Dispersion and Breed were to be kept low, with a range of 1 to 25, with a step of 5, this was not the case for the Slope coefficient, which necessitated a larger range, from 0 to 80, with a step of 20. However, the decision of coefficient range and step is up to the modeler and their own "feel" of the data, so a different modeler would have potentially obtained different calibration results with a slightly different approach to the calibration process.

This is not the case with the GA method of SLEUTH calibration. The three components to be decided upon are the mutation rate, the number of chromosomes in a generation, and the number of generations. All three are chosen beforehand, but many combinations can, and should be tested. In fact, given the UCSB Geography computing constraints, a modeler can afford to test four different GA combinations of mutation rate, population size and number of generations, in the same amount of time of doing one Brute Force calibration.

However, as the GA method of Calibration did not always out-perform the Brute Force method, getting stuck in local maxima, there is room for improvement to the GA. The first area of exploration is the mutation rate. The mutation rate used in this GA was 10%, which is considered very high in the GA community. Both lowering and rising the mutation rate will change the GA, and in doing so may speed up or slow the selection process. Similarly, the GA tends to get stuck on high-performing (elite) chromosomes for many runs (Figure 18.5). This can be improved by allowing for greater mutation and de-emphasizing elitism when such local maxima are reached. The most interesting and obvious developments in the GA for SLEUTH calibration will occur in the modification of the number of generations, possibly creating some stopping criteria, and the population size. This issue will be a balance between the computational speed provided by the GA and casting a wide-enough net in the coefficient space to find the best chromosome.

When comparing the GA and the Brute Force calibrations in terms of computation time, the success of the GA is impressive. For the array size used for Sioux Falls, 486 by 465 pixels, the GA method outpaced the Brute Force by 4 to 5 times as fast, even though the Brute Force calibration was running on a faster computer. The close to 6 h to run the GA for Sioux Falls indicates that Yang and Lo (2003) could shrink their 15-day Brute Force calibration considerably as well.

According to different coefficient choices, Sioux Falls will look either very similar or slightly different in the year 2015. While the model showed growth in all three of the coefficient sets tested, it is not clear which, if any of these are a representation of the truth, as we have 11 years until 2015. The logical continuation of this work is to forecast Sioux Falls urban growth from 1992 to the present, and use contemporary data to validate the different model results. This would allow for a more complete judgment to which method (GA or Brute Force) better predicted the "true" urban growth of Sioux Falls. Validation with present-day data would also assess if the assertion that a higher metric of fit leads to a more valid model run. While it is clear that the GA method of calibration is more efficient computationally, validation could help determine which calibration technique was more effective (validated by the real urban growth of a city). It should be noted, however, that the focus of this work was not to model Sioux Falls' urban growth to 2015. Rather, this effort was to compare using the GA versus the Brute Force calibration methods. The validation of the Sioux Falls forecasts to the present day is a different effort with different challenges, such as the assessment of a spatial fit of a SLEUTH model forecast to contemporary data and the use or disuse of the Monte Carlo simulations.

The above disclaimer notwithstanding, the discrepancies between the forecasts of Sioux Falls' growth in 2015 are possibly due to a number of factors. First, the use of the LeeSallee metric in the metric of fit can be a harsh judge of model fit. The LeeSallee metric judges overlap and any deviations from the overlap are penalized. This may have led to the Best GA coefficient set as being the best reproducer of the 1992 urban extent while not necessarily the ideal coefficient set to be used in forecasting, especially after the coefficients were sent through the Parameter self-modification stage of calibration. In the self-modification stage, the model had to account for a "phase change" in urban form — from the solid globular form of Sioux Falls in 1976 to the speckled 1992 urban layer. The speckled pattern is hard for a model to calibrate to, especially since the earlier data layers were not speckled. The issue of speckled data will have to be resolved, as the use of remote sensing products will surely increase in urban modeling, not decrease.

The two methods of calibrating the SLEUTH model presented here, Brute Force calibration, and a modified GA, illustrate the potential advances of incorporating evolutionary computing methods into spatial modeling. The Brute Force method is just that — it throws the hammer of computation at the problem of searching the complex coefficient space of the SLEUTH coefficients. On the near horizon are computers with enough computational power to manage the large data sets and lengthy routines used in SLEUTH and the complete exploration of the search space will be possible. The recent development of cheap CPUs and homegrown Beowulf parallel computing clusters echoes this sentiment. However, in the present, when there are still computational constraints put on the SLEUTH model user, the GA method of calibration provides a viable and welcome alternative to Brute Force.

The alternative of the GA method posits the question, "how good of a fit really matters for SLEUTH?" Is finding the "best" set of coefficients going to make a difference when running SLEUTH? Is the benefit of the GA dependent on the spatial resolution or the urban region in focus? These questions indicate the path of future exploration of the influence of calibration on the SLEUTH model. Research by Candau (2002) showed that a longer timescale is not necessary for a more accurate calibration of a region. Instead, the most recent urban geospatial data are the most important to obtain in order to calibrate effectively. However, since Sioux Falls was largely formed by 1956, the first control year, it is possible that older historical data would improve the projections.

The primary advantage of the GA is that it slows the arrival of a solution, and therefore prevents the process of getting stuck in a local minimum, as opposed to a more global optimum, as illustrated in this modeling exercise. The strength of the Brute Force method is to carefully explore the coefficient space in a regular way. A hybrid method of calibration could be advantageous. This would entail using the GA to get an estimate of a set of possible solutions, then using the Brute Force approach to consistently "step" around those in a regular manner. This would be an improvement on Brute Force, in terms of speed and accuracy.

This work seeks to illustrate an improvement of a dynamic model of urban growth based on a CA using a GA. While both GAs and CAs have been borne out of the domains of Complexity, Complex Systems, and Chaos Theory (all of which have roots in classical social and physical sciences), these modeling approaches are predisposed to be incorporated into future geospatial techniques. Current methods of incorporation can be seen in many of the chapters in this volume (Benenson and Torrens, Chapter 23, this volume; Xie and Batty, Chapter 19, this volume; Agarwal, Chapter 24, this volume). What they provide is not the ultimate solution or the "be-all and end-all" of models, but an alternative to the static and deductive techniques currently in place. Human and natural phenomena are complicated, complex and multi-faceted. Dynamic models that can embrace this "confusion" rather than simplify it will lead to better models, and in turn, better insight into the consequences of our actions.

ACKNOWLEDGMENTS

The author would like to thank the Santa Fe Institute's Melanie Mitchell for her eloquent introduction of GAs at the Complex System Summer School of 2001. The author would also like to thank Keith Clarke for his programming guidance and insight on SLEUTH. Lora Richards and William Acevedo at the USGS were crucial in the data compilation. Charles Dietzel was integral in the use of his Linux boxes and babysitting the Brute Force calibration. The anonymous reviewers' feedback was very much appreciated and useful. This work has been funded in part by the Lawrence Livermore National Laboratory's SEGRF fellowship program. Prior support was provided by the National Science Foundation Urban Research Initiative grant NSF CMS-9817761 and the UCIME project.

REFERENCES

Arthur, S.T., 2001, A satellite based scheme for predicting the effects of land cover change on local microclimate and surface hydrology. Doctorate Thesis, Penn State, State College.

Balling, R.J., Taber, J.T., Brown, M.R., and Day, K., 1999, Multiobjective urban planning using genetic algorithm. *Journal of Urban Planning and Development-ASCE,* **125**: 86–99.

Bell, C., Acevedo, W., and Buchanan, J., 1995, Dynamic mapping of urban regions: growth of the San Francisco/Sacramento region, Proceedings, Urban and Regional Information Systems Association, San Antonio, pp. 723–734.

Bradley, M.P. and Landy, R.B., 2000, The mid-Atlantic Integrated Assessment (MAIA). *Environmental Monitoring and Assessment*, **63**: 1–13.

Candau, J.T., 2002, Calibration of SLEUTH. Masters Thesis, UCSB, Santa Barbara.

Clarke, K.C. and Gaydos, L.J., 1998, Loose-coupling a cellular automaton model and GIS: long-term urban growth prediction for San Francisco and Washington/Baltimore. *International Journal of Geographical Information Science*, **12**(7): 699–714.

Clarke, K.C., Hoppen, S., and Gaydos, L.J., 1996, Methods and techniques for rigorous calibration of a cellular automaton model of urban growth. In *Third International Conference/Workshop on Integrating GIS and Environmental Modeling*. National Center for Geographic Information and Analysis, Santa Fe, New Mexico.

Clarke, K.C., Hoppen, S., and Gaydos, L., 1997, A self-modifying cellular automaton model of historical urbanization in the San Francisco Bay area. *Environment and Planning B-Planning and Design*, **24**: 247–261.

Colonna, A., Distefano, V., Lombardo, S., Papini, L., and Rabino, G.A., 1998, Learning urban cellular automata in a real world. The case study of Rome Metropolitan Area. In: S. Bandini, R. Serra, and F. Suggi Liverani (Eds.), *Cellular Automata: Research Towards Industry: ACRI'98-Proceedings of the Third Conference on Cellular Automata for Research and Industry, Trieste, 7–9 October 1998*. Springer, London; New York.

Davis, L., 1991, *Handbook of Genetic Algorithms*. Van Nostrand Reinhold, New York.

Gigalopolis, 2004, Project Gigalopolis. NCGIA. www.ncgia.ucsb.edu/projects/gig/

Goldberg, D.E., 1989, *Genetic Algorithms in Search, Optimization, and Machine Learning*. Addison-Wesley Pub. Co., Reading, MA.

Goldberg, D.E. (Ed.), 2002, The design of innovation: lessons from and for competent genetic algorithms. *Genetic Algorithms and Evolutionary Computation 7*. Kluwer Academic Publishers, Boston, MA.

Goldberg, D.E. and Kalyanmoy, D., 1991, A comparative analysis of selection schemes used in genetic algorithms. In: G.J.E. RAWLINS (Ed.), *Foundations of Genetic Algorithms*. M. Kaufmann Publishers, San Mateo, CA, pp. 69–93.

Goldstein, N.C., Candau, J., and Clarke, K.C., 2004, The approaches to simulating the March of Bricks and Mortar. *Computers, Environment and Urban Systems*, **28**: 125–147.

Herold, M., Goldstein, N.C., and Clarke, K.C., 2003, The spatio-temporal form of urban growth: measurement, analysis and modeling. *Remote Sensing of Environment*, **86**: 286–302.

Holland, J.H., 1975, *Adaptation in Natural and Artificial Systems*. University of Michigan Press, Ann Arbor.

Leao, S., Bishop, I., and Evans, D., 2001, Assessing the demand of solid waste disposal in urban region by urban dynamics modelling in a GIS environment. *Resources Conservation and Recycling*, **33**: 289–313.

Lee, D. and Sallee, G., 1970, A method of measuring shape. *Geographical Review*, **60**: 555–563.

Liepins, G.E. and Potter, W.D., 1991, A genetic algorithm approach to multiple-fault diagnostics. In: L. DAVIS (Ed.), *Handbook of Genetic Algorithms*. Van Nostrand Reinhold, New York, pp. 237–250.

Pal, N.R., Nandi, S., and Kundu, M.K., 1998, Self-crossover — a new genetic operator and its application to feature selection. *International Journal of Systems Science*, **29**: 207–212.

Silva, E.A. and Clarke, K.C., 2002, Calibration of the SLEUTH urban growth model for Lisbon and Porto, Portugal. *Computers, Environment and Urban Systems*, **26**: 525–552.

Syswerda, G., 1989, Uniform crossover in genetic algorithms. In: J.D. SCHAFFER (Ed.), *Proceedings of the Third International Conference on Genetic Algorithms, George Mason University, June 4–7, 1989*. M. Kaufmann Publishers, San Mateo, CA, pp. 2–9.

Ucime, 2001, Urban Change — Integrated Modelling Environment. UCSB Geography. www.geog.ucsb.edu/~kclarke/ucime/

Wong, S.C., Wong, C.K., and Tong, C.O., 2001, A parallelized genetic algorithm for the calibration of Lowry model. *Parallel Computing*, **27**: 1523–1536.

Yang, X. and Lo, C.P., 2003, Modelling urban growth and landscape changes in the Atlanta metropolitan area. *International Journal of Geographical Information Science*, **17**: 463–488.

CHAPTER 19

Integrated Urban Evolutionary Modeling*

Yichun Xie and Michael Batty

CONTENTS

19.1 INTRODUCTION

Urban areas have long been recognized as displaying nonlinear, dynamic properties with respect to their growth (Crosby, 1983). Capturing their dynamics, however, is one of the most delicate problems in urban modeling. Only very recently have the conceptual and mathematical foundations for substantive inquiry into urban dynamics been made possible due to our growing understanding of open systems and the way

* First presented at *GeoComputation 2003*, held at the University of Southampton, 8–10 September 2003; see http://www.geog.soton.ac.uk/conferences/geocomp/

0-8493-2837-3/05/$0.00+$1.50

human decision processes feed back into one another to generate the kinds of nonlinearity that characterize urban growth and change. Applications have been made possible by fundamental advances in the theory of nonlinear systems, much of it inspired by theories of dissipative structures, synergetics, chaos, and bifurcation in the physical sciences. In fact, many of the originators of these new approaches have seen cities as being a natural and relevant focus for their work. Prigogine's work on dissipative structures, for example, has been applied to urban and regional systems by Allen (1997) while Haken's work on self-organization has been implemented for city systems by Portugali (2000) and Weidlich (2000). Many of these applications have built around traditional aggregate static approaches to urban modeling pioneered in the 1950s and 1960s, and were motivated as part of the effort to make these models temporally dynamic and consistent with new ideas in nonlinear dynamics (Wilson, 2000).

The development of complexity theory has proceeded in parallel where the concern has been less on spatial simulation per se but more on the way complex systems are composed of many individuals and agents whose behavior drives change at the most local level. Ideas about how life can be created artificially have guided many of these developments and in this context, highly disaggregate dynamic models based on cellular change — cellular automata (CA) — have become popular as a metaphor for the complex system. CA models articulate a concern that systems are driven from the bottom up where local rules generate global pattern, and provide good icons for the ways systems develop in which there is no hidden hand in the form of top–down control. Again cities are excellent exemplars (Holland, 1975). Despite the hype, CA has recently been proposed as a "new science," articulated as the basis for taking a fresh look at a number of different fields of scientific inquiry (Wolfram, 2002).

In fact, the embedding of nonlinear dynamics into traditional cross-sectional static urban models has not led to a new generation of operational land use transportation models for policy analysis, despite the fact that this kind of dynamics is consistent with the way those models are formulated. What has happened is that CA models have found much more favor but these models have all but abandoned the focus on socioeconomic activity in favor of simulating physical change at the level of land use and development. Consequently, most CA models do not treat the transportation sector in any detail whatsoever and, thus, their use in policy analysis is limited. This lack of an explicit transportation dimension is largely due to the way such CA models are structured. CA focuses on physical processes of urban systems and simulates land use changes through rules usually acting upon immediate neighboring cells or at best some wider set of cells which still restrict the neighborhood of spatial influence (White and Engelen, 1993; Xie, 1996; Batty, 1998; Clarke and Gaydos, 1998; Wu and Webster, 1998; Batty et al., 1999; Bell et al., 2000; Li and Yeh, 2000; Wu, 2002). Insofar as transportation enters these models, it is through notions of local diffusion that do not map well onto actual physical movements of short duration such as those characterizing trip-making in cities.

Though many innovative ideas such as genetic algorithms, neural network methods, and stochastic calibration for determining weights and parameters have been proposed and successfully developed (see Goldstein, Chapter 18, this volume), such CA models are still essentially heuristic and simplistic. The origins of CA modeling in urban systems also dictate some of their limitations. Raster-based digital data particularly from remote sensing and GIS software that readily works with such data, has given added weight to models that are composed of cells. The notion too that CA might be used to simulate the evolution of different physical landscapes has influenced their form and structure while the fact that many of the groups developing such models have been remote from policy, has not focused on the effort on real planning applications.

Currently, several profound challenges to CA-based urban simulation models exist. First, both physical and socioeconomic processes interact with each other and their surroundings in complex, nonlinear, and often surprising ways. These processes have subsystem elements that, in turn, may be complex and operate in different ways but in precisely the same geographical space. Different urban elements working in different ways contribute to the emergent properties of the entire system (Wilson, 2000). Each component of a complex urban system may itself be complex. However, current CA models of cities are, to a large degree, limited to physical processes and land development. They ignore urban activities that comprise such spaces which are usually the focus of policy analysis. Second, cells defined as the basic unit of land development are often characterized by a binary state of developed or undeveloped land, or by a land use type within a cell which is usually restricted to only one such use per cell. The basic land unit does not usually carry attributions such as the number of people or households that reside on it, the behavior of its residents, the value of its property or rent, the amenity of the surrounding neighborhood, or the quality of its environment. The exclusion of such socioeconomic features is a serious limitation to realistic applications that adopt CA models for urban planning and related forms of decision-making.

Third, it is difficult to establish compatibility between a cell and a real urban entity. Despite increasingly higher spatial resolution with finer cell sizes adopted in CA models as increases in computer power has enabled larger and larger systems to be represented; pixel-based cellular dynamics seldom matches area-based socioeconomic phenomena. Scale and longitudinal change in the socioeconomic geography of an area further complicate the calibration and validation of CA models. Fourth, CA models are usually supply-driven with demand entirely a function of supply. There is no feedback between demand and supply to reflect any market clearing. This is an important omission as it suggests that CA models do not react to the economic mechanisms that determine how land is actually developed, once again reflecting the disjunction between socioeconomic models, which form part of the urban economic, regional science, and transportation traditions, and this newer tradition of geographical modeling.

To put these criticisms in perspective, our starting point will be *DUEM*, the *D*ynamic *U*rban *E*volutionary *M*odel, which we have developed in several places in southeastern Michigan and which throws into partial relief the limitations of the cellular approach. This model like most CA models of urban development simulates the growth (and decline) of different land uses in cells representing the supply side which is determined by rules governing physical development. These models do not handle geo-demographics, site location, transportation and so on at all well (see Strauch et al., Chapter 20, this volume). Having set the scene with a description of *DUEM*, we will then outline a new model framework *IDUEM* (*I*ntegrated *D*ynamic *U*rban *E*volutionary *M*odeling), which begins to resolve these problems. We will then examine its conceptual structure, improvements to the way demand and supply for land and housing are handled, links to forecasting models, and issues of detailed land parcel and site representation. Our chapter is a first sketch of an ambitious framework for extending and making much more realistic and policy-relevant the CA approach. As such, this is a work in progress.

19.2 STARTING POINTS: *DUEM* AS A CELLULAR MODEL OF LAND USE DEVELOPMENT

19.2.1 An Outline of the Model

DUEM is somewhat different from the standard cellular model of urban development in that it deals with a comprehensive series of land uses, one of which is the infrastructure associated with transportation. In fact, like most other CA models, what goes on in each cell is physical development and there are no measurable attributes of cells such as population levels, rent, density, etc. *DUEM* is also unique in that it provides a strong life cycle focus to land use development reminiscent of Forrester's (1969) model of *Urban Dynamics* and this emphasis makes it highly suited for simulating long-term evolution at the level of the city system. Finally, the graphical user interface is well developed, putting it into a class in which the model is generic, hence applicable to a wide variety of situations by non-expert users. Nevertheless such generalization is difficult; most users find that even standard software has to be refined for particular circumstances. Part of our current effort to develop more applicable model structures is to address the limitations of the generalized user interface that we are currently working with.

In essence, each land use is classified as belonging to one of three life phases — initiating, mature, and declining — which reflect the life cycle of aging with the assumption that as a land use ages, it becomes increasingly less able to act as a generator of new land uses. In fact, we assume that only initiating land uses spawn new uses while mature land uses simply exist *in situ* with declining uses moving to extinction where they disappear, the land they have previously occupied becoming vacant. In *DUEM*, there is an explicit life cycle which ages these land uses through different stages (but with the possibility that a mature or declining use can revert to an earlier category as indeed sometime occurs). Land uses can also make transitions in that a land use can change its type at any stage, although for the most part, this possibility is more likely the older the use and, thus, once again relates to its life cycle.

Initiating land uses drive the growth of new land uses in the model. They spawn new land uses in their neighborhood, which is a restricted field of cells usually symmetrically arrayed around the origin cell but sometimes with directional distortion. The probability of an initiating use spawning another use in this field is a function of the distance away from the central cell and what cells get developed will depend ultimately on how strong the competition is between different land uses being spawned by the particular land use in question. A land use has the potential to spawn any number of different uses but only one of these will occur in each time period. The spawning process is subject to a series of constraints, some within a narrower traditional CA neighborhood around the cell in question which relates to density and type of uses in the neighboring cells, and also subject to regional constraints, which limit what each cell might be used for.

CA models are difficult to present in a closed form that makes their operation transparent. This is because transitions from one state (land use) to another in any cell are determined by various rules which although uniformly applied across all cells, cannot usually be written in continuous algebraic form. Thresholding and counting, for example, are typical operations that make such rules work. Hence this makes the analysis of the dynamics of such models only possible through simulation. The mathematical structure of *DUEM* has been spelt out in some detail elsewhere (Batty et al., 1999) but we do need some formality in presentation if we are to make clear the limitations of this model and demonstrate how these might be resolved. We will define cells using subscripts i and j, land use states as k and l, life cycle aging as τ where the range of τ is subdivided into three classes — initiating, mature and declining — and time itself as t and $t+1$. A land use of type k in cell i with age τ at time t, $S_i^k(\tau, t)$, and this defines the transition as $S_i^k(\tau, t) \rightarrow S_i^l(\tau+1, t+1)$, where aging and state change are clearly marked through the passage of time from t to $t+1$. However, these transitions are not mainly defined by intrinsic changes within the cells but by changes that are taking places in the rest of the system, particularly in the local neighborhood around the cell in question but also in their wider region.

Change itself is in fact generated by two processes: initiating land use spawning new land uses usually in a different place and existing land uses mutating into others (which might be seen as new), which are usually in the same place. In fact, land uses that are in their declining phase make the transition to vacant land *in situ* at some point in this cycle. The way changes take place for new land uses depends on three different sized regions. Most important is the field or district which is wider then the local neighborhood within which the spawning or initiating land uses sits. The distance from the spawning land use is a determinant of where new land use takes place but within the more local neighborhood around this land use, the composition of other land uses is instrumental in determining any state change. At the level of the region which is at the system level, constraints on what are or are not allowed in terms of cells being developed or not, are imposed. We list the three typical land use transitions as follows:

- $S_i^k(\tau = \text{initiating}, t) \rightarrow S_j^l(\tau = 0, \text{ initiating}, \ t+1)$ where the new use is in its initiating phase,
- $S_i^k(\tau = \text{initiating} - \text{mature}, t) \rightarrow S_i^l(\tau + 1 = \text{initiating} - \text{mature}, \ t+1)$, where the changed use can be at a later stage thus reflecting properties of the old use,
- $S_i^k(\tau = \text{declining}, \ t) \rightarrow S_i^*(* = \text{vacant}, \ t+1)$, where the new use is vacant land ready to come back onto the market at a later time period and thus available for new land uses being initiated from existing ones.

The first set of transitions which determine the growth process are influenced by the region, the field, and the neighborhood; the second simply by the field and the neighborhood, and the last simply by the cell itself.

The dynamics emerging from this process is complex in that it is impossible to predict other than through simulating the total land use, which is generated from this process at any one time. Total land use activity for any type is given as $\sum_{i\tau} S_i^k(\tau, t) = S^k(t)$ and for all types as $\sum_k S^k(t) = S(t)$. Not only are these totals controlled by the land development process which operates from the bottom up and whose total predictive capacity is unknown prior to each simulation but the relative proportion of different land uses $S^k(t)/S^l(t)$ are not controlled in any way and can vary dramatically. In one sense, this is an extremely desirable property of CA models for it means they are in the business of predicting total growth or decline, which is largely absent from land use transportation models where such totals are predetermined. However, this is still problematic because the mechanisms at the bottom level based on the land development process are not designed with such total predictions in mind. In short, this is an ambitious goal but much too ambitious given what little we know about such relationships and the way such features are built into the current generation of models.

The problems with this CA model like many others are manifold but in particular there are three key issues. First, there is no feedback between demand and supply. Supply is imposed from outside in that the rules that are used to determine land use transition, hence growth and decline determine what is supplied, and it is supply that preconditions demand. This leads to the second problem, which we have already noted: there is no control over the demand that the model supplies. Total demand has to be scaled artificially if it is to meet certain external known limits and if this is required, the model does not determine how such totals are generated. As we have implied, we consider it almost impossible to devise models based on bottom-up relationships which would produce feasible and realistic totals, at least at this stage. Third, there is no explicit transportation in the model. In fact, in *DUEM* we do generate streets as a distinct land use; we usually define commercial, industrial, residential, vacant land, and two kinds of streets — junctions and segments. The number and the location of streets determine how many other types of land use can be generated. In short, one needs streets as infrastructure to enable other uses to be put in place and vice versa but apart from the physical infrastructure, there are no explicit interaction models which assign traffic flows, for example, to such streets. Moreover, this is the only form of transportation in the model and, thus, other forms of movement — electronic, by air, by rail, and so on — are excluded.

19.2.2 Pedagogic and Real Applications to Small Towns and Metro Regions

To give an idea of what this model can be used for, we present three brief examples. First, we can use the model to generate hypothetical growth patterns and one of the most useful simulations is to show how capacitated growth occurs and how land uses cycle in time through the urban space. As the space fills up, then land uses age and eventually disappear opening up more space for development. In this way the capacitated system cycles up and down and this provides a very useful diagnostic to see how all the various rules for transition between land uses are balanced in

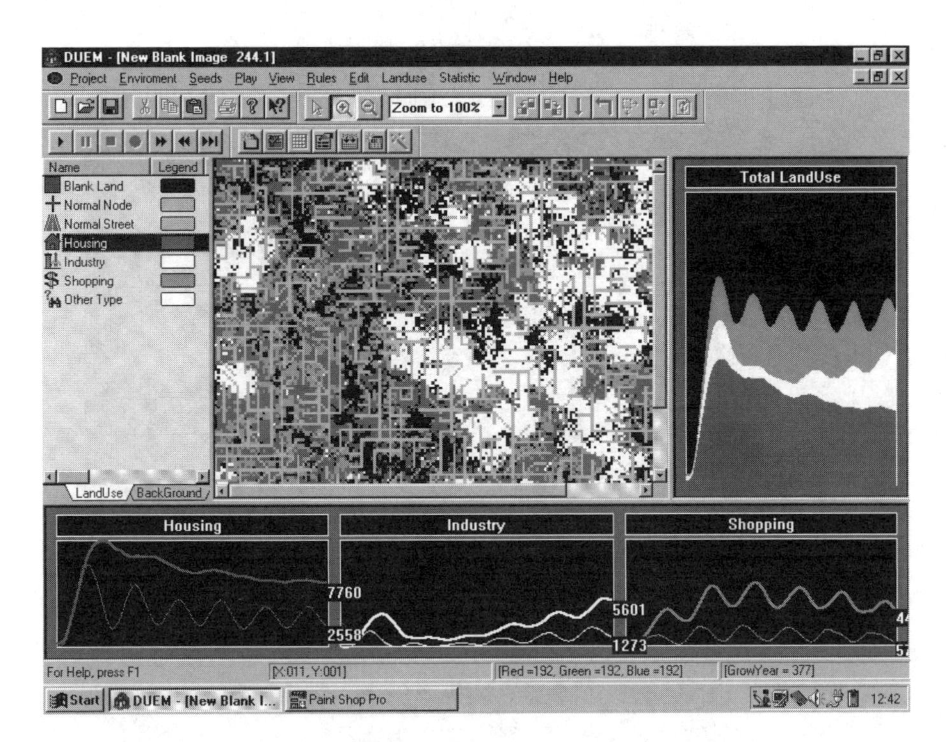

Figure 19.1 Forecasting changes in land use competition using *DUEM*.

the wider simulation. We show such a simulation in Figure 19.1 where the graphs demonstrate how housing, industry, and services oscillate through time and how particular land uses begin to get the upper hand as reflected in the bottom-up rules, which are prespecified. This is an important way of figuring out their plausibility.

In fact, Figure 19.1 provides a clear demonstration of how we do not know in advance how the model predicts the relative ratios of different land uses. As this example is capacitated, when all land is filled, what happens is that we see quite clearly how the ratios of total housing to industry to services change: in Figure 19.1 we see how industry is gradually increasing relative to housing whose proportion is falling with services more or less constant in total, that is $S^{\text{industry}}(t)/S^{\text{housing}}(t) \rightarrow ++$. However what this kind of demonstration does show is that there is no stability in the predictions. We have not run this example for a very large number of time periods but it is entirely possible that in the limit, one land use would dominate and occupy all the space. It might be said that having a simulation device to show this is extremely useful yet all this is actually showing is a limitation of the model, which is undesirable.

We have also applied *DUEM* to the simulation of urban sprawl in Ann Arbor (Figure 19.2A) and to long-term urban growth in the Detroit Metro Region (Figure 19.2B). In terms of the Ann Arbor application, when we run the model with the plausible default rules, we see immediately that housing growth is too focused along street and transport routes. Basically, we find it hard to code into the model rules on clusters which must be in place if the sizes of housing development that

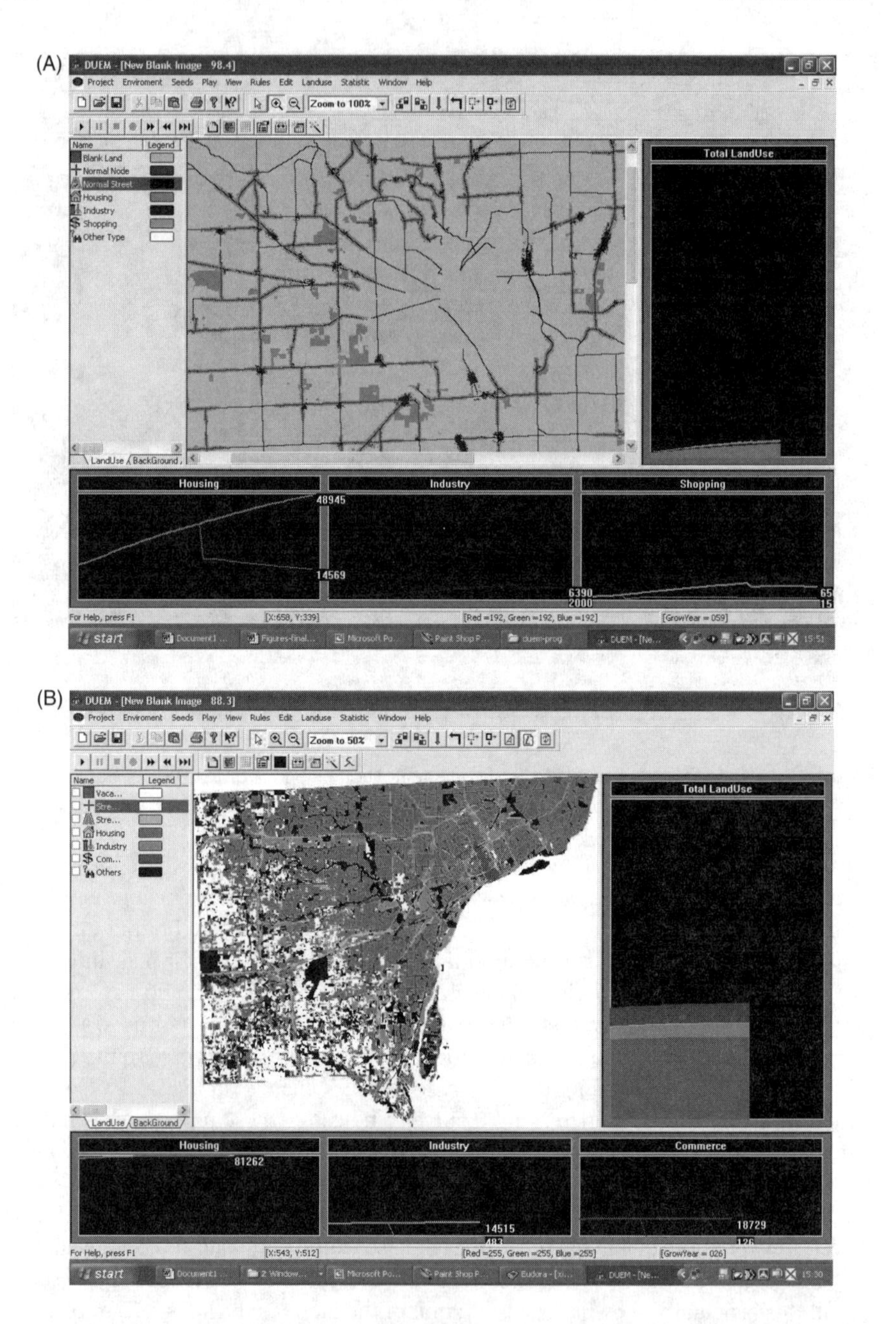

Figure 19.2 **(see color insert following page 168)** *DUEM* simulations in Ann Arbor and Detroit, Michigan: (A) Simulating sprawl in Ann Arbor. Land parcels are hard to assemble into relevant clusters of real development and the simulation predicts growth along roads. The clusters are the existing changes between 1985 and 1990, which spawn the new growth along roads. (B) Simulating long-term housing growth in Detroit. Housing grows in this scenario but in fact Detroit is characterized by decline and abandonment, and it is simply an artifact of the closed space that growth takes place in this fashion.

characterize reality are to be simulated. This again is simply another way of showing how limited the model is. Our application to Detroit also points up the difficulty of this kind of model. In fact, in the area shown in Figure 19.2B, the problem is that the simulations should show massive decline and abandonment in housing but again this is hard to simulate in the model. To develop such features, we need a much better supply side representation and we also need transportation and migration to explicitly represent socioeconomic attributes and magnitudes associated with the relevant populations.

A clear problem with *DUEM* and all cellular models involves the way cells are mapped onto real spatial units. Although the cellular grid is a fairly neutral means of spatial representation in terms of CA models where the cells only contain one land use, this kind of representation is highly abstract. It might be possible to generalize cells into nodes which have differential sizes associated with them but this takes us into representing size in ways that CA models are unable to do. These problems are quite well known and have been documented by O'Sullivan and Torrens (2000) but rather than dwell further on these limitations, we now need to sketch the way we are beginning to extend *DUEM* in integrating it with other models and new models.

19.3 THE DESIGN OF *IDUEM*

We have presented *DUEM* in a number of previous papers (Xie, 1994; Xie, 1996; Xie and Batty, 1997; Batty et al., 1999). From our summary, it is clear that the model is highly physical in nature, built around a process of land supply driving urban development but without any of the detail of the economic decision processes which determine how land is supplied and then balanced with respect to consumer demand. In our extended model — *IDUEM* — we build a central core to the framework around the demand for and supply of residential use (housing and land) but feed the model with data and predictions from more well-established disaggregate and aggregate models involving geo-demographic, geo-economic, transportation, migration and mover processes. The cellular representation is used as the visual interface to the simulation. In the model, we are currently building, which we sketch in this section of the chapter, we will focus on five different issues

- The conceptual structure of *IDUEM*, which will provide the reader with an immediate sense of what we intend.
- Demographic and economic attributes of activities in cells, which will show how the cell structure can be augmented in terms of these kinds of data and characteristics.
- Tight and loose coupling to urban and regional planning models based on micro simulation, agent-based and integrated land use and transportation models.
- Differentiated urban growth, which is marked by the way space is filled and by new subdivision development which characterizes urban growth and sprawl, particularly in U.S. cities.
- Object-based simulation and programming, which lies at the heart of how we are operationalizing and implementing the model as well as the construction of the graphical user interface.

19.3.1 The Conceptual Structure of *IDUEM*

There are two main directions in which we can extend CA models to make them more practically applicable in terms of different modeling styles. The first involves generating a much richer form of disaggregation to the level of the individual or agent and there is considerable momentum at the present time with this type of modeling (Parker et al., 2003). There are several urban applications which show promise (Batty, 2003) but currently these are a long way from practical implementation and in some senses like any CA model, still tend to be pedagogic in nature rather than practical in the policy sense. The second and more conservative strategy is to link our CA models to traditional cross-sectional approaches based on land use and transportation models and more simplistic dynamics such as population and employment forecasting models. As we have indicated, these kinds of models are more pragmatic in structure but are operational and well established. In fact, what we will do is to steer our developments to the latter while at the same time having regard to the former, particularly within the core of the model where we will build around mechanisms to make explicit demand and supply processes governing urban land at the individual household level.

This research core is a model that might be regarded as agent-based but with links directly to more aggregate models. In essence, there are three main types of model reflecting different sectors of the urban system that need to be formally represented. First, overall demand for urban activities, specifically employment and population, can be factored into different kinds of detailed activity such as services, entertainment, population types and so on, as well as their related attributes such as incomes, rents, etc. These can be simulated well using conventional demographic and economic models such as those that are built around population cohort survival, spatial input–output/urban econometric models, and so on. These models provide small area activity forecasts that dimension the more detailed demands which lead to land use change at the level of building blocks or the parcel. The second type of model reflects ways of simulating land supply and these are largely based on land suitability analysis. In a sense, the dictates of the market for land supply are not yet represented in our proposal for at this point we feel that adding issues involving mortgage and capital markets that do clearly influence land supply, is beyond the capability of these models. Land suitability analysis extending to accessibility and related environmental issues is as far as we will go in the current proposal. However, the design philosophy of *IDUEM* (based on the common object model [COM]) will support the extensibility and interoperability needed for integration with types of model derived from mortgage and capital markets in future.

The biggest problem in existing CA models is a lack of a transportation sector but in our proposal because we will be modeling detailed movements within the housing stock, and link these to accounting methods which will be made consistent with discrete choice. These will run within the background but will link to the geo-demographics and economic models and the land suitability analysis by dimensioning, keeping the quantities predicted within reasonable limits. Finally, we see cellular representation as being much relaxed in the overall model in that we will abandon the strict neighborhood-field characterization in favor of tagging individual land parcels and groups of individuals. However, the cellular approach is still useful in terms of

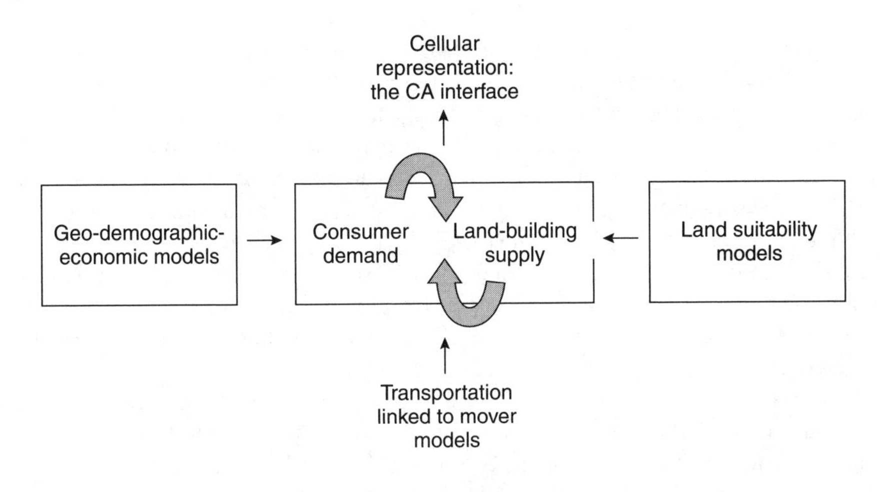

Figure 19.3 The aggregate structure of the *IDUEM*.

visualization and to all intents and purposes, at least superficially, the model will continue to be a CA-like structure. In Figure 19.3, we show the general structure of the model as it is currently developing. We elaborate this in two ways below, in terms of its system architecture and in terms of its detailed submodels. Of course, there are many different ways of looking at such a rich structure but in this chapter, we will not get down to specific ways of implementing the entire structure formally for this is very much a work in progress, whose theoretical structure and applications will be reported as they evolve.

19.3.2 Augmentation of Cell Attributes and Space

According to the classical definition of CA, such a system consists of two basic elements: a cellular space (which is *structural*) and a set of transition rules (which are *functional*) defined over that space. The cellular space is an *infinite* n-dimensional regular Euclidean space on which a neighborhood relation is established. This neighborhood relation specifies a *finite* list of cells, which are called neighbors. In applications, the CA space usually consists of a set of regular grids of the same shape and size, and the CA neighborhood is a subset embracing the same number of cells and displaying a similar structure. A neighborhood usually includes a very small number of cells for two reasons: first, a large neighborhood can lead to tremendous difficulties when formulating CA transition rules; second, local rules where action-at-a-distance is minimal, with cells comprising only first nearest neighbors, for example, give rise to global patterns which are unexpected and have macro structure. This is widely regarded as evidence that such structures are rather good at simulating emergence whose signature is fractal. For instance, von Neumann's construction of a neighborhood only considers four cells around a central fifth cell, while the most used neighborhood (after Moore) includes eight cells surrounding a ninth.

When applying the CA paradigm to spatial systems with policy/planning applications, it is inevitable that the concept of the CA neighborhood be regarded as a crucial bond which connects CA with geographical phenomena. In the context of urban growth, an ideal space unit is likely to be a property parcel as used to represent land in North American legal, real estate, and urban planning applications. A parcel is the smallest cadastral mapping unit. It shows directly the property boundaries associated with land ownership (who owns what and where), property values, its development history, land use type, and the building type and structure, which often occupies the land. It indirectly represents who occupies the plot as well as the occupants' demographic, social, economic, and personal characteristics and behaviors. Furthermore, a parcel is a dynamic commodity circulating on the market as a result of economic development or household change. We have thus decided that the parcel must be formally mapped onto the cell in building new simulation tools to take account of cadastral data. The integration of parcel space involves us in two development phases: first to "cellularize" attribute data based on parcel cadastral data (aggregating parcel demographic, social and economic data to cells), and then to replace cells with parcels in building a new generation of CA with appropriate spatial neighborhoods for conducting dynamic simulation. The latter approach will be implemented and discussed at a future time, but to anticipate how action-at-distance is to be handled, the traditional notion of the restricted physical neighborhood will be much relaxed. It will still exist in *IDUEM* but it will only serve certain obvious functions related to site, whereas situation will be a function of the general kinds of field theory that underlie spatial interaction modeling.

In conventional CA modeling, the state of a cell in the context of urban simulation, is often either binary (developed urban area or open space) or a type of land use as in *DUEM* (residential, industrial, commercial, agricultural, open, etc.). CA in its classic form, simulates cell state changes based on existing cell states and the spatial configurations in their neighborhood (Batty et al., 1999; Li and Yeh, 2000). This is entirely different from traditional modeling practice where urban activities albeit associated with land use, are the objects of simulation. CA models thus miss demographic and socioeconomic attributes and this makes them difficult to root in conventional urban theory. CA's traditional focus on cell states and restricted neighborhood configurations confines them to pedagogic uses, useful as metaphors for spatial exploration or sketch planning tools rather than tools for practical planning or prediction (Batty, 1997).

One of the most significant developments in *IDUEM* is to augment a cell's attribute from a single variable "state," to a comprehensive array of demographic, social, and economic variables. A cell will thus take a new form as "an object." Physically the cell object will have a size dimension (100 m or so), encompassing several parcels, containing several buildings (houses or factories), and household or employment types. The cell object thus represents several sets of attributes, such as household, building (housing), economic, land use, and environmental data. These will in fact be managed by an external database and we envisage that a commercialized relational database management system (RDBMS) such as Microsoft Access will be chosen at this stage so that the *IDUEM* software package is easy to run and to maintain. Cell objects and other GIS datasets will be stored as feature data layers. External models have not been used very much so far in classic CA modeling. The viability of a cell (automaton) is usually determined by its spatial configuration, not its characteristics but our augmentation of cell attributes will transform this traditional notion of CA. The dynamics of cell objects in *IDUEM* will be truly determined by CA attributes. Moreover, CA attributes are direct data inputs to urban and regional models that will be both tightly and loosely coupled in *IDUEM*.

One of the technical breakthroughs in *IDUEM* is the seamless integration between the cellular space, which is the model infrastructure of CA or agent-based models, and geographical space of areas, on which aggregate socioeconomic models are built. This technology makes possible the augmentation of cell attributes in the context of traditional modeling applications. As Figure 19.4 illustrates, the upper left panel of *IDUEM* (*DUEMPro*) Application (the current pilot model) is the content window of vector data layers, representing area-based socioeconomic models. The lower left panel is the content window of raster data layers, typically called housing–building modeling (HBM) data layers in *IDUEM*. This kind of juxtapositioning has not been possible in traditional CA models which have tended to work quite literally at the cell/pixel level. *IDUEM* breaks with this tradition as much because data is hard to force into a cellular representation and the way the land market works requires cells to be configured as plots in that land assembly for development is critical in enabling realistic allocations to be simulated in residential sector which is at the core of this framework.

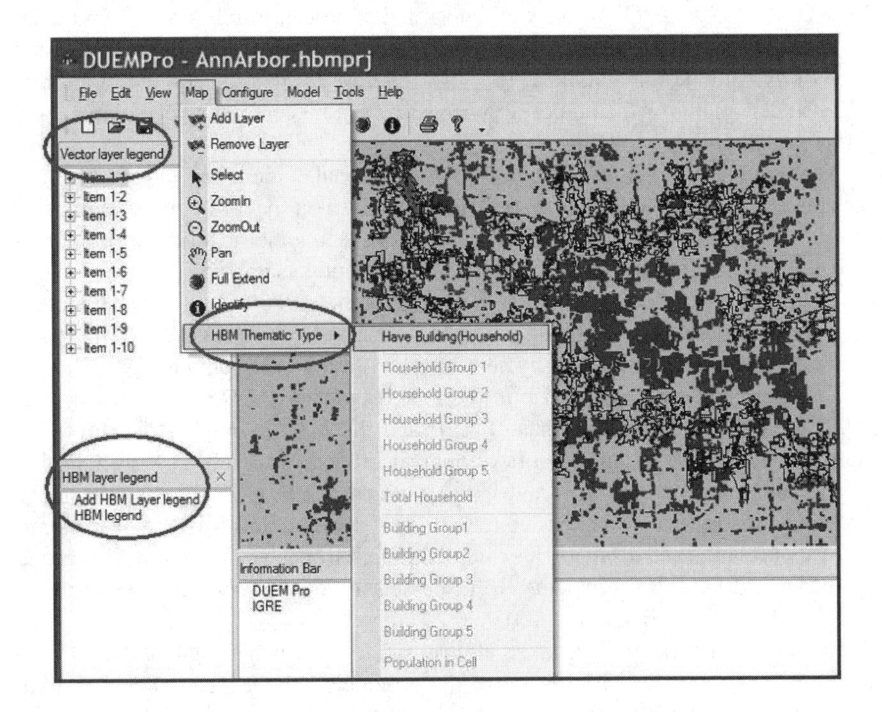

Figure 19.4 A snapshot of the vector and raster/cellular layers reconciled within the *IDUEM* simulation for Ann Arbor, Michigan.

19.3.3 Tightness of Coupling with Urban and Regional Planning Models

One of the major issues in modern software design involves the extent to which software developed for one purpose that might be entirely compatible with another can be linked to that other software in the most effective and seamless way. The loosest such coupling simply involves transferring files traditionally using manual means but more recently through automated desktop and network systems. However, tighter coupling is usually more desirable especially if functions traditionally in one software package are to be split up between many. For example, in *IDUEM* the core demand–supply model leads to the movement of households, firms, and shops, etc. and in turn sets up changes to the transportation flows that are involved, traditionally such flows being estimated using aggregate models that do not interface easily with migrations of any kind over, however, a short scale. In augmented and extended CA modeling of this kind, we thus need to consider such linkages and this suggests at least that in terms of the transportation models that we will build, these will be quite strongly coupled into the system. This is in contrast to more aggregate models such as population forecasting which can produce predictions at higher levels that can be easily factored into separate software packages as control totals.

External models are also used to produce simulation parameters or constraints on the control growth rate, location, or pattern of CA simulations, although there are many efforts reported in literature to integrate external models with CA (Xie, 1996; Batty et al., 1999). *IDUEM* takes an integrated approach to rely on data and data-driven models to answer the question of why growth happens and what is the driving force of dynamic urban automata. In short, *IDUEM* accepts the common notion in urban studies that growth and development is driven by economic development and associated demographic change.

There are three traditional types of cross-sectional static urban model that have been developed over the last 50 years. These are based on spatial-interaction, discrete choice, and spatial input–output analysis in regional econometric form. From an operational point of view, the most popularly referenced models in North America include generalized urban models of the Lowry vintage namely: the *DRAM/EMPAL* models developed by Putman (1983), and the spatial input–output *TRANUS* and *MEPLAN* models, developed, respectively by de la Barra (1989) and Echenique (1994); urban economic oriented structure such as the *CATLAS* (and later *METROSIM* and *NYMTC-LUM*) models developed by Anas (1982), and the *MUSSA* models developed by Martínez (1992); pragmatic land development models with substantially GIS-like functionality such as the California Urban Futures (*CUF*, *CUF-2*) Model (Landis, 1994); and the more comprehensive (and more recent) model structures incorporating discrete choice and disaggregate micro-simulation such as *UrbanSIM* (developed by Waddell, 2000). These models are discussed in detail in several recent reviews (Schock, 2000; Guhathakurta, 2003).

These model structures are taken into consideration when designing the *IDUEM* structure. The interface with several of these models is built into *IDUEM*. A tight coupling with urban and regional models can take advantage of rich demographic, social, and economic data that exists at a micro level for this is an important reflection

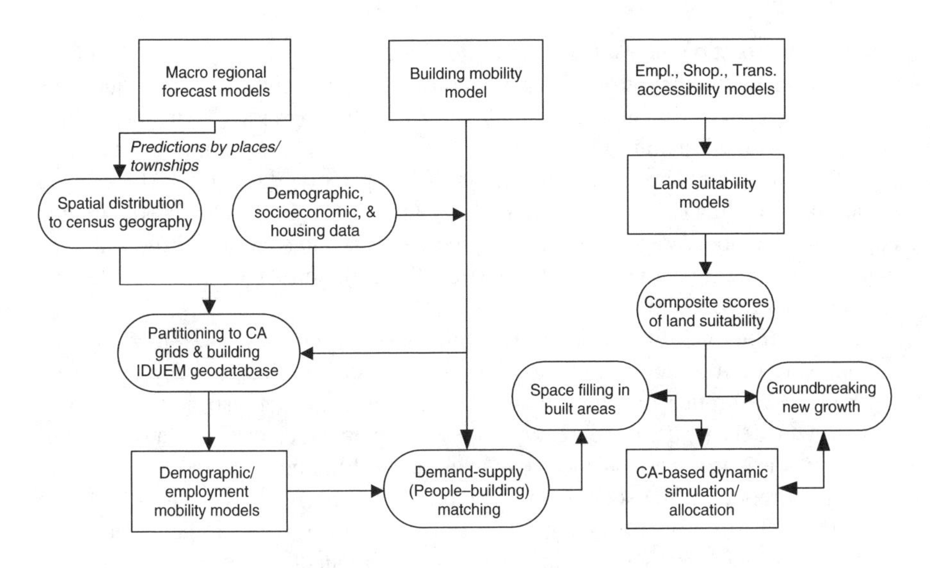

Figure 19.5 The full model structure for *IDUEM*.

of a cell object. These data decide the dynamics of each cell objects and the consequent simulation processes. Another important consideration is the "stimulus" that cell dynamics gives to growth and change. Modeling a cell's viability for change or its "mobility" is an important function in *IDUEM*, which is missing in current CA models. Therefore, *IDUEM* focuses very strongly on mobility type models which determine how urban growth is activated. More details than in Figure 19.3 are shown in Figure 19.5.

Population predictions are generated from macro regional socioeconomic models and are reported over cities, or townships, or minor civil divisions. The technique of Iterative Proportional Fitting (IPF) is implemented in *IDUEM* to partition area-based predictions over the grid space and over time (Beckman et al., 1995). The Household Mobility Model (or the Employment Mobility Model) simulates households (or jobs) which involve decisions to move from current locations. Multinomial logit models are applied to census data to determine movement probabilities as in *UrbanSIM* (Waddell, 2000). Once a household or a job decides to move, it no longer has a location in the study area, and is placed in a temporary allocation set or pool. The space it formerly occupied is made available for "CA space-filling," which we describe briefly Section 19.3.4. This household (or job) in motion will be placed by CA space-filling into what we call "groundbreaking construction" (of a new subdivision) simulation (discussed in Section 19.3.4). A similar approach is taken in implementing the Building Mobility Model.

19.3.4 CA Space-Filling and New Subdivision Development

Urban change takes place in many forms across time and space. Inner city decline is often associated with suburban sprawl. Progressive change in the building stock through deterioration, renovation, demolition, and new construction in existing built areas is accompanied by groundbreaking construction of new subdivisions and new urban centers. *IDUEM* supports simulation of both progressive dynamics which is a more detailed staging of the life cycle effects of *DUEM* with groundbreaking growth which is largely determined on the supply side through GIS-based land suitability modeling (Figure 19.3 and Figure 19.5). The regional macro demographic and economic models determine growth predictions for an area as a whole while the population and employment mobility models determine the number of people who are going to move.

The building mobility model provides the answers with respect to how existing buildings accommodate such moves. The CA simulation allocates people to existing buildings (through matching their socioeconomic characteristics) or determine new structures through groundbreaking construction in which they are housed. Additional persons who are not balanced by the usual equilibrating movements of demand and supply represent the sources for development of new subdivisions. GIS-based land suitability models are executed to rank sites with respect to their suitability. Such suitability models will include sets of tools for calculating accessibility scores for transportation convenience, employment opportunities, and shopping choice; assigning weights and ranks to relevant and available GIS data layers of interest; and composing final scores of suitability for sites available for new development. The suitability scores will determine the order of available sites in terms of the way the CA allocations take place.

The way in which demand and supply is reconciled within *IDUEM* is still under discussion. Because demand is conceived in terms of households and supply in terms of houses/dwelling (buildings) on plots, then the model operates at a much finer scale than *DUEM* where supply was predicted as $S_i^k(\tau, t)$ and demand was simply assumed to be always equal to supply $S_i^k(\tau, t) = D_i^k(\tau, t)$. This supply and demand can be considered as being measured in terms of households and dwellings, although in *IDUEM* one of the processes is a complex balancing that takes place within the temporal structure of the model with any imbalance — where the market has failed to clear — being left until the next time period. The alternative approach which is attractive is to reduce the length of the time interval to a sufficiently small unit to ensure that any imbalance need not be dealt with until the next time period, assuming that such imbalances were a realistic feature of the system (as they probably are). In this way housing market equilibrium would be an ever shifting target.

19.3.5 An Object-Based Simulation and Programming Approach

The last foray into outlining this model structure will briefly note some programming details. As being emphasized, this chapter is a progress report written to present the overall structure of *IDUEM* as well as to indicate how this structure is implemented. *IDUEM* is a new generation of CA model coupling movement (demographic and employment) with building and land use development and as such, an object-oriented approach dominates the new design. First, CA (cells now and parcels later) in *IDUEM* are *objects representing basic urban units*. The urban objects are characterized by two groups of properties: first, the physical properties that are characteristic of building stock, land/property values, land availability scores, environmental amenities, accessibilities to development stimuli, and adjacencies to existing development; and second, the socioeconomic properties including the attributes of population, the number and size of households, age composition, economic situation, employment status, travel time to work, and recent changes in these demographic and socioeconomic characteristics.

The socioeconomic subsystem interacts with the physical subsystem following the classic equilibrium of demand and supply. The socioeconomic subsystem takes into consideration demographic and socioeconomic changes, predicts probable movements of people in terms of matching buildings with preferences, and determines the demand for housing and land development. The physical subsystem simulates the supply of buildings and probable locations of future development based on land suitability. *IDUEM* is thus the first object-oriented CA model to explore the interactions of two most important phenomena of urban growth, detailed migration patterns with respect to housing and subsequent land development. These are in fact the drivers of urban sprawl in particular and urban growth in general whose understanding and prediction remains the rationale for this kind of model, at least in the first instance.

Urban entities and their physical and socioeconomic properties are analyzed and processed in the style of objects. The match (equilibrium) between demand and supply (between household migration, building construction, and land supply) is realized through multiple variance analysis of object properties. One simple illustration in the context of residential growth simulation is that households are classified according to the median household income, while housing is categorized by the type. The median household income is classified into five object types: very low, low, middle, upper middle, and high, according to the natural breaks within the data. Housing is divided into five types: apartment (AP), duplex (DUPLEX), low amenity single family (SF1), middle amenity single family (SF2), and high amenity single family (SF3). Probabilities of transition between types and between incomes are generated through intersections between the Block-Group Data Layer (containing the median household income data) with the Land Use Data Layer (including different types of housing information). The matching probability set is then used to determine where the household might move according to its household income. The CA simulation routine will finally place the household in a location based on the CA rules and the physical and socioeconomic properties of a land cell object.

IDUEM is being coded as a suite of object-oriented programs following the COM. The database management is being implemented as ActiveX database access objects. The user interface and graphic visualization are coded as VC++ multiple documents and VC++ DLLs. The models coupled with *IDUEM* are being slowly integrated through programming as either VC++ DLLs or Java Serverlets. In Figure 19.6, we show the typical user interface for generating an allocation of households to houses (as illustrated earlier in Figure 19.4), which illustrates the degree of control the user has over the simulation.

At Step 1, we load both grid-based HBM data, and area-based land suitability data. The HBM data layer is a composite data layer through several preprocesses. They include grouping and categorizing residence housing data, grouping and categorizing

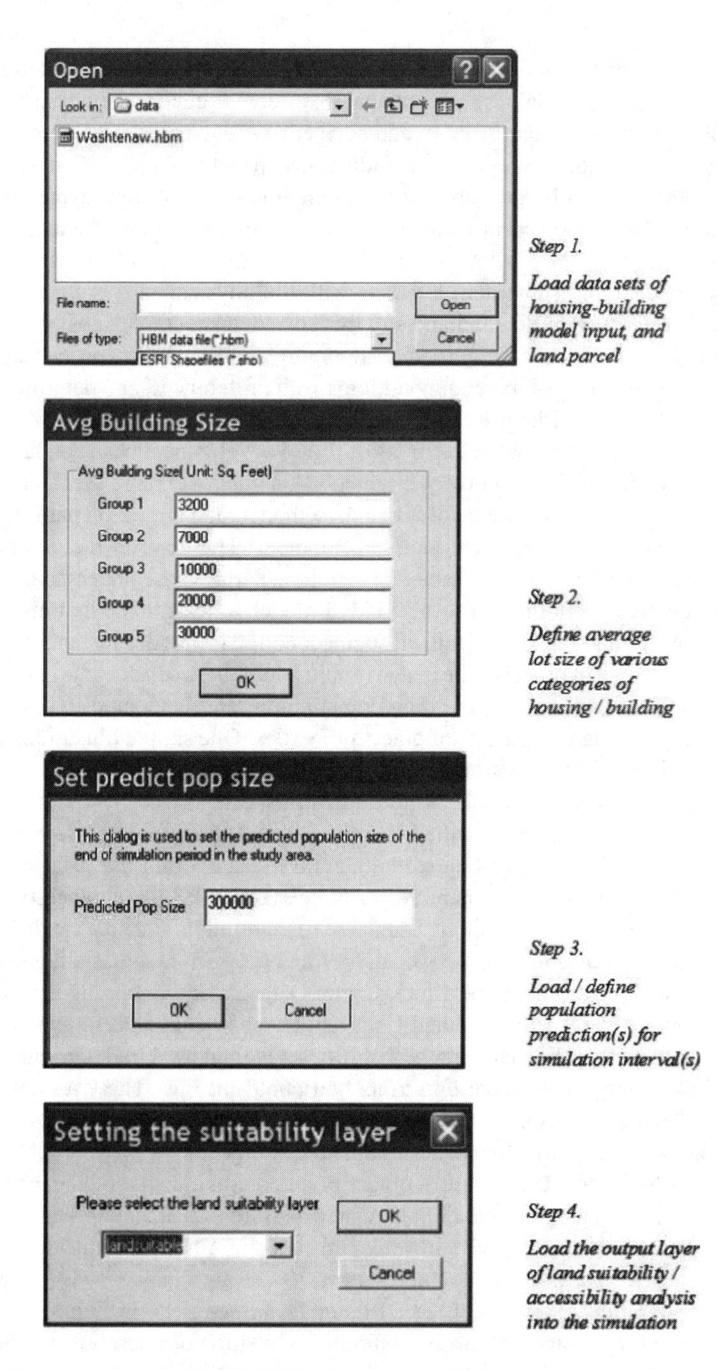

Figure 19.6 **(see color insert following page 168)** Screenshots of the *IDUEM* simulation for the town of Ann Arbor, Michigan.

socioeconomic data from Census Block Group Data, matching through multiple variance analysis. The land suitability data is the union outcome of land accessibility analyses and land suitability analysis. The tools of objectizing, categorizing, and matching housing–building with household are not packaged in *IDUEM* at this moment, but will be in the future. The two pull-down menus (in nonactivated mode) in Figure 19.4 are prototype tools for viewing and re-configuring the matching table of housing–building with households.

Step 2 shows how to parameterize housing–building size constraints in the simulation. Step 3 illustrates an integration of population prediction in the simulation. Step 3 supports more complex situations, the predictions over multiple intervals and over subdivisions. A file browser will be opened for selecting a file that contains the prediction data in this complex simulation. After the specification of predicted population, the predicted value(s) will be partitioned over time and space by an IPF routine and then a probability matrix of mobility will be calculated to guide the simulation. Step 4 allows users to confirm which layer contains the results of the land suitability analysis. Then *IDUEM* simulation will be launched. One simulation outcome was shown earlier in Figure 19.4 where a detailed analysis shows that already the model is successful in predicting sizeable developments, which account for the physical properties of land parcels, households demand, and the structure of the building process.

19.4 NEXT STEPS: CURRENT DEVELOPMENT AND FUTURE PLANS

The focus of our current work is on building the demand–supply core of the model, linking this to external geo-demographic and geo-economic models and to land supply suitability potentials. Currently, no attempt has been made to handle the transportation component in anything but a cursory and temporary way. All preliminary applications of *IDUEM* are being tested using data from the urban areas of Ann Arbor, Detroit, and Ypsilanti in Michigan where the pilot research has been concentrated and continues to be on predicting the location of population growth and housing development between 1985 and 2020, as illustrated in Figure 19.2. As Figure 19.4 reveals, a consciousness has been exercised on that models such as these should be as tightly coupled as possible with contemporary and proprietary GIS software and data formats and to this end the user interface has many features, which enable links to such external software.

What this chapter has illustrated is that for large-scale urban modeling projects, many different components need to be developed both sequentially and in parallel and at any point in the development of the wider framework, it is incomplete. This chapter has simply sketched the first stages of this model development and offered some snapshots of progress. However, we consider that for CA models to become applicable in the urban planning process specifically and more generally as part of land development, then it is necessary to move away from the literal cellular frame itself and begin to incorporate the detail of the geometry and geography of the real city as well as its linkages through transportation activity. To do this, many different modeling traditions need to merge and this suggests that integrating different models, establishing consistency between them, and making them work together should be at the forefront of this variety of urban simulation.

REFERENCES

Allen, P. M., 1997, *Cities and Regions as Self-Organizing Systems: Models of Complexity* (London: Taylor and Francis).

Anas, A., 1982, *Residential Location Markets and Urban Transportation: Economic Theory, Econometrics, and Policy Analysis with Discrete Choice Models* (New York: Academic Press).

Batty, M., 1997, Cellular automata and urban form: a primer. *Journal of the American Planning Association*, **63**, 266–274.

Batty, M., 1998, Urban evolution on the desktop: simulation using extended cellular automata. *Environment and Planning A*, **30**, 1943–1967.

Batty, M., 2003, Agents, Cells and Cities: New Representational Models for Simulating Multi-Scale Urban Dynamics. A paper presented to the Conference on *Framing Land Use Dynamics*, University of Utrecht, The Netherlands, April 16–18, 2003.

Batty, M., Xie, Y., and Sun, Z., 1999, Modeling urban dynamics through gis-based cellular automata. *Computers, Environments and Urban Systems*, **233**, 205–233.

Beckman, R. J. et al., 1995, *Creating Synthetic Baseline Populations* (Washington, D.C.: Transportation Research Board Annual Meeting).

Bell, M., Dean C., and Blake, M., 2000, Forecasting the pattern of urban growth with PUP: a web-based model interfaced with GIS and 3D animation. *Computers, Environment and Urban Systems*, **24**, 559–581.

Clarke, K. and Gaydos, L., 1998, Loose-coupling a cellular automaton model and GIS: long-term urban growth prediction for San Francisco and Washington/Baltimore. *International Journal of Geographical Information Science*, **12**, 699–714.

Crosby, R. W., 1983, Introduction and asking better questions. In R.W. Crosby (Ed.), *Cities and Regions as Nonlinear Decision Systems* (Boulder, CO: Westview Press), pp. 1–28.

de la Barra, T., 1989, *Integrated Land Use and Transport Modelling* (Cambridge, U.K.: Cambridge University Press).

Echenique, M. H., 1994, Urban and regional models at the martin centre. *Environment and Planning B*, **21**, 517–534.

Forrester, J. W., 1969, *Urban Dynamics* (Cambridge, MA: MIT Press).

Guhathakurta, S., 2003, *Integrated Land Use and Environmental Models* (Berlin: Springer-Verlag).

Holland, J. H., 1975, *Adaptation in Natural and Artificial Systems* (Ann Arbor, MI: University of Michigan Press).

Landis, J. H., 1994, The California urban futures model: a new generation of metropolitan urban simulation models. *Environment and Planning B*, **21**, 399–420.

Li, X. and Yeh, A. G. O., 2000, Modelling sustainable urban development by the integration of constrained cellular automata and GIS. *International Journal of Geographical Information Science*, **14**, 131–152.

Martínez, F., 1992, The bid-choice land use model: an integrated economic framework. *Environment and Planning A*, **24**, 871–885.

O'Sullivan, D. and Torrens, P. M., 2000, Cellular models of urban systems. In S. Bandini and T. Worsch (Eds.), *Theoretical and Practical Issues in Cellular Automata: ACRI'2000:*

Proceedings of the Fourth International Conference on Cellular Automata for Research and Industry (London: Springer-Verlag), pp. 108–117.

Parker, D. C., Manson, S. M., Janssen, M. A., Hoffman, M. J., and Deadman P., 2003, Multi-agent systems for the simulation of land-use and land-cover change: a review. *Annals of the American Association of Geographers*, **93**, 314–337.

Portugali, J., 2000, *Self-Organization and the City* (Berlin: Springer-Verlag).

Putman, S., 1983, *Integrated Urban Models: Policy Analysis of Transportation and Land Use* (London: Pion Press).

Schock, S., 2000, *Projecting Land Use Change* (Washington, D.C.: EPA/600/R-00/98, National Exposure Research Laboratory, EPA).

Waddell, P., 2000, A behavioral simulation model for metropolitan policy analysis and planning: residential and housing market components of UrbanSIM. *Environment and Planning B*, **27**, 247–263.

Weidlich, W., 2000, *Sociodynamics: A Systematic Approach to Mathematical Modelling in the Social Sciences* (Amsterdam, The Netherlands: Harwood Academic Publishers).

White, R. and Engelen, G., 1993, Cellular automata and fractal urban form: a cellular modeling approach to the evolution of urban land-use patterns. *Environment and Planning A*, **25**, 1175–1189.

Wilson, A. G., 2000, *Complex Spatial Systems: The Modelling Foundations of Urban and Regional Analysis* (Harlow, Essex, U.K.: Prentice Hall).

Wolfram, S., 2002, *A New Kind of Science* (Urbana, IL: Wolfram Media, Inc.).

Wu, F., 2002, Calibration of stochastic cellular automata: the application to rural-urban land conversions. *International Journal of Geographical Information Science*, **16**, 795–818.

Wu, F., and Webster, C. J., 1998, Simulation of land development through the integration of cellular automata and multi-criteria evaluation. *Environment and Planning B*, **25**, 103–126.

Xie, Y., 1994, *Analytical Models and Algorithms for Cellular Urban Dynamics*, unpublished Ph.D. dissertation, State University of New York at Buffalo, Buffalo, NY.

Xie, Y., 1996, A generalized model for cellular urban dynamics. *Geographical Analysis*, **28**, 350–373.

Xie, Y. and Batty, M., 1997, Automata-based exploration of emergent urban form. *Geographical Systems*, **4**, 83–102.

CHAPTER 20

Linking Transport and Land Use Planning: The Microscopic Dynamic Simulation Model ILUMASS

Dirk Strauch, Rolf Moeckel, Michael Wegener, Jürgen Gräfe, Heike Mühlhans, Guido Rindsfüser, and Klaus-J. Beckmann

CONTENTS

20.1 INTRODUCTION

All cities in Europe struggle with the problems of urban sprawl and traffic congestion, yet mostly with little success. There is a growing awareness that market forces will continue to lead to ever more dispersed, energy-wasteful urban settlement patterns

0-8493-2837-3/05/$0.00+$1.50

and that only a combination of land use policies, such as the promotion of higher-density, mixed-use urban forms, and of transport policies to promote public transport and contain the automobile can free metropolitan areas from their increasing auto-dependency. It is therefore necessary to develop modelling approaches in which the two-way interaction between transport and land use is modelled (Alvanides et al., 2001).

Today, there is a new interest in integrated models of urban land use and transport provoked by the environmental debate. In the United States and in Europe the number of integrated urban land use transport models that can be used for assessing environmental impacts of land use and transport policies is increasing (Wegener, 1998).

20.2 OVERVIEW OF THE JOINT RESEARCH PROJECT ILUMASS

The project ILUMASS (Integrated Land Use Modelling and Transportation System Simulation) is an example for the development of an integrated model as stated in the Section 20.1.

ILUMASS aims at embedding a microscopic dynamic simulation model of urban traffic flows into a comprehensive model system that incorporates changes of land use, the resulting changes in activity behaviour and in transport demand, and the impacts of transport and land use on the environment (Figure 20.1, see also Wegener, 1998; Strauch et al., 2002). The results of the policy scenarios will contribute to knowledge about feasible and successful policies and policy packages to achieve sustainable urban transport (Claramunt et al., 2000).

The ILUMASS project aims at developing, testing, and applying a new type of integrated urban land-use/transport/environment (LTE) planning model. Urban LTE models simulate the interaction between urban land-use development, transport demand, traffic, and environment (Figure 20.1).

The distribution of land use in the urban region, such as residences, workplaces, shops, and leisure facilities, creates demand for spatial interaction, such as work, shopping, or leisure trips. These trips occur as road, rail, bicycle, or walking trips

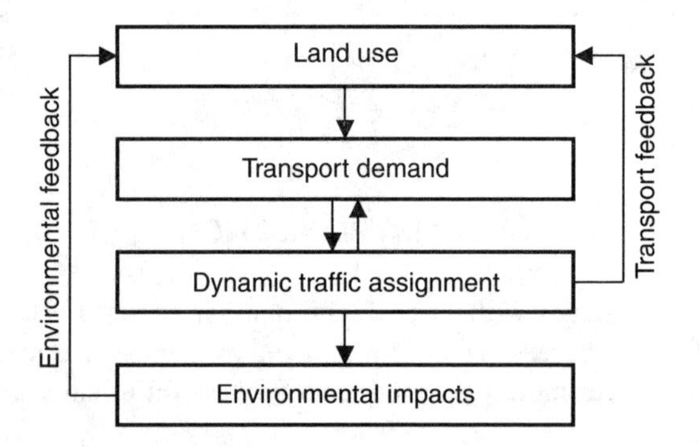

Figure 20.1 Feedbacks in LTE models.

over the transport network in the region, and they have environmental impacts. There are two important kinds of feedback: the accessibility provided to locations in the region by the transport system influences the location decisions of developers, firms, and households. Firms and households also take environmental factors, such as clean air and absence of traffic noise, in location decisions into account.

20.2.1 Project Organisation and Main Objectives of ILUMASS

ILUMASS is conducted by a consortium of German research institutions consisting of the German Aerospace Center (DLR) in Berlin, the Institute of Spatial Planning at the University of Dortmund (IRPUD) together with Spiekermann & Wegener, Urban and Regional Research (S&W), Dortmund, the Institute of Urban and Transport Planning at the RWTH Aachen University (ISB), the Institute of Theoretical Psychology at the University of Bamberg (IfTP), the Centre of Applied Computer Science at the University of Cologne (ZAIK), and the Institute of Sustainable Infrastructure Planning at the University of Wuppertal (LUIS) under the co-ordination of DLR.

The work programme of ILUMASS consists of six interrelated work packages:

- Microsimulation of land use changes (IRPUD/S&W).
- Microsimulation of activity patterns and travel demand (ISB/IfTP).
- Microsimulation of traffic flows by dynamic traffic assignment (ZAIK).
- Simulation of goods transport (DLR).
- Microsimulation of environmental impacts of transport and land use (LUIS).
- Integration and co-ordination (DLR).

The main components of ILUMASS are shown in Figure 20.2.

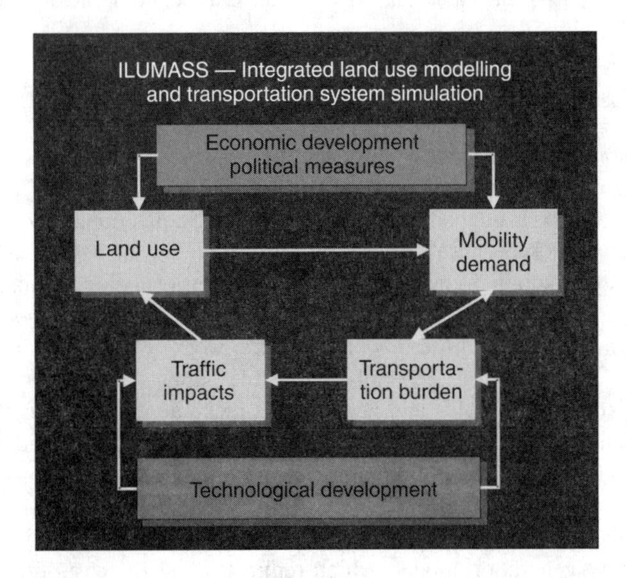

Figure 20.2 Main components of *ILUMASS*.

The *land use* component of ILUMASS is based on the land use parts of the existing urban simulation model developed at the Institute of Spatial Planning at the University of Dortmund (IRPUD) but is microscopic like the transport parts of ILUMASS. Microsimulation modules include models of demographic development, household formation, firm lifecycles, residential and non-residential construction, labour mobility in the regional labour market, and household mobility in the regional housing market. The *transport* part of ILUMASS models daily activity patterns and travel and goods movements based on state-of-the-art models of household activity patterns and the resulting mobility behaviour of individual household members and on a microscopic simulation model of travel flows developed by a team of German universities in earlier projects. The *environment* modules of ILUMASS calculate the environmental impacts of transport and land use, such as greenhouse gas emissions, air pollution, traffic noise, barrier effects, and visual impairment by transport and selected emissions.

The ILUMASS approach takes account of deficiencies of existing urban land use/transport models which are too aggregate in their spatial and temporal resolution to model aspects that are crucial for achieving sustainable urban transport, such as

- Multipurpose unimodal and intermodal trip chains and time of day of trips,
- The interaction between activity and mobility patterns of household members,
- New lifestyles and work patterns, such as part-time work, telework, and teleshopping,
- The interaction between travel demand, car ownership and residential and firm location,
- The interaction between land use and built form and mobility behaviour,
- Environmental impacts of transport such as traffic noise and exposure to air pollution,
- Feedback from environmental impacts to the behaviour of firms and households.

20.2.2 The Microscopic Approach in ILUMASS

The innovation of the ILUMASS approach is a continuous microscopic transformation of land use, activity and transport demand, and environmental impacts. The design of the land use model takes into account that the collection of individual micro data (i.e., data which because of their micro location can be associated with individual buildings or small groups of buildings) or the retrieval of individual micro data from administrative registers for planning purposes is neither possible nor, for privacy reasons, desirable. The land use model, therefore, works with synthetic micro data, which can be retrieved from generally accessible public data. The synthetic population consists of households and persons that make activities, firms that provide workplaces and that offer goods or services, and buildings for residential, commercial, or public use. Since the synthetic micro data are statistically equivalent to real data a microsimulation model can run with synthetic data (Moeckel et al., 2003).

The activity generation model, which replicates and forecasts time dependent O-D-matrices (input for the traffic flow model), is based on the microsimulation of the individual activity scheduling process. For each simulated person — one person stands for a defined number of people of the synthetic population — the daily and weekly sequence of different activities and trips is generated. In a first step for each person, an individual activity repertoire is generated, which contains a set of activities and their characteristic attributes for execution, for example, duration, frequencies, priorities, and period of time (preferred start and end time) including an individual set of possible locations. In a second step, based on a skeleton schedule (routine or habitual activities), the different activities of the repertoire are put together in an individual activity programme. The modelling of this activity scheduling process underlies a lot of decisions (long-, mid- and short-term), about which activity has to be scheduled next, how to perform the activity, and how to solve conflicts which may occur between different activities and trips during the scheduling process. Therefore, an empirical database is built up, which contains initial information on different activity attributes on time, space, and mode as well as parameters describing the planning related attributes such as flexibility, variability, and routines. The activity generation model is integrated in an iterative modelling process and linked with information about accessibility of locations and travel times and therefore it is directly connected to the land use and traffic flow simulation (Schäfer et al., 2001; Thill, 2000). The microscopic traffic flow model establishes the connection between the infrastructure of the city and the individual activity behaviour. In that step of the model, the planned trips are realized taking their interaction into account. As a result, information about the practicability of the planned trips is available. That information is used in an iteration process in which plans are rescheduled leading to an equilibrium situation in which all plans are feasible. In addition to this short-term feedback the environmental impact of traffic is assessed and used to influence long-term planning of the simulated individuals.

The result is a comprehensive model system incorporating changes of land use, the resulting changes in activities and in transport demand, and the impacts of transport on the environment.

20.3 STUDY AREA

The study region for tests and first applications of the model is the urban region of Dortmund (Figure 20.3). The area consists of the city of Dortmund and its 25 surrounding municipalities with a population of about 2.6 million. The area is subdivided into 246 statistical zones. However, the spatial resolution of 246 zones is not sufficient for microsimulation of transport and land use and for modelling environmental impacts such as air quality and traffic noise. These types of models require a much higher spatial resolution. Therefore, raster cells of 100 by 100 m in size are introduced in the modelling system and are used as addresses for activities. In order to bridge the data gap between zones and raster cells, GIS-based techniques are used to disaggregate zonal data to raster cells. Figure 20.4 is a detailed map of the city centre of Dortmund (the small square in the centre of Figure 20.3) showing the built-up area, the zone boundaries and the raster cells. In total, about 207,000 raster cells cover the study area.

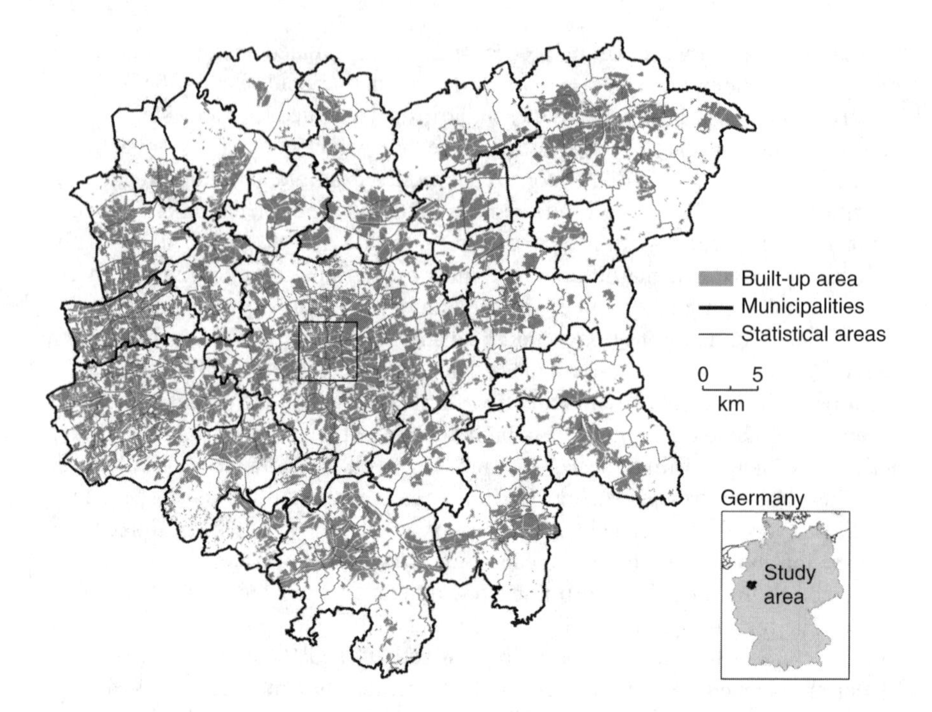

Figure 20.3 The study region of Dortmund and its 25 surrounding communities.

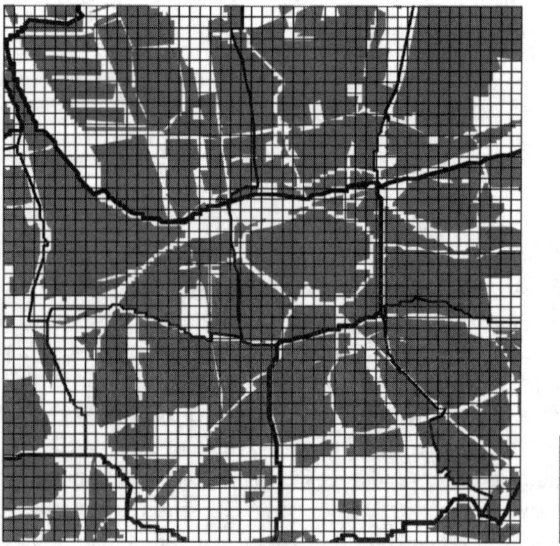

Figure 20.4 The Dortmund city centre with raster cells.

20.4 MICROSIMULATION MODULES IN ILUMASS

20.4.1 Microsimulation of Changes in Land Use

The module *Microsimulation of changes* in *land use* was developed by the Institute of Spatial Planning at the University of Dortmund (IRPUD) and the partner Spiekermann & Wegener (S&W), Urban and Regional Research, Dortmund.

Major input data are synthetic "populations" of households, dwellings, non-residential buildings, and firms in the base year as well as the road and public transport networks (Figure 20.5). Households and household members, firms and workers, and residential and non-residential buildings are aged every simulation period and undergo changes by choices, transitions or policies occurring during the simulation period. For each forecasting year, the distributions of households, persons, firms, and workers are passed to the microsimulation modules forecasting travel and freight transport demand and dynamic traffic assignment. The traffic flows, link loads, and travel times and costs so generated are fed back to the land use model in which they, through accessibility, affect the behaviour of developers, households, and firms (transport feedback). In addition they serve as input to the environmental modules which calculate the resulting environmental impacts of transport and land use. These, in turn are fed back to the land use models and affect the location decisions of developers, households, and firms (environmental feedback) (Figures 20.1 and 20.5).

Work on the synthetic population of households, persons, and firms is ongoing (Moeckel et al., 2003). Population, household, labour, employment, and housing data were collected for the 246 zones of the study region and disaggregated to raster cells of 100×100 m size using GIS-based land use data as ancillary information (Spiekermann and Wegener, 2000). In addition, data on schools, universities, car ownership, land prices, household income, and parking facilities were collected and disaggregated to raster cells using kriging and other spatial interpolation methods. Every simulation period the synthetic population consisting of households, persons, dwellings, firms, and non-residential floorspace is updated by events such as households move or buy a car, persons can marry, get divorced, get children, or increase their income, new dwellings are constructed and others are upgraded, firms move or changes their vehicle fleet, and non-residential floorspace is built or demolished. The integrated model will be calibrated using data from household activity and travel surveys conducted in the study region and validated using aggregate time series data of population, housing, and employment as well as data from traffic counts in the study region. The model will then be used to study the likely impacts of various policy alternatives in the fields of land use and transport planning. Scenarios might cover land use planning alternatives, such as policies promoting high-density mixed-use inner-city development or policies fostering decentralised polycentric regional development, or transport infrastructure changes, such as new motorways or rail lines, or regulatory policies, such as area-wide speed limits, or monetary policies, such as road pricing, higher petrol taxes, or changes in rail fares or parking fees.

The definition of policy scenarios together with local planners will be a test of the policy relevance of the models. The results of the policy scenarios will contribute

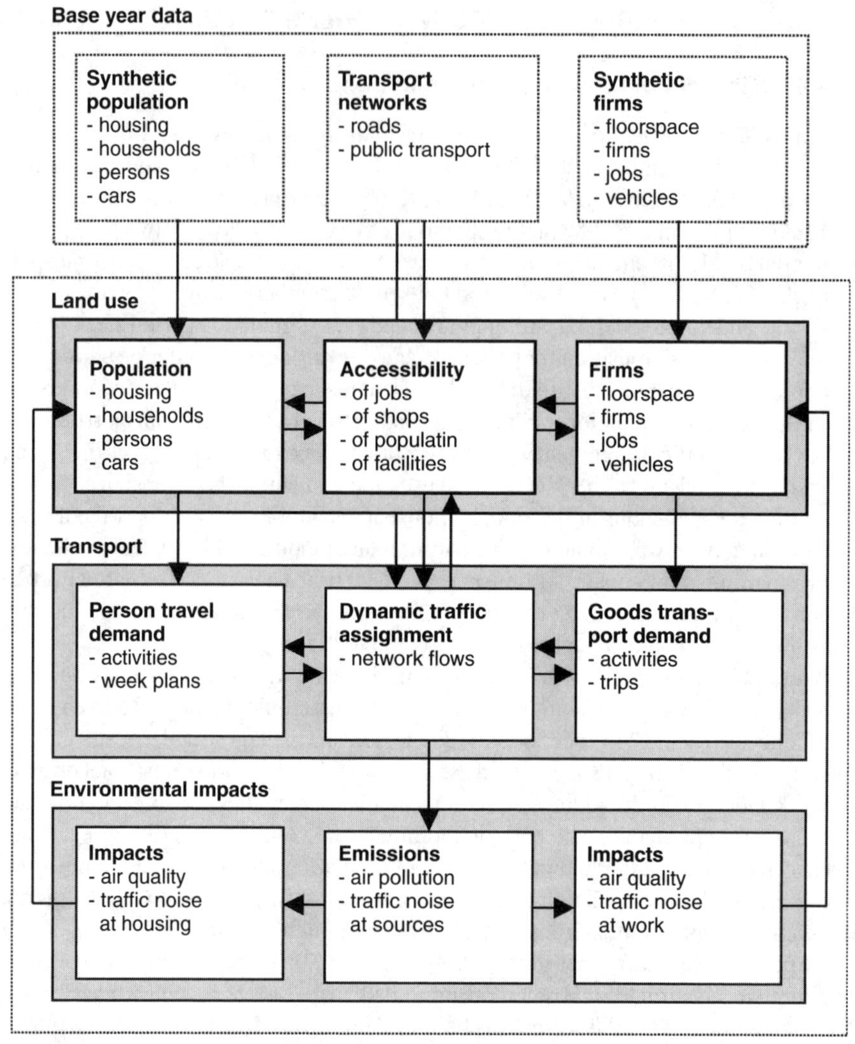

Figure 20.5 The *ILUMASS* model and the integration of the land-use model.

to knowledge about feasible and successful policies and policy packages to achieve sustainable urban transport and may be used for finalising the new land-use plan and mobility master plan of the city of Dortmund.

20.4.2 Microsimulation of Activity Pattern and Travel Demands

The microscopic module of activity pattern and travel demands was developed by the Institute for Urban and Transport Planning — RWTH Aachen University.

The main objective, therefore, is to improve the empirical basis of a scheduling approach integrated in a microsimulation framework of individual behaviour. In ILUMASS this new approach to tracing these underlying activity scheduling decision processes is developed. The goal was to utilize a computerized hand-held survey instrument that allows the gathering of information from subjects at regular intervals as close in time to real decision points. The computerized platform also enables the instrument to automatically trace and/or prompt for certain attributes of the decisions process, such as the sequence of decision inputs, thereby reducing respondent burden.

The project's main interest focused on the scheduling behaviour of individuals related to out of home activities (and derived travel), especially the sequences of the planning of distinct activity and travel attributes. The latter implying a desire to trace how the timing, location, involved persons, mode, etc. attributes of activities are differentially planned. Given this focus, a self-administered and computer-based instrument was considered to have the best chance of capturing the various variables of interest. The successful applications of CHASE (Computerized Household Activity Scheduling Elicitor) and the experiences with a CHASE survey done in Aachen, Germany, in 1999 (Mühlhans and Rindsfüser, 2000) support that view. CHASE was the first computer aided self-interview of activity scheduling behaviour as it occurred in reality in the household. CHASE was developed by Doherty and Miller (Doherty and Miller, 2000).

The EX-ACT survey was developed with these concerns in mind, and represented a unique solution to the problem. It builds upon these experiences in several key ways:

- Usage of hand-held computers or Personal Digital Assistants (PDAs) instead of laptop computers as in CHASE. This is a major improvement in terms of the flexibility of the interviewees and a situational data entry.
- More in-depth tracking of activity scheduling decisions, including the tracking of entering, modifying, or deletion of activities and travel (as with CHASE), but also tracking of how distinct attributes of activities (timing, location, involved persons, mode) are differentially planned. This represents a significant increase in detail, and conceptually is more behaviourally realistic as people often plan different attributes of activities on different time and space horizons.
- Other general improvements in instrument design, concerning form layout, instrument structure, user friendliness, and preliminary database setup.

The EX-ACT survey was completely administered on the PDA, and included the following main components:

- pre-interview
- exploration of an individual activity-repertoire
- an initial data-entry (already planned activities for the following survey period)
- a multi-day main scheduling exercise
- and a post-interview.

Figures 20.6 to 20.8 show some of the forms used in EX-ACT. For a more detailed description see Rindsfüser et al. (2003). Figure 20.6 shows two examples

Figure 20.6 Two screenshots from the Pre-Interview (*Source:* Rindsfüser et al., paper presented in *10th International Conference on Travel Behaviour Research*, 2003).

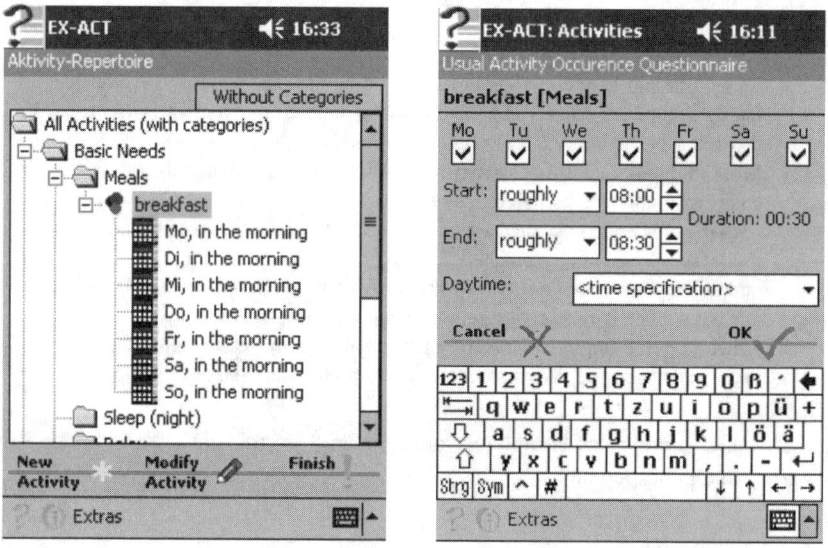

Figure 20.7 Two screenshots from the Activity Repertoire (*Source:* Rindsfüser et al., paper presented in *10th International Conference on Travel Behaviour Research*, 2003).

of the type of forms used in the Pre-Interview to capture public transit information, and household member information. Additional forms in the Pre-Interview were directed to prior decision of the household and their members with major influence on behaviour.

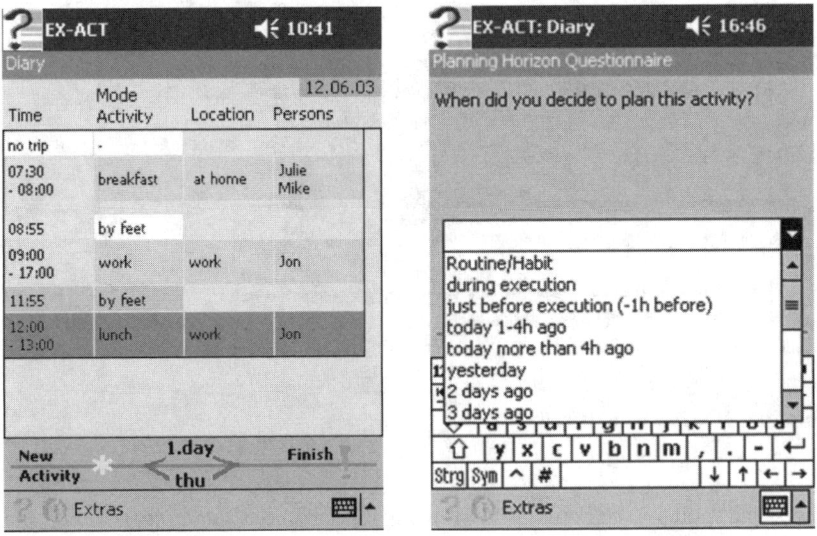

Figure 20.8 Two screenshots from the Activity Planning Diary (*Source:* Rindsfüser et al., paper presented in *10th International Conference on Travel Behaviour Research*, 2003).

To capture the activity repertoire, a tree-style view is presented to respondents showing nine main activity categories with multiple sub-categories of possible activity type choices, as shown in Figure 20.7.

Two screenshots from the scheduling diary are displayed in Figure 20.8. The screenshot at left depicts the main view of the diary wherein data entry begins. Activities for a single day are listed in sequence according to start times, with all activities having the same height (as opposed to height corresponding to the duration of the activity as in a typical dayplanner). This allowed for a much more compact display required of the PDA. Activity attributes (time, mode, location, involved persons) are displayed in columns. Colour-coding is used to indicate several key attributes of the schedule, including the degree of exactness to which each attribute has been specified. On the right-hand side of Figure 20.8 the prompt appearing every time an entry is made "after the fact" (the activity is already undertaken) concerning when the activity attribute decision was actually made is displayed.

In combination, these various instrument components captured the following information

- Socio-demographic characteristics of the household and the household members.
- Details of available and used transport modes (e.g., car fleet, season tickets owned).
- Interviewees activity-repertoire, assessed through a series of questions concerning the types of activities they typically perform along with their attributes (such as normal frequency, durations).
- One-week continuous activity planning decisions.
- Two days worth of in-depth planning characteristics.
- Resulting/realized activity-travel patterns (as in a traditional diary).

Figure 20.9 shows the survey cycle and different survey parts. The goal was to gather information about a variety of activities, including typical (routine or habitual),

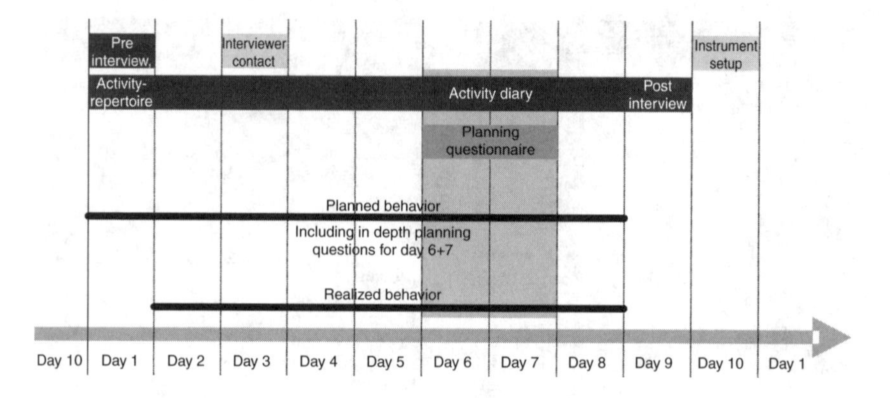

Figure 20.9 Cycle and parts of the EX-ACT survey.

planned and spontaneous activities. Following this, an initial data-entry period is spent describing activities already planned for the week to follow. At this stage, one or more of the attributes of a planned activity can be left undecided (e.g., "have not thought about the location yet"), or only partially planned (e.g., the planned day might be Monday or Tuesday). The pre-interview, the activity-repertoire, and the initial data-entry are all completed with the assistance of an interviewer (see also more detailed sections below). During the week to follow, subjects are instructed to continue adding activities (one-week diary) they have planned for future days (using same format as initial data-entry), but also to make modifications/deletions to activities as they change, and update undecided or partially planned activity attributes. They are instructed to do so whenever a decision about one or more activity-attributes or activities as a whole occurs.

All the while, the program tracks automatically the sequence of decisions made, and prompts the user for supplemental information on certain decisions, such as the reasons for the modifications, or when exactly a particular decision has been made (this is especially important for tracking impulsive decisions that are entered into the program after-the-realisation). To reduce respondent burden, these supplemental prompts are asked only for a sample of decisions, weighted more heavily towards the end of the survey (since this allows several days in advance to capture scheduling decisions as they evolve). The final component of the survey is an interviewer assisted post-interview, in which additional information on activities may be asked along with an assessment questionnaire concerning the software and hardware usability.

Although the initial application will result in the equivalent of a week long activity planning diary plus the tracing of the scheduling decisions, the software has been developed with special settings that allow the researcher to pare down the software so that it may serve as an activity or trip diary alone, or be used for a varying number of days. In addition there are many other settings (including a switch between German and English text version), controlled via a database, which allow specific adjustments in the frequency of supplemental prompts and language settings.

Application of the survey on 402 individuals took place from November 2002 to February 2003 in the study area of ILUMASS, the city of Dortmund (see Section 20.3).

The instrument was implemented on a COMPAQ iPaq 3850 (PDA) with Windows CE 3.0 operating system and programmed using Microsoft Visual Embedded Tools in Visual Basic. The software automatically starts upon turning on the PDA. The software was coded by Interactive Instruments (Bonn, Germany) and the survey data collection was conducted by SOKO-Institut (Bielefeld, Germany).

The data interpretation and the database provided through EX-ACT has just started and is still ongoing. So a conclusion can only be tentative. In general, the survey was successful in terms of handling the devices and handling the instrument EX-ACT. Here, it must be stated that the brief overview which is described above supports a first success of the concept. There are many questions arising while getting deeper insight to the data. These analyses are the subject of the current work. First results of EX-ACT were published in Rindsfüser et al. (2003).

20.4.3 Further Microsimulation Sub-Modules in ILUMASS

There are further sub-modules in ILUMASS in progress, they will be specified below.

The module *Microsimulation of traffic flows* is developed by the Centre of Applied Computer Science (ZAIK) at the University of Cologne. The interfaces processing the road and public transport networks prepared by the working group IRPUD Dortmund were completed and tested and the classification of cars and commercial vehicles defined (Figure 20.5). The interfaces linking the modules calculating travel demand with the dynamic traffic assignment were defined. Alternative methods to model public transport route choice behaviour of travellers were compared and integrated into the existing multimodal route planner. The concept for modelling individual mobility was finalised and integrated into the existing dynamic assignment algorithm.

The module *Psychological Actor Model of Individual Intentions and Decisions* (PSI) is developed by the Institute of Theoretical Psychology (IFTP) at the University of Bamberg. The work has focused on the integration of the PSI-model into the weekly activity planner AVENA. The interfaces to link the activity model with the land use models and the AVENA activity planner are largely completed.

The module *urban goods transport* is developed by German Aerospace Center (DLR). The complex work on modelling urban goods transport resulted in the decision to develop a simplified goods transport model in ILUMASS.

The module *Microsimulation of environmental impacts of transport and land use* is developed by the Institute of Sustainable Infrastructure Planning at the University of Wuppertal (LUIS). The work on environmental impacts has focused on defining the main groups of moving and fixed emission sources and establishing the methodology of estimating emissions and spatial dispersion models of greenhouse gases, pollutants, and traffic noise and defining the interfaces between the land use and transport modules and the environmental sub-models.

20.5 INTEGRATION AND MODULE STRUCTURE IN ILUMASS

The ILUMASS project aims at integrating modules to a complete modelling system (Figure 20.5). The complete model ILUMASS consists of 6 sub-modules to characterise the complex interactions between urban development, general social-political conditions and mobility.

The first step involves the independent development of all sub-modules, and abstracts initially from the task of linking them with other models. The linkage of the modules and, in particular, the enabling of backward-linkages between the modules is a research topic that, to date, has received little analytical treatment.

The simulation procedure consists initially of processing a hierarchical chain of individual modules in a time interval. The output of the model then forms a subset of the input for the following model. A time interval is defined here by the longest typical simulation period of an individual module (e.g., for Dortmund this would be one year). The outputs of all the sub-modules serve in the next time interval as a new input data set. Such a system would be a bottom-up simulation, however, iterative backward linkages have to be included. In the ILUMASS model system, there are both uni-dimensional cause–effect relationships as well as strongly linked connections between the activity reports of the surveyed individuals, the daily schedules, and the trip times in the network that are calculated by different sub-modules. Consistency between the data set, thus, necessitates consideration of the interactions between the modules. For example, following the calculation of concrete travel times over the network, the time that an individual has to conduct a certain activity during the day does not generally correspond to the original assumptions. The solution of this problem, therefore, requires an iterative approach.

For this purpose each individual module has to be integrated into a standardised operating system. In this step, the modules are not visualised, but the results of the simulation could be imported and processed in a conventional Geographic Information System. The data communication within the program system will result via Input and Output data files. The coordination of the program system will be assumed by a control algorithm, which is currently under development by DLR.

The main tasks of this control algorithm are:

- successive running of the modules (program parts)
- waiting for respective results before editing and running the next (further) module (algorithm steps) (see Figure 20.10).

For the above reason the new program has the ability to run a complete scenario (simulation). Then the analysis of the several results will be carried out by each project member. With the adoption of data bases, the Module Integration is more flexible, however, more complex. Therefore, it is necessary to establish the data communication with data bases and not only via data files (Etches, 2000).

There are some advantages by using this approach. Potential users (e.g., municipalities, planning bureaus) are able to adopt different scenarios direct from the data base with a comfortable (graphical) user interface.

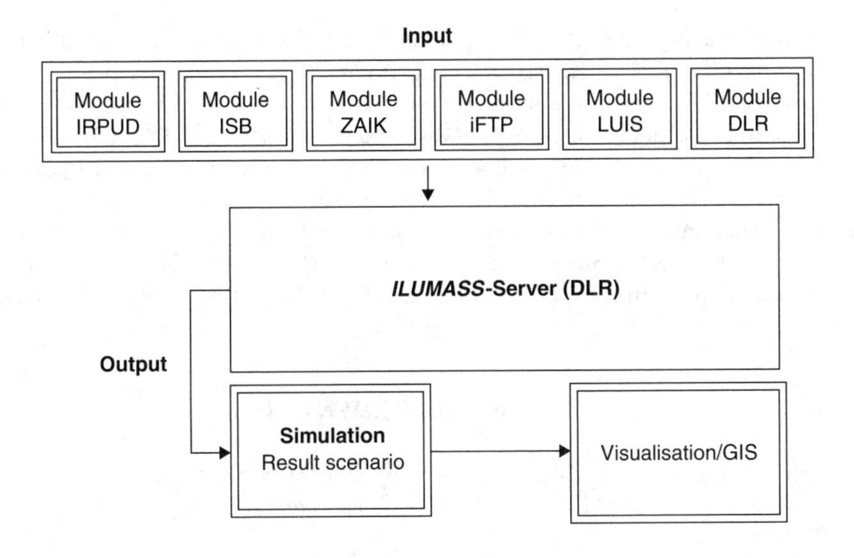

Figure 20.10 Integration and Module Structure in *ILUMASS*.

20.6 FUTURE ASPECTS OF LAND USE/TRANSPORT/ ENVIRONMENT SIMULATION MODELS

The ILUMASS model is completely disaggregate and deals with micro locations and movements of individual agents and destinations (households, firms, and persons) on a surface of pixel-like grid cells combining a microscopic land use model with a microscopic activity based travel demand model and microscopic environmental impact models in one unified modelling framework. It remains to be asked whether the movement towards ultimate disaggregation in content, space, and time is the most suitable approach.

From a technical point of view, the prospects are excellent. More powerful computers will remove former barriers to increasing the spatial, temporal, and substantive resolution of models. The wealth of publicly available high-spatial resolution data will reduce aggregation errors in spatial models. Geographic Information Systems will become the mainstream data organisation of urban models (Clarke, 2003). Spatial disaggregation of land use and transport network data in raster GIS will permit the linkage between land use transport models and dispersion (emission–immission) air quality and noise propagation models. Multiple representation of spatial data in raster and vector GIS will combine the advantages of spatial disaggregation (raster) and efficient network algorithms (vector) (Dietzel, 2003). It will be possible to replace aggregate probabilistic approaches (e.g., entropy maximising) by disaggregate microsimulation approaches.

When completed, the integrated ILUMASS model will be the only European counterpart to the growing number of large North-American modelling projects utilising advanced microsimulation approaches for the integrated planning of sustainable land use, transport, and environment in urban regions, such as the California Urban Futures

(CUF) Model at the University of California at Berkeley (Landis and Zhang, 1998a, b), the Integrated Land Use, Transport and Environment (ILUTE) model at Canadian universities led by the University of Toronto (Miller, 2001), the Urban Simulation (UrbanSim) model at the University of Washington, Seattle (Waddell, 2000), and the models of the Transport and Land Use Model Integration Program (TLUMIP) of the Department of Transportation of the State of Oregon, U.S. There are no efforts of comparable size in Europe. There are a few national projects, such as the Learning-Based Transportation Oriented Simulations System (ALBATROSS) of Dutch universities (Arentze and Timmermanns, 2000) or the ILUMASS-Project in Germany described in this paper.

ACKNOWLEDGMENT

The joint research project ILUMASS is supported by a grant from the German Ministry of Education and Research (Bundesministerium für Bildung und Forschung — BMBF).

REFERENCES

Alvanides, S., Openshaw, S., and MacGill, J., 2001, Zone design as a spatial analysis tool, In Tate, N. and Atkinson, P.M. (Eds.), *Modelling Scale in Geographical Information Science*, London, pp. 141–157.

Arentze, T. and Timmermanns, H., 2000, ALBATROSS — a learning based transportation oriented simulation system. European Institute of Retailing and Services Studies, Eindhoven.

Claramunt, C., Jiang, B., and Bargiela, A., 2000, A new framework for the integration, analysis and visualisation of urban traffic data within geographic information systems, In Thill, J.C. (Ed.), *Geographic Information Systems in Transportation Research*, Oxford, 3–12.

Clarke, K., 2003, The limits of simplicity: toward geocomputational honesty in urban modelling, In *Proceedings of the 7th International Conference on GeoComputation*, Southampton, U.K., September 2003 (CD-ROM).

Dietzel, C., 2003, Spatio-temporal difference in model outputs and parameter space as determined by calibration extent, In *Proceedings of the 7th International Conference on GeoComputation*, Southampton, U.K., September 2003 (CD-ROM).

Doherty, S.T. and Miller, E.J., 2000, Interactive methods for activity scheduling processes, In Goulias, K. (Ed.), *Transportation* **27**, 75–97.

Etches, A., 2000, A temporal geo-spatial database in support of an integrated urban transportation system, In ZAGEL, B. (Ed.), *GIS in Transport und Verkehr*, Heidelberg, 33–44.

Landis, J. and Zhang, M., 1998a, The second generation of the California urban futures model. Part 1: Model logic and theory, *Environment and Planning B: Planning and Design* **25**, 657–666.

Landis, J. and Zhang, M., 1998b, The second generation of the California urban futures model. Part 2: Specification and calibration results of the land use change module, *Environment and Planning B: Planning and Design* **25**, 795–824.

Miller, E.J., 2001, Integrated Land Use, Transportation, Environment (ILUTE) Modelling System, *http://www.ilute.com/* (accessed 22 September 2004).

Moeckel, R., Schuermann, C., Spiekermann, K., and Wegener, M., 2003, Microsimulation of land use, In *Proceedings of the 8th International Conference on Computers in Urban Planning and Urban Management (CUPUM)*, Sendai, Japan, Centre for Northeast Asian Studies (CD-ROM).

Moeckel, R., Spiekermann, K., and Wegener, M., 2003, Creating a synthetic population, In *Proceedings of the 8th International Conference on Computers in Urban Planning and Urban Management (CUPUM)*, Sendai, Japan, Centre for Northeast Asian Studies (CD-ROM).

Mühlhans, H. and Rindsfüser, G., 2000, Computergestützte Erhebung und Analyse des Aktivitätsplanungsverhaltens, *Stadt Region Land* (*SRL*), **68**, Institut für Stadtbauwesen und Stadtverkehr, RWTH Aachen, pp. 69–78.

Rindsfüser, G., Mühlhans, H., Doherty, S.T., and Beckmann, K.J., 2003 (in press), Tracing the planning and execution of activities and their attributes — design and application of a hand-held scheduling process survey, Paper presented at the *10th International Conference on Travel Behaviour Research*, August 10–15, 2003, Lucerne, Switzerland.

Schäfer, R.-P., Strauch, D., and Kühne, R., 2001, A Geographic Information System (GIS) for the Integration of Heterogenous Traffic Information, In GHASSEMI, F. (Ed.), *MODSIM 01. — Proceedings of the Conference MODSIM 01*, Canberra/Australia, December 2001 (Publications of The Australian National University, Canberra), 2075–2080.

Spiekermann, K. and Wegener, M., 2000, Freedom from the tyranny of zones: towards new GIS-based models, In FOTHERINGHAM, A.S. and WEGENER, M. (Eds.), *Spatial Models and GIS: New Potential and New Models*, GISDATA **7**, London, pp. 45–61.

Strauch, D., Hertkorn, G., and Wagner, P., 2002, Mikroskopische Verkehrssimulation, Flächennutzung und Mobilität — Entwicklung eines neuen Planungsinstrumentariums im Verbundprojekt *ILUMASS*, In Möltgen, J. and Wytzisk, A. (Eds.), *GI-Technologien für Verkehr und Logistik*, IfGI prints **13**, Univ. Münster, Inst. f. Geoinformatik, Münster, pp. 133–146.

Thill, J.-C., 2000, Geographic information systems for transportation in perspective, In Thill, J.-C. (Ed.), *Geographic Information Systems in Transportation Research*, Oxford, pp. 167–184.

Waddell, P., 2000, A behavioural simulation model for metropolitan policy analysis and planning: residential location and housing market components of Urban Sim, *Environment and Planning B: Planning and Design*, **27**, 247–263.

Wegener, M., 1998, Applied models of urban land use, transport and environment: state-of-the-art and future developments, In Lundqvist, L., Mattson, L.-G., and Kim, T.J. (Eds.), *Network Infrastructure and the Urban Environment: Recent Advances in Land Use/Transportation Modelling*, Berlin, Heidelberg, New York, 245–267.

CHAPTER 21

Simulating Urban Dynamics in Latin American Cities

Joana Barros

CONTENTS

21.1 INTRODUCTION

Third World cities are known for their inherent chaotic and discontinuous spatial patterns, and rapid and unorganised development process. The present chapter focuses on the dynamics of development of Latin American cities, focusing on a specific kind of urban growth that happens in these cities, called "peripherisation." "Peripherisation" can be defined as a kind of growth process characterised by the formation of low-income residential areas in the peripheral ring of the city. These areas are incorporated into the city by a long-term process of expansion in which some of the low-income areas are recontextualised within the urban system and occupied by a higher economic group while new low-income settlements keep emerging on the periphery.

0-8493-2837-3/05/$0.00+$1.50

The present study explores these ideas with a number of agent-based simulation experiments to increase the understanding of the relation between the dynamics and the resultant morphology of rapid urbanisation. We use urban modelling techniques to unfold the problem of urban growth of Latin American cities through their dynamics. We will show from very simple simulation exercises that some of the most important issues regarding urban dynamics of Latin American cities can be investigated.

We will present the Peripherisation Model (Barros and Alves Jr., 2003), a simulation model that explores a specific mode of urban growth that is characteristic of Latin American city morphology and "City of Slums" (Barros and Sobreira, 2002) that focuses on the role of spontaneous settlements in the global spatial dynamics of Third World cities. We will then relate the results of these experiments with examples from reality through some maps of two Latin American cities: São Paulo and Belo Horizonte, Brazil.

21.2 URBAN GROWTH IN LATIN AMERICA

While the problem of urban growth in Europe and North America has been formulated in terms of sprawl, in the Third World and, more specifically, in Latin America the main focus has been the rapidity of growth of cities. During the period between 1950 and 1980 growth rates were very high (Hall, 1983; Valladares and Coelho, 1995) and, based on these data, studies anticipated continuing high rates of growth. It was believed that many areas would double in population and a few would triple, creating urban areas that by the year 2000 would be without parallel in history (Hall, 1983). Latin American countries went from being predominantly rural to predominantly urban in few decades, with high concentrations of urban population in cities with more than one million inhabitants (UNCHS, 1996). This rapid urbanisation produced various kinds of social problems as housing stock and urban infrastructure were not enough to house all the migrants that arrived to the cities.

However, this population change has shown marked signs of change since 1980. After decades of explosive urbanisation, urban growth rates have slowed, the rate of metropolitan growth has fallen, and fertility rates have declined (Valladares and Coelho, 1995). Moreover, rural to urban migration has come to have a much smaller role in urban population growth and, most recently, the pace of urban expansion has been maintained by births in the cities. These new trends have been detected in the period between 1980 to 1990, and have been confirmed by recent censuses.

The changed demographic trends have had impacts on urbanisation patterns, which can be defined mainly by a process of population deconcentration: a fall in the overall rate of population growth; a reduced concentration of population in the core of metropolitan areas coupled with significant growth of small and medium-sized municipalities; and a declining rate of demographic growth in regional capitals and major urban centres.

The main problem of urban growth in Latin American cities is no longer the high rates of population growth and rural–urban migration. Rather, it is the spatial pattern of growth, the peripherisation process, which enlarges the peripheral rings of cities and metropolis despite the reduction in the overall urban growth rates. In the

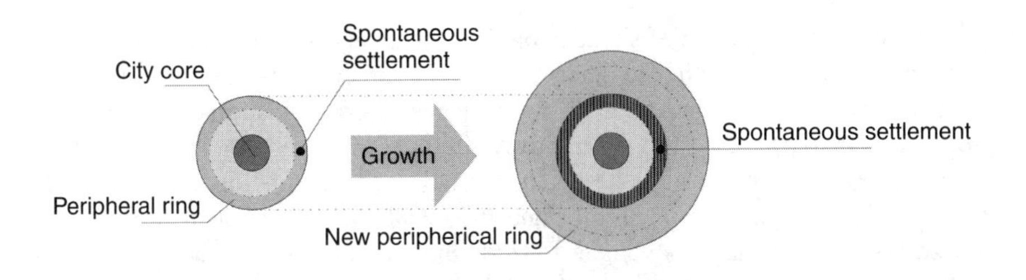

Figure 21.1 Schema of recontextualisation of spontaneous settlements.

larger cities of Latin America, the demographic growth rate has slowed right down, migration has taken second place to natural increase, and the bulk of the housing stock now consists of upgraded (or in process of upgrading) low-income residential areas, with a large number of spontaneous settlements.

The phenomenon of peripheral growth can now be considered as an established process of growth of most Latin American cities. Peripherisation can be defined as a kind of growth process characterised by the expansion of borders of the city through the massive formation of peripheral settlements, which are, in most cases, low-income residential areas. These areas are incorporated to the city by a long-term process of expansion in which some of the low-income areas are recontextualised within the urban system and occupied by a higher economic group while new low-income settlements keep emerging on the periphery (see schema in Figure 21.1).

Peripherisation is an urban spatial problem which has strong effects in social and economic terms, a problem that is unlikely to fade away without a well-informed planning action. The peripheral ring of Latin American cities consists mostly of low-income housing including large spontaneous settlements, and which usually lack urban services of any kind. As such, peripherisation clearly constitutes a social problem. However, it is not a problem only in the sense of the extreme social inequalities that appear in the city in a very concrete spatial form. Rather, the problem is the perpetuation and continuity of such a process in space and time.

To better understand the process of peripheral growth, it is necessary to study the spatial elements that compose it and the dynamics of its formation, growth, and consolidation. The study of spontaneous settlement in the global context of urban growth is important not only for the purposes of planning but also for understanding the nature and scope of urban problems in developing countries. The need to discuss the dimensions of urban growth in developing countries along with the implications of the corresponding mushrooming of spontaneous settlements has been addressed by UNCHS (1982, 2003).

As a static phenomenon, spontaneous settlements are seen only as a housing problem; from a dynamic perspective, however, they present a problem of urban development. So far the point of view of statics has been the approach for spontaneous settlements, by both, the research community and planners. The connection between spontaneous settlements and urban growth has attracted little, if any, attention. The high rates of growth have been seen as the main cause of the formation of spontaneous settlements in Third World cities, together with the inability of the state to deal with rapid urbanisation.

Although the formation of spontaneous settlements is considered a consequence of the high rates of growth, the persistence of this process despite the recent slow down of these rates in Latin American cities suggests a consolidation of this dynamic process as the normal mode of urban growth in those cities, and the core–periphery as their spatial pattern. It must be noted that although the topological structure of location (core–periphery) remains the same, the actual spatial location of the periphery is modified in a constant movement towards the city's borders. New spontaneous settlements keep emerging on the urban peripheral rings, expanding the urban frontiers as a consequence of the core's development. The spatial structure can be considered as "a pattern in time" (Holland, 1995), since it is a dynamic phenomenon in which the spatial pattern is constantly being reproduced.

In terms of urban planning policies the phenomenon is also seen from a static point of view. The focus of the governments' interventions is still on the local or housing scale, upgrading settlements and providing housing tracks for the low-income groups. There has been no focus either on the dynamics of the process or on the linkage between the local and the global scales, that is, the overall growth of the cities, which has been seen as a mere result of a demographic phenomenon. Peripherisation, like urban sprawl, is a suburbanisation phenomenon. Whilst urban sprawl has been studied in detail and its main features seem to be broadly understood, in Latin American's case the understanding of the peripherisation process remains a central issue. On the contrary of sprawl, which is an inherently spatial problem, urban peripherisation is essentially a social problem with spatial character. From a social point of view, peripherisation is not a simple problem to solve, neither is it in the hands of planners to attempt it. As a spatial problem, and more specifically as an urban development problem the phenomenon still needs to be further investigated.

Like urban sprawl, peripherisation is fragmented and discontinuous development. It also presents problems related to urban sustainability, transportation, and the cost of infrastructure and urban services. Studies from the 1970s suggest that the lowest densities in Latin American cities are found in high-income residential areas, the highest densities are in middle-class areas, and the densities of spontaneous settlements are somewhere between these two (Amato, 1970). Finally, an interesting difference between urban sprawl and peripherisation is related to the fact that, while urban sprawl is directly related to the preference of people for suburban settings, peripherisation is not a direct consequence of locational preference. On the contrary, people who move to the city's border do not wish to live there but are impelled to.

21.3 MODELLING URBAN DYNAMICS

Urban dynamics have been traditionally studied by using modelling techniques. Since the 1960s, a number of models have been developed and, more recently, with advances and popularisation of computer tools, the possibilities to explore urban systems and dynamic processes from this viewpoint have increased considerably.

The computer became an important research environment in geography and techniques and tools have been developed ever since. In urban applications, the use of automata-based models, more specifically cellular automata (CA), has replaced traditional transport and land use models, shifting the paradigm of urban models towards a complexity approach. CA proved to be an inherently dynamic simulation tool, which has been widely developed and proved to be useful for a number of different urban applications. CA models have been successfully applied as operational models, that is, for real-world applications (White and Engelen, 1997; Almeida et al., 2002; Xie and Batty, Chapter 19, this volume), as have heuristic descriptive models, with their more theoretical objectives (see Batty, 1998; Portugali, 2000). Agent-based models have been used to study a number of different aspects of urban phenomena, from pedestrian movement (Schelhorn et al., 1999; Batty et al., 2002) to spatial change (Parker et al., 2001; Benenson and Torrens, Chapter 23, this volume), including studies about cognitive aspects of micro-scale human behaviour in urban space (Agarwal, Chapter 24, this volume).

Differently from CA models, which explores only landscape–landscape interactions through fixed neighbourhood relationships, agent-based models allow modellers to work with three distinct layers or interactions: Agent–agent, agent–landscape, and landscape–landscape. However, due to such a detailed framework, agent-based modelling benefits can, sometimes, "exceed the considerable cost of the added dimensions of complexity introduced into the modelling effort" (Couclelis, 2001).

In Section 21.4, we will present some experiments with an agent-based model, which has been used to investigate different aspects of urban dynamics in Latin America. For the purposes of the present chapter, we will assume that an agent-based model is an automata-based model like cellular automaton, in which the transition rules of the CA are replaced by actual decision-making rules. CA models explore only the spatial layer of the city (landscape–landscape) and, although transition rules often were representations of human decision-making, this representation is not explicit. Like in CA models, the choice of increasing the degree of complexity of the model or keeping it simple depends entirely on the researcher and the purposes of the model in hand. Agent-based models came to meet the understanding that human decision-making plays a major role in urban processes and change, and we argue that the fact that agent–landscape interactions are explicit in the model opens up an avenue for analysis of dynamic processes that link spatial development with social issues. This kind of analysis is of fundamental importance when dealing with cases of strong social differentiation, as the case of urban dynamics in the Third World.

21.4 PERIPHERISATION MODEL

The aim of our simulation model is to develop heuristic-descriptive models on the decentralised process underlying the process of growth in Latin American cities. The model was elaborated in such a way that the behaviour rules were as simple as possible. It is totally based on the agent–landscape relationship and we do not explore neither landscape–landscape nor agent–agent relationships. The model was built on a Starlogo platform, that is a user-friendly parallel programming tool developed by the Epistemology and Learning Group of the Massachusetts Institute of Technology (Resnick, 1994). The model represents a process of succession and expansion by simulating the locational process of different economic groups in an

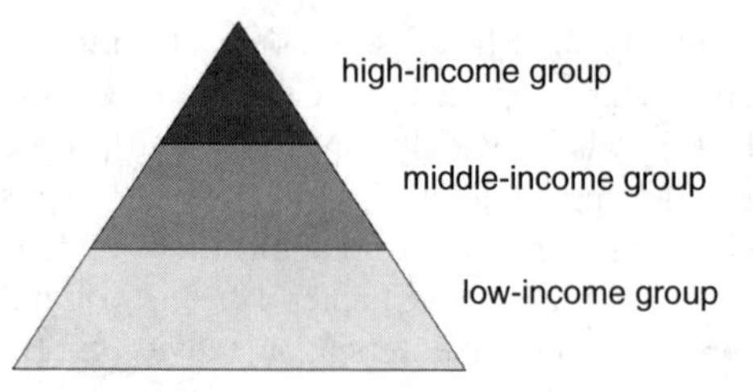

Figure 21.2 Pyramid model of distribution of population into economic groups in Latin America.

attempt to reproduce the residential patterns of these cities. In the model, the population is divided into distinct economic groups according to the pyramidal model of distribution of income in Latin American countries, which suggests that the society is divided into three economic groups (high, medium, and low income) where the high-income group are a minority on the top of the triangle, the middle-income group are the middle part of the triangle, and the low-income group is on the bottom of the triangle (see Figure 21.2).

Our simulation model assumes that, despite the economic differences, all agents have the same locational preferences, that is, they all want to locate close to the areas that are served by infrastructure, with nearby commerce, job opportunities, and so on. As in Third World cities these facilities are found mostly close to the high-income residential areas. In the model's rules, agents look for a place close to a high-income group residential area. What differentiates the behaviour of the three income groups are the restrictions imposed by their economic power. Thus, the high-income group (represented in the model in light grey) is able to locate in any place of its preference. The medium-income group (in white) can locate everywhere except where the high-income group is already located; and, in turn the low-income group (in dark grey) can locate only in otherwise vacant space (see flowchart of agent's behavior rules in Figure 21.3).

In the model there are agents divided into three economic groups in a proportion based on the division of Latin American society by income. All the agents have the same objective, that is, to be as close as possible to the high-income places; but each income group presents a different restriction to the place they can locate. Since some agents can occupy another agent's cell, it means that the latter is "evicted" and must find another place to settle.

The focus of this study is the generic phenomenon of peripherisation and the dynamics that shape this process. The phenomenon is studied here independently of the scale, even though the scale of the simulation can be considered as metropolitan.

In what follows, we will present some of the experiments in which we have tested the behaviour of the model with different parameters. In simulation models, sensitivity analysis is a necessary step towards the evaluation of the model. A sensitivity analysis consists on the study of the relationship between input and output of a model or, in other words, the study of the effects that the change of parameters has on the output which, in this case, is the spatial pattern. This study will allow us to understand the behaviour of the model and identify the role of each parameter, which will permit us to draw informed conclusions about reality from the results of the model and test our hypothesis more effectively.

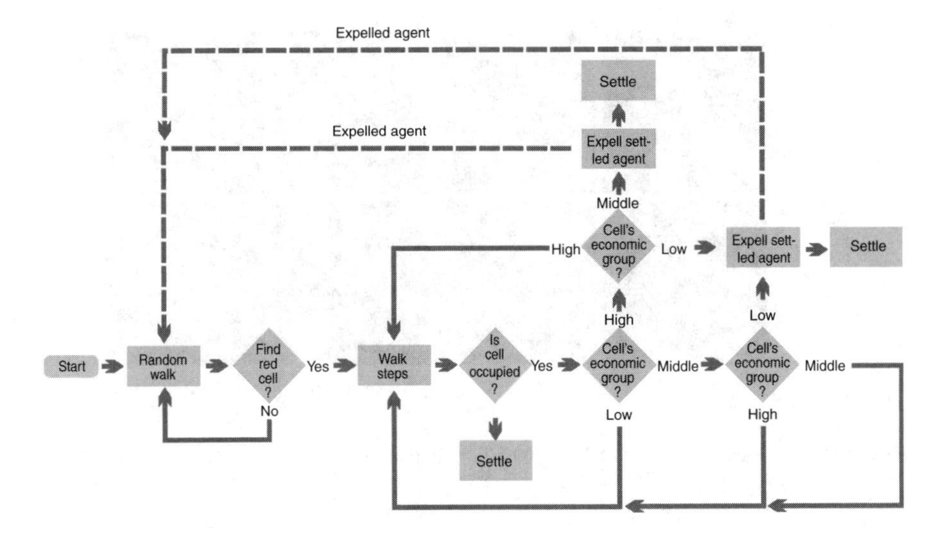

Figure 21.3 Flowchart of the agent's rules.

The results and experiments presented here are of exploratory nature and consist of the first round of tests with the model that will be part of a complete sensitivity analysis. The figures below show representative runs of the tests with parameters, which are analysed qualitatively through observation of the spatial patterns and relating it to the Latin American city spatial pattern described in the literature review. The next step of this work is to carry out a quantitative analysis of these patterns, using landscape and fractal metrics.

Two main parameters define the behaviour of the peripherisation model: "step" and "proportion of agents per income group." Step is the number of pixels that the agent walks before trying to settle in a place (cell). This parameter represents how far people are willing to settle from their ideal location.

In Figure 21.4, we can see the sequence of snapshots of the peripherisation model with different values for the parameter "step." For the experiments below, all parameters were fixed except the one is being tested. The sequence of snapshots were taken with the same initial condition, that is, one seed only on the coordinates (0, 0) and the same proportion of agents per income group, that is, 10% light grey, 40% white, and 50% dark grey. It is interesting to notice that the different values for "step" determine a different spatial development with the same number of timesteps. Each timestep represents an iteration, that is, the time that a agent takes to go through the procedure presented in the flowchart on Figure 21.3. The first set of snapshots (step = 1) presents a more spread spatial development than the second experiment (step = 2), with less homogeneity. Also, it tends to develop more rapidly in space than the first experiment. This is due to the fact that the larger the step, the more empty spaces are left between cells, making the search virtually easier, that is, the agents find an appropriate place to settle faster.

We have also tested the behaviour of the model with different proportions of agents per income group. All of them respect the pyramidal model, which is acknowledged as the distribution of population per economic group in Latin America. It is important to

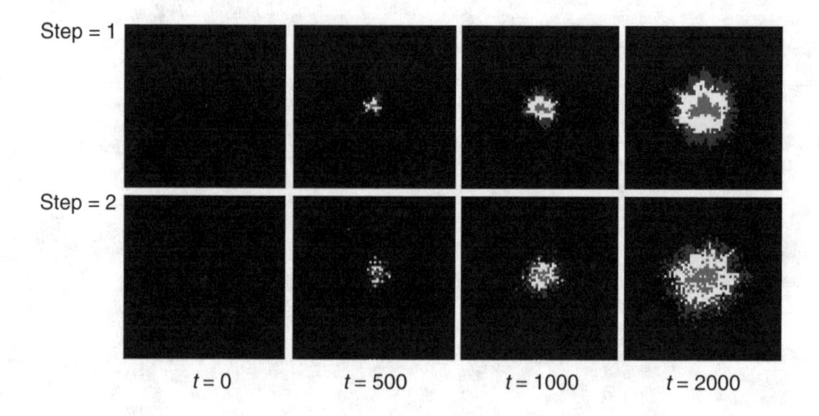

Figure 21.4 Sequences of snapshots from Peripherisation model using different values for parameter "step".

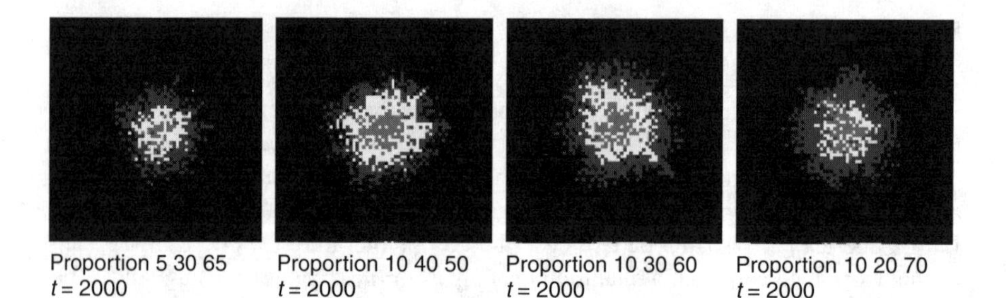

Figure 21.5 Snapshots from peripherisation model using different proportions of agents per income group.

note that the proportion of agents per economic group differs from country to country and even from city to city and that the proportion in the model represents a relative proportion only, as there is no definition of economic group implied in the model. In the following experiment (Figure 21.5), we have tested three proportions 5 30 65 (5% of high-income agents in light grey, 30% of middle income agents in white, and 65% of low-income agents in dark grey, respectively), 10 40 50, and 10 30 60. The number of steps was fixed and is equal to 2 (step = 2).

We can see that the spatial development occurs at different speeds according to the proportion of agents per income group. This means that different income groups develop spatially at different speeds, as can be observed in Figure 21.5, where we can see that the first sequence (proportion 5 30 65) presents much slower spatial development than the other three experiments. This is because there are fewer high-income agents (in dark grey) in this experiment and the high-income agents settle faster than the other two because they can settle anywhere. Also, according to the model's rules, the more high-income cells (light grey) there are, the faster the other two groups of agents will settle.

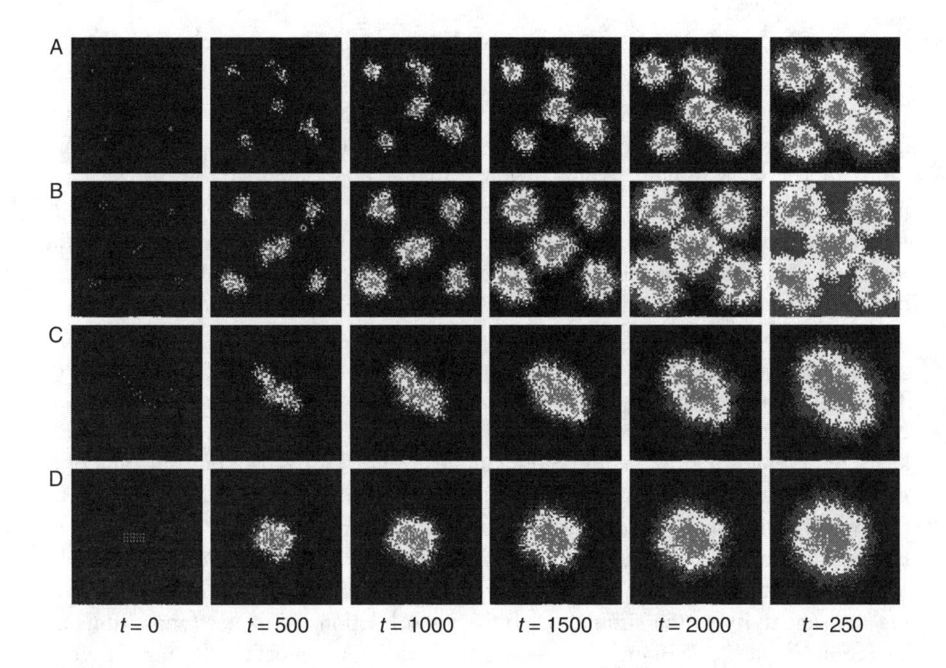

Figure 21.6 Sequences of snapshots from peripherisation model testing different initial conditions: multiple seeds (A and B), path or road (C), and grid of colonial city (D).

We also carried out some tests with different initial conditions to explore the idea of path dependence in the model's behaviour. All experiments present fixed parameters (step = 2 and proportion of agents per income group = 10 40 50). The idea behind these experiments is to verify the behaviour of the model with multiple seeds and how the spatial pattern produced varies according to the initial distribution of these seeds. Figure 21.6 shows some of these experiments. Experiments A and B explored the idea of multiple seeds. This is the case of metropolitan areas, which are the result of the combination of several cities or villages that ended up as a single spatial area because of their spatial proximity and growth. The initial condition of the sequence C uses as initial condition a line of seeds, which represent roughly a path or a road, which is a very common situation in the real world. Sequence D presents as an initial condition an attempt to resemble a grid of colonial Portuguese and Spanish cities, typical in Latin America. It is interesting to note that the spatial development starts with a very mixed structure, and as time passes, the core–periphery structure emerges. These experiments show that the initial conditions do not change the general spatial pattern produced by the model. In Figure 21.6A, B, for instance, we can see how each seed develops into a small centre that merge together, creating a "middle income buffer zone" separating high- and low-income groups, and thus reproducing the spatial pattern of Latin American cities described by Johnston (1973). As in reality, the spatially segregated pattern consolidates in the model and as the simulation runs, the spatial development expands maintaining the core–periphery structure.

21.5 CITY OF SLUMS

City of Slums (Barros and Sobreira, 2002) was built upon the peripherisation model by combining the original peripherisation logic to a consolidation rule. This rule refers to a process in which spontaneous settlements are gradually upgraded, and, as time passes, turn into consolidated favelas or, in other words, spontaneous settlements that are harder to evict. As a result of the introduction of the consolidation logic, the city of slums model generates a more fragmented landscape than the homogeneous concentric-like spatial distribution of classes in which consolidated spontaneous settlements are spread all over the city.

The consolidation process is built into the model through a "cons" variable. This cons variable has its value increased at each iteration of the model and, at a certain threshold ("cons limit"), the low-income cell turns into the consolidation state, represented by the medium grey colour in Figure 21.7. If a high-income or a medium-income agent tries to settle on the low-income cell in a stage previous to the consolidation threshold, the low-income cell is replaced by the respective new occupant's economic group. Otherwise, consolidated cells are "immune" to eviction.

We have run the same kind of experiments using different parameters, also testing the sensitivity of the simulation to the consolidation parameter (cons limit) as shown in Figure 21.7. Figure 21.8 shows two snapshots sequences testing different initial conditions. As can be observed, at the beginning of both simulations there are no consolidated spontaneous settlements in the virtual urban landscape. After some iterations, medium grey cells appear, following a similar distribution of spontaneous settlement in actual cities.

In recent years, a great deal of effort in pure and applied science has been devoted to the study of nontrivial spatial and temporal scaling laws which are robust, that is, independent of the details of particular systems (Batty and Longley, 1994; Bak, 1997; Gomes et al., 1999). Spontaneous settlements tend to follow these scaling laws in both scales, local and global (Sobreira and Gomes, 2001; Sobreira, 2002). This multiscaling order is analysed here by a fragmentation measure which is related to the diversity of sizes of "fragments" (built units) in these systems. The spatial pattern of distribution of spontaneous settlements produced by the model was analysed and then compared with the distribution of spontaneous settlements in real cities by using

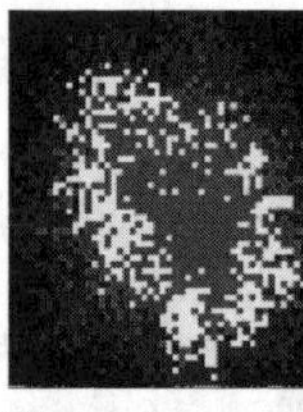

Step = 2
Time = 2000
Conslimit = 4

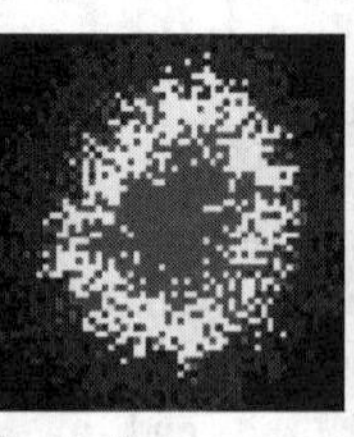

Step = 2
Time = 2000
Conslimit = 5

Step = 1
Time = 2000
Conslimit = 4

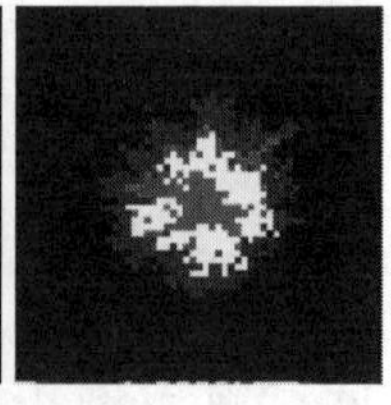

Step = 1
Time = 2000
Conslimit = 5

Figure 21.7 Variations of step and consolidation threshold parameters.

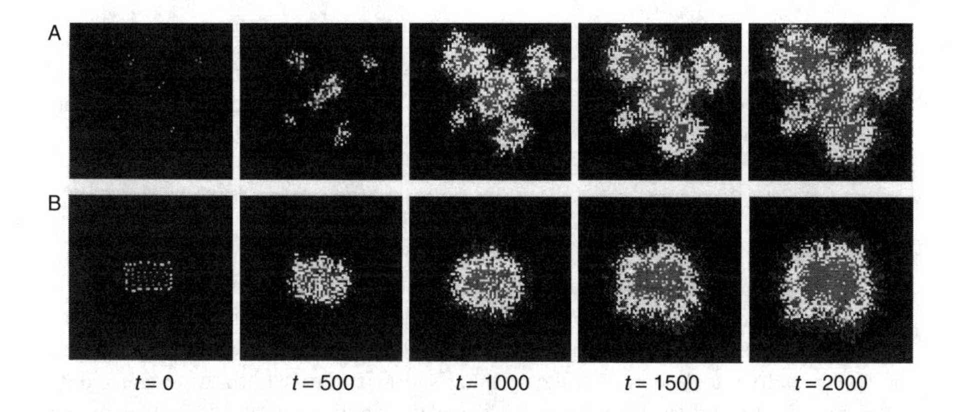

Figure 21.8 Experiment with different initial conditions, polycentric (A) and colonial grid (B).

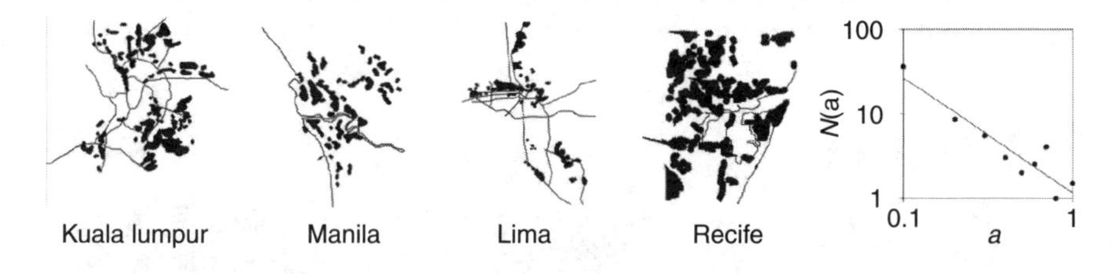

Figure 21.9 Fragmentation pattern of settlements in three Third World cities: Kuala Lumpur, in Malaysia; Manila, in Philippines; and Lima, in Peru.

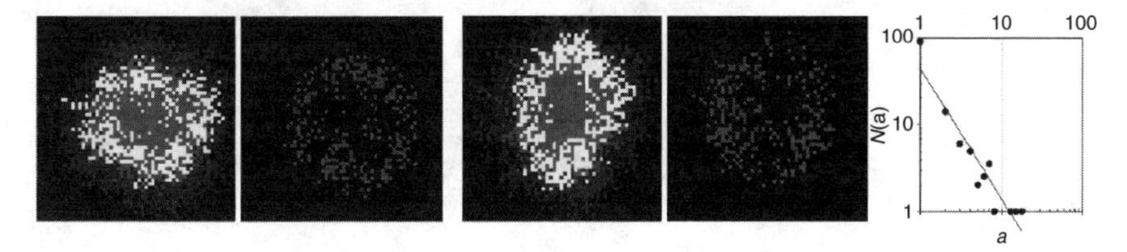

Figure 21.10 **(see color insert following page 168)** Distribution of settlements in the city of slums model with fragmentation graph.

a fragmentation measure (Sobreira, 2002), which is related to the diversity of sizes of "fragments" in systems.

In Figure 21.9, the fragmentation pattern is analysed through the size distribution of settlements in three Third World cities and compared with the size distribution of settlements in the City of Slums simulations in Figure 21.10. In particular, the settlements in each city were grouped according to their area, and the relation between number of settlements ($N(a)$) and respective size interval (a) were plotted in a log–log graph. As one can observe from Figure 21.9, the scaling law which describes the settlement's size distribution in the real cities falls in the same statistical fluctuation of the law which describe the size distribution of the City of Slums simulations. Figure 21.10 describe the same scaling relation $N(a) \sim a - \alpha$, where $\alpha = 1.4 \pm 0.2$.

Both global scale fragmentation patterns (real and simulated) are statistically the same. The negative exponent α indicates a non-linear scaling order, in which there is a great number of small units (settlements), a small number of big units, and a consistent distribution between them (Barros and Sobreira, 2002).

21.6 COMPARISON WITH REALITY

In what follows, we show some simple maps built from the Census 2000 dataset for two Latin American cities: São Paulo and Belo Horizonte. Although these maps are static representations of patterns of income concentration, together with the simulation model they help us to demonstrate the peripherisation pattern in Latin American cities.

Figure 21.11 shows the maps of income distribution in two Brazilian cities: São Paulo and Belo Horizonte. The city of São Paulo has a population of over 10.4 million inhabitants and occupies an area of 1509 km^2, out of which 900 km^2 are urbanized. Its metropolitan region is comprised of 39 autonomous cities with

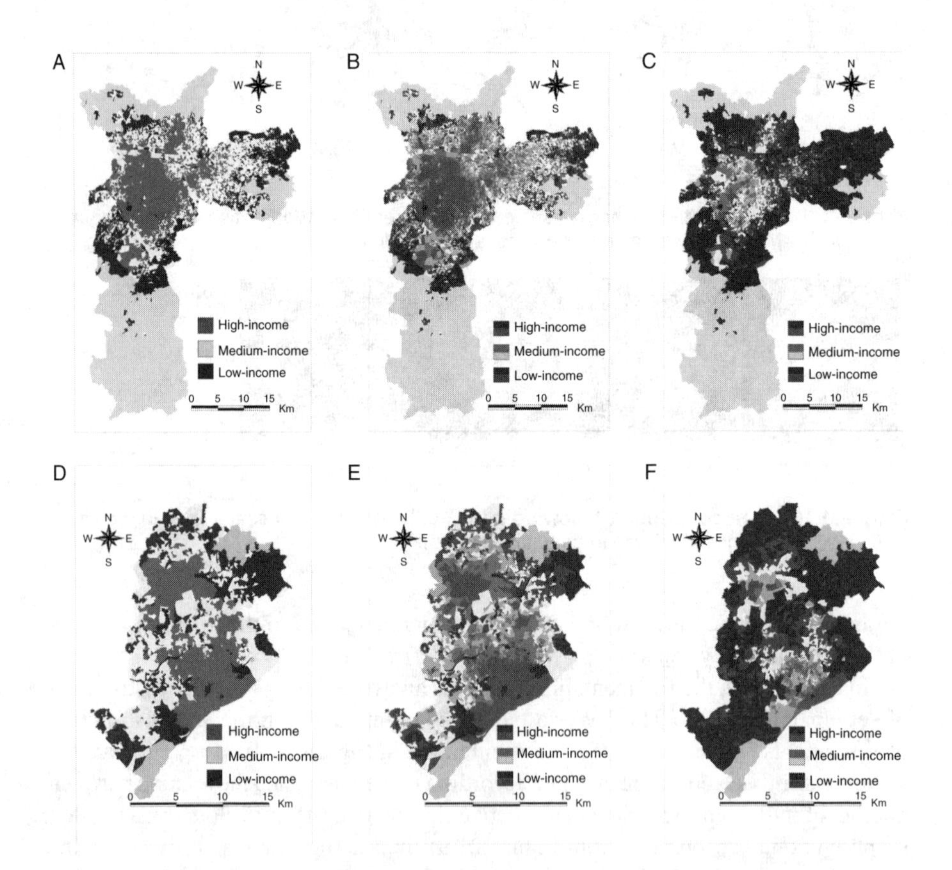

Figure 21.11 **(see color insert following page 168)** Maps of São Paulo (A to C) and Belo Horizonte (D to F) showing distributions of income in the urban area. Maps A, B, D, and E use quantile breaks and maps C and F use natural breaks.

a resident population of more than 17.8 million inhabitants occupying an area of 8501 km^2. Belo Horizonte is today the third largest metropolitan area of Brazil and is comprised of 33 municipalities that have a population of 4.8 million inhabitants over an area of 9179.08 km^2. Belo Horizonte city itself is far more compact than its metropolitan region with a population of 2.2 million inhabitants in area of 330.23 km^2.

The maps show the distribution of income per census sector in the urbanized area of São Paulo (Figure 21.11A to C) and Belo Horizonte (Figure 21.11D to F). The limits of the maps are the administrative boundaries of the cities and, therefore, the metropolitan areas (Great São Paulo and Great Belo Horizonte) are not included in those maps.

The data used here are the average of the head of household monthly income per census sectors (enumeration district/census block) and is part of the Census 2000 dataset provided by the Brazilian Institute of Geography and Statistics (IBGE). This variable was chosen because of its similarities to the rules of the Peripherisation Model, which is based on the division of agents into economic groups.

For the maps of São Paulo city we used a total of 12,428 census sectors out of the 13,278 sectors available and for Belo Horizonte's maps we used a total of 2,549 sectors out of 2,564. Each census sector contains an average of 250 households or 750 inhabitants. The remaining sectors represent either nonurbanised areas or sectors without available information for the population variable.

The aggregated data per urban census sector were normalised by the number of householders in each sector and then classified into three ranges (Figure 21.11A, D) or six ranges (Figure 21.11B, C, E, F). The maps use medium grey for the higher income groups, white for middle-income groups, and dark grey for the lower-income groups, as in the simulation model to aid comparison. As in the images produced by the model, we can easily identify a concentric pattern in Figure 21.11A, D, in which the high-income groups are concentrated towards the centre of the urban area and the concentration of high-income groups decreases towards the urban periphery. The graduation is more easily observed in Figure 21.11B, E where the same data were graduated into six classes, showing a decrease in the income distribution towards the edge of the city.

Figure 21.11C, F show a different classification of income, where the number of dark grey areas is fewer in comparison to the two other maps. In these maps, the classification was done according to the natural groupings of the data (income) values, and, what the map actually shows is that there are very few people that belong to a high-income group and a lot of people that belong to the low-income group. It should be noted, however, that we have not used established definitions of income groups either in the simulation model or in the maps shown above, and our focus is on the relative locational pattern of these groups within the city only. As such, the actual number in each income group is not relevant for the present study.

Of course, the spatial pattern in the maps is not as concentric as the patterns produced by the simulation model. This is due to various aspects such as initial conditions, topography, the presence of water bodies, etc. In particular, the topography of these areas has strong influences on the spatial development of these cities. Belo Horizonte, for example, was built in a valley, which constrained its development to the east that forced its development to the north and south.

It is also important to mention that the maps shown here do not encompass the metropolitan area of those two cities, but they are restricted to their administrative boundaries. This means that São Paulo and Belo Horizonte are actually part of polycentric urban areas like the ones showed in Figure 21.6A, B (on Section 21.4) and, therefore, the analysis of their urban form is restricted.

21.7 CONCLUSIONS AND FUTURE WORK

Latin American cities present a specific mode of urban growth, which differs in many aspects from urban growth in Western countries. Yet, most theories about urban growth, morphological patterns, and inner-city change have treated the city as a generic entity and, therefore, the knowledge on the specificities of the dynamics of different kind of cities has not been greatly explored to date.

The chapter has presented two simulation experiments exploring different features of the urban growth phenomenon in Latin American cities and then related these experiments with examples from reality through maps of Latin American cities and fragmentation measurements.

It seems clear that the actual development process of Latin American cities consists of socio-economic inequality that is reproduced in space by the locational process. The peripherisation process was caused initially by the high rates of urban growth in these countries but it is now consolidated as the normal process of development of these cities. The result is an emergent pattern of spatial segregation characterised by stark differences between core and periphery, which is consolidated as the residential spatial pattern of Latin American cities. The perpetuations of both process and spatial pattern reinforce the social inequality, which was their cause in the first place, working as a vicious cycle.

The dynamic modelling exercises presented in this chapter allowed us to develop an understanding of the rapid urbanisation process and investigate its dynamics, changing our perspective of the problem from a demographic viewpoint to a dynamic and morphological one. This research has taken a step in the direction of bringing new perspective to an old problem. However, the peripherisation model is at a very preliminary stage of development and is still a rough simplification of the phenomenon. It is possible to refine the model and encompass a more complex behaviour as well as spatial constraints.

Thus, the next steps consist on the further development of the model, introducing spatial constraints such as slopes and water bodies and adding inner-city processes such as filtering and movement of high-income groups to the outskirts. Then, a systematic sensitivity analysis must be carried out, together with the use of landscape and fractal metrics of the generated spatial pattern. These measures will also allow us to compare results of the model with the development of actual Latin American cities spatial patterns using census data.

The need for an increased understanding of urban spatial phenomenon in cities of the Third World is essential in order to provide a basis for future planning actions and policies. The approach outlined in this study is part of an ongoing research programme and will be further developed. Nevertheless, we believe it provides evidence that urban modelling tools together with GIS and fine scale data can provide an appropriate basis for research on Latin American urban process. We also believe it is necessary to approach the problem relating morphology and dynamics, for which dynamic modelling provide the appropriate means.

ACKNOWLEDGEMENTS

This research is supported by the Brazilian Government Agency CAPES. The author would like to thank Sinesio Alves Jr. and Fabiano Sobreira for their contribution in the work presented in this chapter.

REFERENCES

Agarwal, P., Chapter 24, this volume.

Almeida, C.M., Monteiro, A.M.V., Camara, G., Soares-Filho, B.S., Cerqueira, G.C., Pennachin, C.L., and Batty, M., 2002, Empiricism and stochastics in cellular automaton modeling of urban land use dynamics. *CASA Working Papers*, available on-line at http://www.casa.ucl.ac.uk/working_papers/Paper42.pdf.

Amato, P., 1970, A comparison: population densities, land values and socioeconomic class in four Latin American cities. *Land Economics*, **46**, 447–455.

Bak, P., 1997, *How Nature Works: The Science of Self-Organized Criticality* (Oxford: Oxford University Press).

Barros, J. and Alves Jr., S., 2003, *Simulating Rapid Urbanisation in Latin American Cities* (London: ESRI Press).

Barros, J. and Sobreira, F., 2002, City of slums: self-organisation across scales. *CASA Working Papers Series*, available on-line at http://www.casa.ucl.ac.uk/working_papers/Paper55.pdf.

Batty, M., Desyllas, J., and Duxbury, E., 2002, The discrete dynamics of small-scale spatial events: agent-based models of mobility in carnivals and street parades. *CASA Working Papers*, available on-line at http://www.casa.ucl.ac.uk/working_papers/Paper56.pdf.

Batty, M., 1998, Urban evolution on the desktop: simulation using extended cellular automata. *Environment and Planning A*, **30**, 1943–1967.

Batty, M. and Longley, P., 1994, *Fractal Cities: A Geometry of Form and Function* (London: Academic Press).

Benenson, I. and Torrens, P., Chapter 23, this volume.

Couclelis, H., 2001, *Why I No Longer Work with Agents* (Louvain-la-Neuve: LUCC Focus 1 Office).

Gomes, M., Garcia, J., Jyh, T., Rent, T., and Sales, T., 1999, Diversity and Complexity: Two sides of the same coin? *The Evolution of Complexity*, **8**, 117–123.

Hall, P., 1983, *Decentralization Without End? A Re-Evaluation* (London: Academic Press).

Holland, J.H., 1995, *Hidden Order: How Adaptation Builds Complexity* (Reading, MA: Addison-Wesley).

Johnston, R.J., 1973, Towards a general model of intra-urban residential patterns: some cross-cultural observations. *Progress in Geography*, **4**, 84–124.

Parker, D., Berger, T., and Manson, S.M., 2001, *Agent-Based Models of Land-Use and Land-Cover Change* (Irvine, CA: LUCC).

Portugali, J., 2000, *Self-Organization and the City* (Berlin: Springer).

Resnick, M., 1994, *Turtles, Termites, and Traffic Jams: Explorations in Massively Parallel Microworlds* (Cambridge, MA: MIT Press).

Schelhorn, T., O'Sullivan, D., and Thurstain-Goodwin, M., 1999, STREETS: an agent-based pedestrian model. *CASA Working Paper Series*, available on-line at http://www.casa.ucl.ac.uk/working_papers.htm.

Sobreira, F., 2002. *The Logic of Diversity: Complexity and Dynamic in Spontaneous Settlements* (Recife: Federal University of Pernambuco).

Sobreira, F. and Gomes, M., 2001, The geometry of slums: boundaries, packing and diversity. *CASA Working Paper Series, 30*, available on-line at www.casa.ucl.ac.uk.

UNCHS, 1982, *Survey of Slum and Squatter Settlements* (Dublin: Published for UNCHS (Habitat) by Tycooly International Pub.).

UNCHS, 1996, *An Urbanizing World: Global Report on Human Settlements* (Oxford: Oxford University Press for the United Nations Centre for Human Settlements (Habitat)).

UNCHS, 2003, *The Challenge of Slums: Global Report on Human Settlements* (London: Earthscan).

Valladares, L. and Coelho, M.P., 1995, *Urban Research in Brazil and Venezuela: Towards an Agenda for the 1990s* (Toronto: University of Toronto).

White, R. and Engelen, G., 1997, Cellular automata as the basis of integrated dynamic regional modelling. *Environment and Planning B: Planning and Design*, **24**, 235–246.

Xie, Y. and Batty, M., Chapter 19, this volume.

CHAPTER 22

Error Propagation and Model Uncertainties of Cellular Automata in Urban Simulation with GIS

Anthony Gar-On Yeh and Xia Li

CONTENTS

0-8493-2837-3/05/$0.00+$1.50

22.1 INTRODUCTION

GIS databases are approximations of real geographical variation with very limited exceptions (Goodchild et al., 1992). Understanding of errors and uncertainties of GIS is needed in the implementation of GIS techniques in most situations. There are two main types of GIS errors: (a) data source errors that exist in GIS databases; and (b) error propagation through the operation performed on the data by using GIS functions.

Recently, a series of urban Cellular Automata (CA) have been developed for modeling complex urban systems with the integration of GIS (Batty and Xie, 1994; Wu and Webster, 1998; Li and Yeh, 2000). The application of CA in urban modeling can give insights into a wide variety of urban phenomena. Urban CA have better performance in simulating urban development than conventional urban models that use mathematical equations (Batty and Xie, 1994). Urban CA have much simpler forms, but produce more meaningful and useful results than mathematical-based models. Temporal and spatial complexities of urban development can be well simulated by properly defining transition rules in CA models. CA is capable of providing important information for understanding urban theories, such as the emergence and evolution of forms and structures. They are also used as planning models for plan formulation.

Although many studies have been reported in urban simulation, the errors and uncertainties of CA have not attracted much attention so far. This issue should be important because a huge volume of geographical data is usually used in urban simulation, especially in modeling real cities. Spatial variables are usually retrieved from GIS and imported to CA modeling processes. It is well known that most GIS data are affected by a series of errors. Like many GIS models, urban CA simulation is not without problems because of the inherent data errors and model uncertainties. These errors will propagate in CA simulation and affect the simulation outcomes. This requires the evaluation of the influences of source errors and error propagation on simulation results. Although there are many studies on error types and error propagation in GIS, very little research has been carried out to examine the issue in CA simulation. This chapter attempts to examine the influences of errors and uncertainties on urban simulation by carrying out some experiments. This can help urban planers to understand more clearly the meanings and implications of simulation outcomes.

22.2 ERRORS AND UNCERTAINTIES IN URBAN CA

Spatial modeling with GIS is an important topic in researches and applications in geography. It uses GIS powerful functions to model a variety of resource and environmental problems. In recent years, a class of dynamic spatial modeling has been developing rapidly by the integration of CA with GIS. CA is dynamic spatial models which have powerful capabilities in modeling complex systems in physics, chemistry, biology, and geography. Particularly, CA and GIS have been used to simulate urban systems for testing urban theories (Wu and Webster, 1998; Webster and Wu, 1999) and formulating development plans for urban planning (Yeh and Li, 2001, 2002).

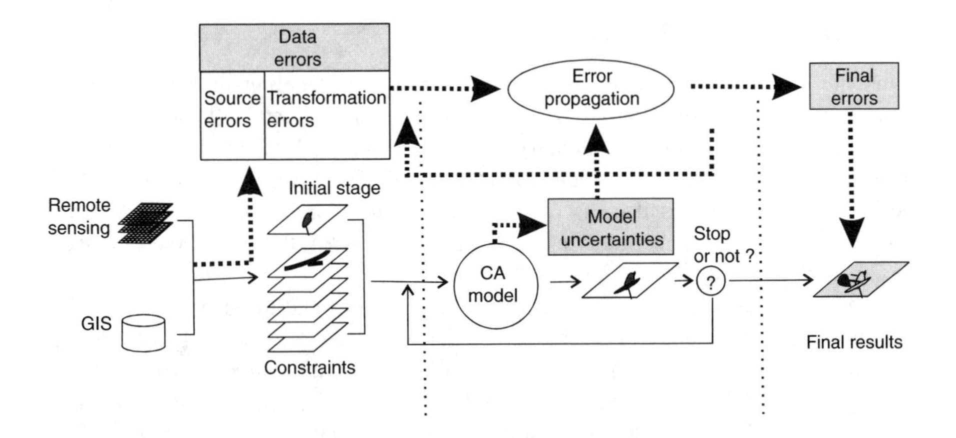

Figure 22.1 Data errors, model uncertainties, and error propagation in cellular automata.

A major concern for urban CA models is their errors and uncertainties if they are applied to real cities. CA models for geography and urban simulation are significantly different from Wolfram's deterministic CA models (Wolfram, 1984). Wolfram's models have strict definitions and use very limited data. This allows CA models to produce stable outputs without any error and uncertainty. However, urban CA models usually need to input a large set of spatial data for realistic simulation. The outcome of CA models will be affected by a series of errors and uncertainties from data sources and model structures (Figure 22.1).

22.2.1 Errors from Data Sources

When spatial data from GIS are used in urban CA, two major types of source errors can be identified:

22.2.1.1 Positional Errors

Positional errors in GIS can affect the accuracy of urban simulation. Such errors can cause mistakes in estimating conversion probability, which is related to proximity variables. Positional accuracy has been widely discussed in many GIS studies (Goodchild, 1991; Veregin, 1999). The positional errors for points can be measured by the discrepancy between the actual location and recoded location. Euclidean distance can be used for the measurement. The spatial error for a set of points has been commonly represented by root mean squared error (RMSE), which is computed as the square root of the mean of the squared errors. It is calculated in both x- and y-directions.

The position errors for lines can be represented using some variant of the epsilon band (Veregin, 1999). There is a certain probability of observing the "actual" line within the band. The simplest one is to assume the band and the distribution are uniform. However, recent studies show that both the band and distribution might be non-uniform (Caspary and Scheuring, 1993; Veregin, 1999).

22.2.1.2 Attribute Errors

Attribute errors convey that something is wrong for labeling at each location. Conventional surveying maps have errors that are associated with human errors (e.g., reading errors) and instrumental errors (e.g., unstable conditions). For example, a site labeled as vegetables on a map may turn out to be grass on the ground. A DEM derived from contours is also susceptible to the errors of interpolation. These errors also contribute to the uncertainty of urban simulation when these attributes are used in CA models.

There are concerns on how data source errors contribute to uncertainties in the results of GIS operations and computational models. Yet, it is only relatively recently that much attention has been paid to the problems of data source errors and error propagation in GIS (Heuvelink et al., 1989). There are attempts to use quantitative models to determine the magnitude of error propagation in GIS. For example, Heuvelink et al. (1989) present detailed methods to derive error propagation equations using Taylor series. The advantages of quantitative models are that they are able to yield analytical expressions of error propagation and the computation is not intensive. Another method to analyze error propagation is Monte Carlo simulation, which has been widely used in many applications. The advantages include easy implementation and generally applicability, but the disadvantage is the lack of an analytical framework.

Unlike Wolfram's traditional CA models, urban CA models are usually implemented in heterogeneous space by importing spatial information from GIS. The simulation of real cities needs to use many spatial data and the simulation results are very sensitive to data errors. Spatial data quality should be a major concern in urban simulation. There are a number of errors that may be present in source data. Although GIS data are stored in digital formats, they are not error-free. GIS databases are often created by means of digitizing paper maps. Errors will be introduced during the transfer of the source map to the digital database. Most GIS databases do not have information about the error of the source maps. When no accuracy information is kept, it becomes extremely difficult to evaluate the true accuracy of the final results of modeling (Heuvelink, 1998).

22.2.2 Transformation or Operation Errors

Besides data source errors, there are also new errors from common GIS transformations. Some standard GIS operations have to be carried out to generate additional or specific information that is not already stored in GIS as inputs to CA models. GIS databases only contain basic data, such as land use, terrain features, and transport information. These data can be used to derive user-specific information that is unavailable in the original data sets. For example, proximity variables can be calculated by applying GIS operations to these data. These operations may include:

- Vector–raster transformation
- Raster–raster transformation (e.g., resampling)
- Overlay or buffer operations
- Other complex operations (e.g., classification).

There are two major types of GIS data formats — vector and raster. The conversion between vector and raster format is a common task in GIS operations. Urban CA models are implemented using a raster format — cells. Therefore, the inputs of GIS data to CA models should be prepared in raster format. Vector data have to be converted into raster data first before spatial data can be handled by most of urban CA. It is apparent that the conversion of vector data into raster data will result in a loss of spatial detail.

Even for raster data, raster–raster transformation is required for two purposes — registration of different layers of data and conversion of data from one spatial resolution to another. Registration of different sources of raster data is an important procedure for using geographical data. Geo-referencing of maps is usually done by using affine transformation or polynominal transformation. The transformation will resample data by using the method of nearest neighbor, bilinear interpolation, or cubic convolution. It is possible that new errors may be created by the mistakes of registration or resampling. Conversion of data by changing cell size can allow them to be comparable. However, when raster data are converted from a higher spatial resolution to a lower spatial resolution, there is a loss of information.

The transformation related to GIS overlay can be implemented by "cartographic algebra" (Burrough, 1986). Sometimes, multicriteria evaluation (MCE) may be required when a number of spatial factors are involved in urban simulation (Wu and Webster, 1998). These operations can generate new errors during the process of data handling. GIS operations are in effect a computational model which is merely an approximation to reality (Heuvelink, 1998). Model errors can be introduced in GIS database when such operations are carried out.

Environmental factors or constraints are usually incorporated in an urban CA. This type of information is obtained by using ordinary GIS operations, such as overlay analysis or transformation. For example, constrained CA models may be developed to simulate planned urban development (Li and Yeh, 2000). The purpose is to prohibit uncontrolled urban development according to the constraint information provided by a GIS. A series of resource and environmental factors can be retrieved from the GIS database and imported to CA models as site attributes. These factors may include topography, land use types, proximity and agricultural productivity (Li and Yeh, 2000). Constraint scores can be calculated by using GIS linear or non-linear transformation functions. However, there are uncertainties in defining the forms of transformation functions.

Errors can also be produced during proximity analysis or buffer analysis in GIS. In urban simulation, a common procedure is to calculate urban development probability. Urban development probability decides whether land development can take place during the simulation process. Urban development probability is estimated based on the attractiveness for urban development. It is more attractive for urban development if a site has greater accessibility to major transport networks or facilities. Some distance variables are used to represent the attractiveness, including various distances to roads, railways, town centers, hospitals, and schools. These variables can be calculated conveniently from GIS layers of corresponding points and lines. A major problem is that there may be positional errors in representing points and lines in GIS layers. These errors can originate from human errors (e.g., mis-registration) or model errors (e.g., limitations of pixel size). These positional errors can cause uncertainties in urban simulation.

Other operations on spatial data can also bring about uncertainties. An example is that attribute errors may come from the classification of remote sensing data. Remote sensing classification is mainly based on spectral characteristics. Sensors' noises, atmospheric disturbance, and limitations of classification algorithms are all liable for classification errors. For example, some pixels may be misclassified for their land use types by employing classification techniques to remote sensing data. These errors can generally be measured by comparing ground data with classification results. A confusion matrix can be constructed to indicate the percentages of correctly and wrongly classified points.

The existence of mixed pixels also leads to the uncertainty in remote sensing classification. It is well known that remote sensing and other raster data are subject to the errors caused by spatial resolution limitations. Remote sensing images are made up of pixels. Each pixel corresponds to a basic sampling unit which records ground information. Conventional remote sensing classification assumes the following conditions (Fisher and Pathirana, 1990):

- All pixels are purely occupied by a land use type
- Any one single pixel is just entitled to one land use type
- Different land use types should generate a distinct signature.

In reality, these assumptions are not realistic because of the existence of mixed pixels. A mixed pixel indicates that there is more than one type of land use occupying a single pixel. General methods may have errors in classifying mixed pixels. There are uncertainties when these data are stored in GIS and further used for urban simulation. For example, initial urban areas for urban simulation may be obtained from classification of remote sensing images. Classification errors can significantly influence the simulation of urban growth because the errors can propagate through the simulation process.

22.2.3 Model Uncertainties in Urban CA Modeling

The error problems of CA models are further exacerbated by taking into account model uncertainties. There are other types of errors that are not produced physically during the process of data capture. These errors come from models themselves due to limited human knowledge, complexity of nature, and limitation of technology. In CA simulation, not only do input errors propagate through the simulation process, but model errors as well. Like any computer models, CA models could disagree with reality even when the inputs were completely error-free. CA models are only an approximation to reality. Most of the existing CA models are just loosely defined and a unique model does not exist. Various types of CA models have been proposed according to individuals' perception and preference, and requirements of specific applications. The simulation results are hard to repeat when different CA models are used.

A series of inherent model errors can be identified for CA models. They are related to the following aspects:

- Discrete entities in space and time
- Neighborhood definitions (types and sizes)
- Model structures and transition rules
- Parameter values
- Stochastic variables.

22.3 EVALUATION OF ERRORS AND UNCERTAINTIES OF URBAN CA

22.3.1 Error Propagation in CA Modeling

Assessment of error propagation in CA modeling is important for understanding the results of simulation. In urban simulation, initial conditions, parameter values, and stochastic factors play important roles in influencing simulation results. Unexpected features can emerge during CA simulation because of the interplay of various local actions. CA simulation may become meaningless if the behavior of the automation is completely unstable and unrepeatable. Fortunately, it is found that CA simulation can produce stable results at the macroscopic level (Benati, 1997). The general shape of CA simulation remains the same, although the configuration may be changed. However, the behaviors of CA simulation are unpredictable to a certain extent at the microscopic level.

Error and uncertainty can propagate through the modeling process. There are many studies to show how such errors propagate in GIS manipulation, such as the common overlay operation (Veregin, 1994). The original errors may be amplified or reduced in the modeling process. All the errors inherent in individual GIS layers can contribute to the final errors of the output during the overlay of these layers.

Error propagation in CA models is different from that of GIS overlay operations. In GIS operations, mathematical expressions can be given to calculate the errors presented in simple overlay using the logical *AND* and *OR* operators. CA models adopt relatively complicated configurations by using neighborhood and iterations. The simulation is a dynamic process in which very complex features can arise according to transition rules. The transition of states is influenced by the states in a neighborhood. It is almost impossible to develop strict mathematical equations for the error propagation in the dynamic process. It can be seen from Figure 22.1 that error propagation in CA models is quite complicated because of the use of dynamic looping.

A convenient way to examine error propagation in urban simulation is to perturb spatial variables and assess the error terms in the outcome of simulation. Sensitivity analysis has been used to establish the effects of error in a database on analytical outcomes in general GIS analysis (Lodwick, 1989; Fisher, 1991). Monte Carlo simulation is often used to sensitize spatial data, and then the sensitized data are used to determine the accuracy of outcomes. Fisher (1991) has presented two algorithms to perturb categorically mapped data, as exemplified by soil map data, and to assess the error propagation.

The Monte Carlo method seems to be most suitable for the study of error propagation in CA simulation. Standard error propagation theory cannot be used in some models which involve complicated operations (Heuvelink and Burrough, 1993). The Monte Carlo method is a convenient way to study error propagation when mathematical models are difficult to define. Although the Monte Carlo method is very computationally intensive, increasingly this is less problematic because of the advancement of computer technology. When the Monte Carlo method is used, perturbations will be inserted in spatial variables so that the sensitivities of the perturbations in urban simulation can be examined. The Monte Carlo simulation should have more advantages because explicit mathematical equations cannot be built for urban CA models.

A simple realization of noise is to use the uncontrolled perturbation, which assumes no knowledge about the errors. The perturbation can be carried out to simulate attribute errors for the following spatial data that are used as the main inputs to urban CA models: (a) land use types; (b) initial urban areas; (c) suitability analysis.

The Monte Carlo method can be used to assess the influences of attribute errors of cells on the simulation results. Urban simulation is based on the attributes of each cell. These attributes may or may not be correct due to data source errors. Initial land use types for urban simulation are usually obtained from the classification of remote sensing data. Classification errors have effects on the final outcome of urban simulation. For example, some of the initial urban areas may be wrongly located because of misclassification. This means that the original states have errors and the errors can be propagated in urban simulation. The final results can be influenced by the errors inherent in the data sources. Since CA simulation is based on neighborhood functions, the functions should control the process of error propagation. The error propagation in CA should be examined to ensure the appropriate application of CA in urban simulation.

The following experiment was to evaluate the impacts of the attribute errors on the simulation results. The initial images had two major types of land use — urban areas and non-urban areas. It is expected that the initial image may be subject to classification errors for these two land use types. There is only some general information about the classification errors in most situations. The accuracy of land use classification from satellite remote sensing usually falls within the range of 80 to 90% (Li and Yeh, 1998). However, the detailed locations of classification errors are not available in most situations.

The first step of the experiment was to perturb the classified satellite remote sensing images with some errors. 20% errors were randomly generated in the classified remote sensing images since there is no prior knowledge about the spatial locations of the errors. Then, a very simple urban CA was used to examine the error propagation. The use of too complicated CA cannot isolate the effects of model uncertainties. The model is based on the following rule-based structure (Batty, 1997):

IF any cell $\{x \pm l,\ y \pm l\}$ is already developed

THEN $N\{x,y\} = \sum_{ij \in \Omega} D\{i,j\}$ ($D\{i,j\} = 1$ for a developed cell, Otherwise $D\{i,j\} = 0$)

&

IF $N\{x,y\} > T_1$ and $R > T_2$

THEN The cell$\{x,y\}$ is developed

where $N\{x, y\}$ is the total number of the developed cell in the neighborhood, T_1 and T_2 are the threshold values, R is a random variable, and cell $\{i, j\}$ are all the cells which from the Moore neighborhood Ω including the cell $\{x, y\}$ itself.

An experiment, which examines error propagation during urban simulation, was carried out in Dongguan in the Pearl River Delta. It simulated the land development in 1988 to 1993 when rapid urban expansion took place in the region. In this study, the parameter l for the neighborhood size was set to 3. The threshold values of T_1 and T_2 determine how many cells can be converted into urban areas at each time step (iteration). Lower values of T_1 and T_2 allow a larger number of cells to be developed. If the objective is to use the same amount of land consumption, lower values of T_1 and T_2 will require fewer time steps to finish the simulation. Therefore, the values of T_1 and T_2 can be defined according to the amount of land consumption and the time steps to complete the simulation.

The experiment was to examine the influences of source errors on simulation outcomes. The values of T_1 and T_2 were set to 10 and 0.90, respectively. The model was used to simulate 23,330.5 ha. of land development in Dongguan in 1988 to 1993. The land area of Dongguan is 2,465 km^2. The model used a grid of 709 $\times$ 891 cells with a resolution of 50 m^2 on the ground for the simulation. The model runs two times — one with the input of original initial urban areas, and the other with the input perturbed by 20% errors. The baseline test was to simulate urban growth using the initial urban areas without error perturbation. The simulation was compared with that of 20% errors perturbed to the initial urban areas. The errors were computed by comparing the simulated results with the actual land development obtained from remote sensing.

Figure 22.2 shows the error propagation during the simulation. The simulation without error perturbation also has errors for the simulation results. Higher simulation

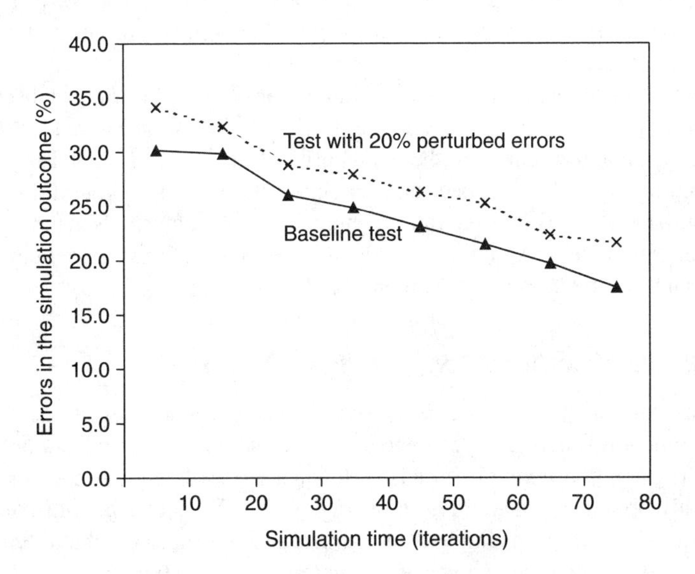

Figure 22.2 Error propagation of CA with 20% error perturbed to original initial urban areas.

errors were obtained when the initial urban areas were perturbed with 20% errors at original land use types. However, the increased errors are much less than 20%. The increased errors only amount to about 5%. This means that the perturbed errors (20%) was significantly reduced during the simulation process. It is because CA adopt neighborhood functions which have averaging effects to reduce the errors. The analysis also indicates that all the errors will be reduced with time. This is because land available for development will be reduced as the urban areas grow. The simulation will then be subject to more constraints which minimize the chance of producing errors.

22.3.2 Model Uncertainties in CA Modeling

22.3.2.1 Discrete Space and Time

CA models are implemented using discrete space and time. Cells, which are in the form of discrete space, are the basic unit of CA models. However, discrete cells are only the approximation to the continuous space with loss of spatial detail. There are questions on how to choose proper cell size and cell shape. A large cell size may be preferable for reducing data volume, but it may reduce spatial accuracy. Uniform cells are commonly used because they are simple for calculation. However, irregular cells may be more suitable under particular circumstances (O'Sullivan, 2001). An example is to use irregular cells to represent land parcels or planning units.

CA models adopt discrete time (iterations) to represent actual time in simulation. There are problems on how to decide the interval of discrete time (the total number of iterations). The larger the interval of the discrete time is, the smaller the number of iterations becomes. The discrete simulation time of CA is different from the continuous real time. The outcome of simulation from 100 iterations is not the same as that from 10 iterations. There is a need to assess the influences of discrete time on CA simulation. Temporal errors can be introduced in CA because of the use of approximate discrete time.

Figure 22.3 is the experiment results which clearly show the effect of discrete time on urban simulation. Figure 22.3A uses only 10 iterations to generate the simulation result. It is much different from the actual urban form that is obtained from remote sensing in Figure 22.3D. It is because local interactions are important for generating realistic urban forms. Too few iterations cannot allow spatial details to emerge during the simulation process. An increase in the number of iterations can help to generate more accurate simulation results (Figure 22.3B, C).

22.3.2.2 Neighborhood Configuration

The most important notion of CA is the so-called neighborhood function. The neighborhood function is defined to estimate the conversion probability from one state to another. The neighborhood function is based on a series of neighborhood operations. It is usually obtained by summing or averaging the values of input attributes at the neighborhood. A simple example is to estimate the conversion probability based on the summation of the total number of a state (e.g., land use type) in a 3×3 window.

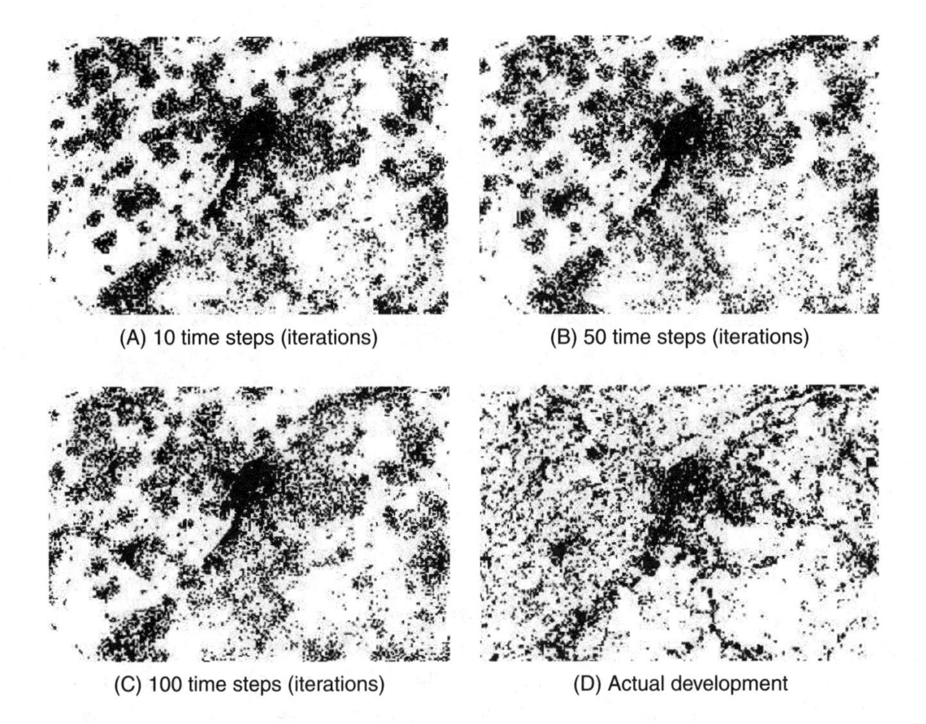

Figure 22.3 The influences of discrete time steps on simulation accuracies.

It is easy to perceive that the original data errors will be reduced for a large size of neighborhood. However, this will also reduce spatial detail due to the averaging effects.

Neighborhood configuration affects the results of CA simulation. There are two common types of neighborhoods — von Neumann and Moore neighborhood. A way to examine neighborhood effects is to see how cities grow under different neighborhood influences. The Moore neighborhood will lead to exponential urban growth, which is different from actual growth patterns. The von Neumann neighborhood can be used to reduce the growth rate. However, the two neighborhoods are generally in a rectangle form which has side effects in urban simulation.

22.3.2.3 Model Structures and Transition Rules

It is expected that different model forms will have impacts on the outcome of CA simulation. A variety of urban CA models have been proposed to tackle specific problems in urban simulation. Model variations are usually related to individual preferences and the requirements of applications. It is essential to define transition rules, which are the core of CA models. However, it is quite *ad hoc* to determine transition rules because there is no unique way to do so. Substantially different methods have been proposed for defining transition rules. They include:

- Simulating the births, survivors, and death of cells with the notion of the game of Life (Batty and Xie, 1994).

- Estimating development probability based on the analytical hierarchy process (AHP) of multicriteria evaluation (Wu and Webster, 1998).
- Defining transition rules with fuzzy sets (Wu, 1999).
- Calculating transition potentials using a predefined parameter matrix (White and Engelen, 1993).
- Simulating land use changes using five controlling factors — *Diffusion, Breed, Spread, Slope, Road* (Clarke et al., 1997).
- Simulating urban conversion using "gray-values" (Li and Yeh, 2000).
- Incorporating planning objectives in urban simulation (Yeh and Li, 2001).
- Calibrating and simulating urban development with neural networks (Li and Yeh, 2001).

Development probability is the function of a series of spatial variables. These spatial variables are usually measured using GIS tools. There is also a controversial issue on how to choose variables. When a series of variables are present, there is difficulty to judge which variable should be selected or removed from CA models. The selection of variables is a matter of experience. The use of more or less number of variables will affect the outcome of CA simulation. Moreover, the ways to measure and standardize these variables will also affect simulation results. An example is to obtain proximity variables using GIS functions. GIS functions are used to calculate the influences of a source (center). For example, a closer distance to a utility (market center) will have a higher score of attractiveness for urban development. The attractiveness of a center will decrease as the distance increases. It is straightforward to use the Euclidean distance to indicate the influences of centers. However, a transformed form (e.g., a negative exponential index) may be more appropriate to represent the actual influences of centers. It can more appropriately represent the situation that the influences from centers do not decrease in a linear form as distance increases. The problem is that there are uncertainties in defining parameter values for the negative exponential function.

22.3.2.4 Parameter Values

CA model errors are also introduced by mistakes in assigning parameter values. It is a debatable issue on how to define parameter values. CA models need to use many spatial variables and thus many parameters. For example, White et al. (1997) present a CA model to simulate urban dynamics. Their models need to determine as many as $21 \times 18 = 378$ parameter values. Parameter values should be defined before CA models can be executed. Parameter values have critical influences on the outcome of CA simulation (Wu, 2000). It is quite tedious to define proper parameter values when the number of variables is large. A very simple method to find suitable parameter values is to use the so-called visual test (Clarke et al., 1997). It is based on the trial and error approach in which the impact of each parameter is assessed by changing its value and holding other parameters constant. Wu and Webster (1998) provide another method that uses AHP of multicriteria evaluation (MCE) techniques to decide parameter values. The pairwise comparison was used to recover weight vector by which suitability of the land can be computed. However, the comparison will become much difficult when there is a large set of variables. Moreover, the weights cannot be properly given when there are relevant variables.

These above methods have uncertainties because parameter values are decided with subjective influences. Objective methods should be used to remove the uncertainties. There is some limited work for finding optimal parameter values using exhaustive computer search. Clarke and Gaydos (1998) develop a relatively robust method to find suitable parameter values based on computer search algorithms. It tests various trials of parameter combinations and calculates the difference between the actual data and simulated results for each trial. The parameter values can be found according to the best fit of the trials. The computation is extremely intensive as the possible combinations are numerous. It usually needs a high-end workstation to run hundreds of hours before finding the best fit. It is practically impossible to try all the possible combinations. Computation time will even increase exponentially when there are a larger number of parameters. A more robust way is to train neural networks and find the parameter values of CA by using the observation data of remote sensing (Li and Yeh, 2001). This can help to reduce the uncertainties in defining the parameter values of CA.

22.3.2.5 Stochastic Variables

Most urban CA are not deterministic for simulating complex urban systems. Deterministic models may have problems in representing many geographical phenomena. These phenomena have manifested some unpredictable features which cannot be explained by independent variables due to the complexity of nature. It is almost impossible to forecast exact future patterns by using any kind of computer models. Frequently, urban CA models have to incorporate stochastic variables to represent the uncertainty of nature. Some "noises" are artificially added to urban CA models by using controlled stochastic variables to produce "realistic" simulation (White and Engelen, 1993). In the transition rules, calculated development probability is compared with a random number to decide whether the transition is successful or not (Wu and Webster, 1998). This can allow a certain degree of randomness to be inserted in urban simulation. However, there are questions when these models are used for urban planning. It is because each simulation will generate different results although the inputs are the same. A planner may be in a dilemma as to which result is suitable for planning. There is a concern on the repeatability of urban simulation using stochastic variables and the use of urban simulation results in preparing land use development plans.

An experiment was carried out to examine the uncertainty of stochastic CA. It repeatedly ran the CA model ten times and then examined the overlapping of the outcomes from an overlay analysis (Figure 22.4). In the overlay analysis, the urban areas are coded with 1, and non-urban areas are coded with 0. If CA are deterministic, the urban areas and non-urban areas in the ten different simulations should be the same. They should be 100% overlapping in the overlay. The overlay will only yield two values — 10 for urban areas and 0 for non-urban areas. However, the stochastic CA will not generate the same simulation results. The 10 simulations will not be totally overlapping and the overlay will yield 11 values in the hit count. The hit counts of 10

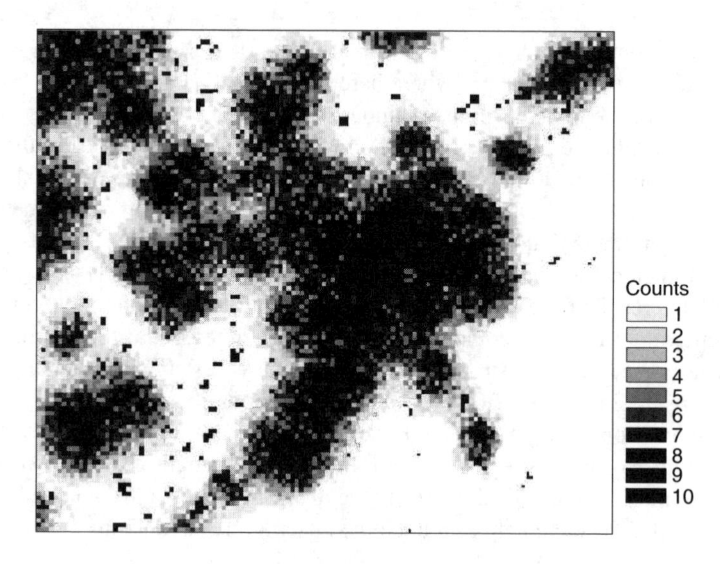

Figure 22.4 Overlay of the simulation results by repeatedly running the stochastic CA ten times.

and 0 in the overlay represent the urban areas and nonurban areas respectively with 100% confidence. The hit counts from 1 to 9 represent urban areas in some simulations while nonurban areas in others. Therefore, the hit counts of 1 to 9 represent the areas with uncertainty. However, the cells with a larger value of hit count (e.g., 9) have higher confidence to be urban areas in the simulation.

The areas with uncertainty in the simulations should be within a small percentage of the total simulated urban area. Otherwise, the simulations are meaningless. According to the experiment, it is interesting to see that the uncertainties only mainly exist at the fringe areas of each urban cluster. Stochastic CA can produce consistent simulation results in the core urban areas. This means that stochastic CA can maintain stability at the macro-level while they may have subtle changes at the micro-level for each simulation. This characteristic should be useful for urban planners to understand the implication of CA urban simulation.

22.4 CONCLUSION

Like many GIS models, urban CA have the inherent problems related to data errors and model uncertainties. This study demonstrates that the outcomes of urban CA simulation are affected by the errors of GIS data and the structures of CA models. The errors and uncertainties will in turn affect planning and development decisions when the results of urban CA simulation are applied in the planning and development process. Although there are many studies on data errors and error propagation in GIS analysis, not much research has been carried out to examine these issues in urban CA simulation.

GIS data are the main inputs to most urban CA models. A large amount of GIS data is usually required for producing realistic urban simulation. It is well known that most GIS data are subject to a series of errors. There are many possibilities of creating errors in spatial data as the errors can come from original sources and even be produced in the process of data capture. New errors can also be created during GIS operations. These errors will propagate in CA simulation and affect the simulation results. There is concern whether CA models can produce reliable and repeatable results, especially when it is applied to urban planning. Although some researchers may be aware that errors can propagate through CA simulation, they rarely pay much attention to this problem in practice because of the complexity. When GIS data are used as inputs to CA models, the source errors will propagate and affect the outcomes of simulation. A particular example is the errors in labeling land use types during land use classification. The experiment shows that the errors in the initial land use types can propagate through CA simulation. However, the errors will be much reduced in the simulation outcomes because of the averaging effects of the neighborhood and iterations of CA.

Simulation uncertainty is further increased by model uncertainty. The relationship between errors and outcomes is complicated for dynamic models. CA also have a series of inherent model uncertainties. These uncertainties are related to a number of factors in defining CA models — the neighborhood, cell size, computation time, transition rules, and model parameters. Most CA models have incorporated stochastic variables in urban simulation. This has allowed some unpredictable features to be inserted in the simulation process. There are arguments that uncertainty is necessary for generating realistic urban features, such as the emergence of new urban centers during the simulation process. A simple overlay of two repeated simulations from stochastic CA can reveal the discrepancy between them. Fortunately, the discrepancy only exists in the fringe areas of urban clusters according to the experiments. It means that stochastic CA can generate stable simulation results at the macro-level although there are variations at the micro-level. This characteristic is important to ensure the applicability of stochastic CA in simulating planning scenarios. Therefore, planners should run urban CA a couple of times repeatedly when CA are used for selecting development sites. Planners can then select the simulated development sites with more certainty.

The issues of data errors, error propagation and model uncertainties are important but often neglected in urban CA models. This chapter has examined and addressed some of these issues by carrying out experiments with GIS data. Many model errors are related to model configurations, that is, how to define a proper model to reflect the real process of urban development. This study demonstrates that some, however, are quite unique to CA: (1) the number of iterations can cause different spatial patterns; and (2) the discrepancy of simulation can be found at the edge of simulated urban areas. The above analyses can help to understand the implications of CA simulation for urban planning. However, further work is needed to develop a methodology for reducing the influences of errors and producing more reliable simulation results.

ACKNOWLEDGMENT

This study is supported by the funding from the Hong Kong Research Grants Council (Contract No. HKU 7209/02H), Hong Kong SAR, PR China.

REFERENCES

Batty, M., 1997. Growing cities (Working paper, Centre for Advanced Spatial Analysis, University College London).

Batty, M. and Xie, Y., 1994. From cells to cities. *Environment and Planning B: Planning and Design*, **21**, 531–548.

Benati, S., 1997. A cellular automaton for the simulation of competitive location. *Environment and Planning B: Planning and Design*, **24**, 205–218.

Burrough, P.A., 1986. *Principles of Geographical Information Systems for Land Resource Assessment* (Oxford: Clarendon Press).

Caspary, W. and Scheuring, R., 1993. Positional accuracy in spatial database. Computers. *Environment and Urban Systems*, **17**, 103–110.

Clarke, K.C. and Gaydos, L.J., 1998. Loose-coupling a cellular automata model and GIS: long-term urban growth prediction for San Francisco and Washington/Baltimore. *International Journal of Geographical Information Science*, **12**, 699–714.

Clarke, K.C., Gaydos, L., and Hoppen, S., 1997. A self-modifying cellular automaton model of historical urbanization in the San Francisco Bay area. *Environment and Planning B: Planning and Design*, **24**, 247–261.

Fisher, P.F., 1991. Modeling soil map-unit inclusions by Monte Carlo simulation. *International Journal of Geographical Information Systems*, **5**, 193–208.

Fisher, P.F. and Pathirana, S., 1990. The evaluation of fuzzy membership of land cover classes in the suburban zone. *Remote Sensing of Environment*, **34**, 121–132.

Goodchild, M.F., 1991. Issues of quality and uncertainty. In *Advances in Cartography*, J.C. Müller (Ed.) (Oxford: Elsevier Science), pp. 111–139.

Goodchild, M.F., Sun, G.Q., and Yang, S.H.R., 1992. Development and test of an error model for categorical data. *International Journal of Geographical Information Systems*, **6**(2), 87–104.

Heuvelink, G.B.M., 1998. *Error Propagation in Environmental Modelling with GIS* (London: Taylor & Francis).

Heuvelink, G.B.M. and Burrough, P.A., 1993. Error propagation in cartographic modeling using Boolean logic and continuous classification. *International Journal of Geographical Information Systems*, **7**, 231–246.

Heuvelink, G.B.M., Burrough, P.A., and Stein, A., 1989. Propagation of errors in spatial modeling with GIS. *International Journal of Geographical Information Systems*, **3**(4), 303–322.

Li, X. and Yeh, A.G.O., 1998. Principal component analysis of stacked multi-temporal images for monitoring of rapid urban expansion in the Pearl River Delta. *International Journal of Remote Sensing*, **19**, 1501–1518.

Li, X. and Yeh, A.G.O., 2000. Modelling sustainable urban development by the integration of constrained cellular automata and GIS. *International Journal of Geographical Information Science*, **14**, 131–152.

Li, X. and Yeh, A.G.O., 2001. Calibration of cellular automata by using neural networks for the simulation of complex urban systems. *Environment and Planning A*, **33**, 1445–1462.

Lodwick, W.A., 1989. Developing confidence limits on errors of suitability analysis in GIS. In *Accuracy of Spatial Databases*, M.F. Goodchild and S. Gopal (Eds.) (London: Taylor & Francis), pp. 69–78.

O'Sullivan, D., 2001. Exploring spatial process dynamics using irregular cellular automaton models. *Geographical Analysis*, **33**, 1–18.

Veregin, H., 1994. Integration of simulation modeling and error propagation for the buffer operation in GIS. *Photogrammetric Engineering and Remote Sensing*, **60**, 427–435.

Veregin, H., 1999. Data quality parameters. In *Geographical Information Systems*, P.A. Longley, M.F. Goodchild, D.J. Maguire, and D.W. Rhind (Eds.) (New York: Wiley), pp. 177–189.

Webster, C.J. and Wu, F., 1999. Regulation, land-use mix, and urban performance. Part 2: Simulation. *Environment and Planning A*, **31**, 1529–1545.

White, R. and Engelen, G., 1993. Cellular automata and fractal urban form: a cellular modelling approach to the evolution of urban land-use patterns. *Environment and Planning A*, **25**, 1175–1199.

White, R., Engelen, G., and Uijee, I., 1997. The use of constrained cellular automata for high-resolution modelling of urban land-use dynamics. *Environment and Planning B: Planning and Design*, **24**, 323–343.

Wolfram, S., 1984. Cellular automata as models of complexity. *Nature*, **31**, 419–424.

Wu, F., 1999. A linguistic cellular automata simulation approach for sustainable land development in a fast growing region. *Computer, Environment, and Urban Systems*, **20**, 367–387.

Wu, F., 2000. A parameterised urban cellular model combining spontaneous and self-organising growth. In *GIS and Geocomputation*, P. Atkinson and D. Martin (Eds.) (New York, NY: Taylor & Francis), pp. 73–85.

Wu, F., and Webster, C.J., 1998. Simulation of land development through the integration of cellular automata and multicriteria evaluation. *Environment and Planning B: Planning and Design*, **25**, 103–126.

Yeh, A.G.O. and Li, X., 2001. A constrained CA model for the simulation and planning of sustainable urban forms using GIS. *Environment and Planning B: Planning and Design*, **28**, 733–753.

Yeh, A.G.O. and Li, X., 2002. A cellular automata model to simulate development density for urban planning. *Environment and Planning B*, **29**, 431–450.

CHAPTER 23

A Minimal Prototype for Integrating GIS and Geographic Simulation through Geographic Automata Systems

Itzhak Benenson and Paul M. Torrens

CONTENTS

0-8493-2837-3/05/$0.00+$1.50

23.1 INTRODUCTION

New forms of simulation have come into popular use in geography and social science in recent years, supported by an array of advances both in the geographical sciences and in fields outside geography. These models and simulations can be characterized by a distinctly innovative approach to modeling — the geosimulation approach. Geosimulation is concerned with automata-based methodologies for simulating collectives of discrete, dynamic, and action-oriented spatial objects, combining cellular automata and multi-agent systems in a spatial context (Benenson and Torrens, 2004a, b). In geosimulation-style models, urban phenomena as a whole are considered as the outcome of the collective dynamics of multiple animate and inanimate urban objects.

Geosimulation models are more commonly based on cellular automata (CA) and multi-agent systems (MAS). Applied in isolation, CA and MAS approaches have been used to simulate a wide variety of urban phenomena and there is a natural imperative to combine these frameworks for exploratory and applied simulation in urban geography (Torrens, 2002a, b, 2003, 2004a, b). Nevertheless, the direct amalgamation of CA and MAS eventually suffers from awkward compromises, a function of the necessity for CA partition of urban space into cells. Given cell partition, no matter whether units are identical or vary in size and shape, cells are either granted some degree of 'agency' and are simply reinterpreted as artificial agents (Box, 2001) or MAS are imposed on top of CA and simulated agents respond to averaged cell conditions (Benenson, 1998; Polhill et al., 2001). These frameworks are certainly useful, especially in studying abstract models, but in a real-world milieu are a function of the limitations of the available tools rather than the structure of real urban systems or the laws of their behavior (see Agarwal, Chapter 24, this volume, for a discussion of the limitations of agent-based models).

The dataware for the real-world simulations is usually provided by Geographic Information Systems (GIS), which have an integral role in the development of geosimulation models. Dramatic changes in geographic databases during the last decades of the 20th century have ushered in a new wave of urban modeling (Benenson et al., 2002; Benenson and Omer, 2003). Automated procedures for data collection — remote sensing by spectrometer, aerial photography, scanners, video, etc. — have provided new information sources at fine resolutions, both spatial and temporal (Torrens, 2004a, b). New methodologies for manipulating and interpreting spatial data developed by geographic information science and implemented in GIS have created added value for these data. Information collection is much more pervasive than before (Brown and Duguid, 2000), and high-resolution databases for urban land-use, population, real estate, and transport are now more widespread (Benenson and Omer, 2003). Cellular spaces populated with agents are too limiting an interface between GIS and geosimulation, and there remains much potential for fusing GIS and geosimulation into full-blown, symbiotic simulation systems. In this chapter, we propose a minimal but sufficient framework for integrating GIS, CA, and MAS into what we term *Geographic Automata Systems* (GAS). This approach is intended to be "down to earth", and we will demonstrate the implementation of the approach, in this chapter, with reference to GAS-based software developed at the Environment Simulation Laboratory of the Porter School of Environmental Studies at the University of Tel Aviv — the *Object-Based Environment for Urban Simulations* (OBEUS). This chapter builds on discussions of urban geocomputation (Benenson and Omer, 2000; Torrens and O'Sullivan, 2000) and software environments for GeoComputational research in urban contexts (Benenson et al., 2001), which were partially presented at previous GeoComputation meetings.

23.2 THE BASIC AUTOMATA FRAMEWORK

Put very simply, an automaton is a processing mechanism; a discrete entity, which has some form of input and internal states. It changes states over time according to a set of rules that take information from an automaton's own state and various inputs from outside the automaton to determine a new state in a subsequent time step. Formally, an automaton, **A**, can be represented by means of a set of *states* **S** and a set of *transition rules* **T**.

$$\mathbf{A} \sim (\mathbf{S}, \mathbf{T}) \tag{23.1}$$

Transition rules define an automaton's state, S_{t+1}, at time step $t + 1$ depending on its state, $S_t (S_t, S_{t+1} \in S)$, and *input*, I_t, at time step t:

$$\mathbf{T}: \quad (S_t, I_t) \rightarrow S_{t+1} \tag{23.2}$$

Automata are discrete with respect to time and have the ability to change according to predetermined rules based on internal (**S**) and external (**I**) information.

In terms of urban applications, automata lend themselves to specification as city simulations with myriad states and transition rules. However, to make sense, an individual automaton should be as simple as possible in terms of states, transition rules, and internal information (Torrens and O'Sullivan, 2001). Simplicity is a characteristic of the most popular automata tools in urban geography, Cellular Automata — a system of spatially located and inter-connected automata.

23.2.1 From General to Cellular Automata

Cellular automata are arrangements of individual automata in a partitioned space, where each unit (cell) is considered as an automaton **A**, for which input information **I** necessary for the application of transition rules **T** is drawn from **A**'s *neighborhood* **R**. In urban applications, cells are most commonly used to represent land units with state representing possible land-uses (White et al., 1997). Usually, CA lattices are partitioned as a regular square or hexagonal grid. We can specify

Equations (23.1) and (23.2) to introduce an automaton, **A**, belonging to a CA lattice as follows:

$$\mathbf{A} \sim (\mathbf{S}, \mathbf{T}, \mathbf{R}) \tag{23.3}$$

where **R** denotes automata neighboring **A**.

Although there are direct analogies between land parcels and cells on the one hand and land-uses and cell states on the other, there were no geographical applications of CA models before the 1990s. There were a few examples published in the 1970s (Chapin and Weiss, 1968; Tobler, 1970, 1979; Albin, 1975; Nakajima, 1977), but the field was largely ignored in terms of research until interest was revived in the 1980s (Couclelis, 1985; Phipps, 1989). Beginning in the 1990s, CA modeling became a popular research activity in geography, with pioneering applications in urban geography (Batty et al., 1997; O'Sullivan and Torrens, 2000).

In terms of space, neighborhood relationships are important for rendering CA as *spatial* systems. In basic CA, neighborhoods have identical form for each automaton, for example, Moore or von Neumann (Figure 23.1A). In the last decade it became clear, however, that reliance on regular partitions of space is largely superficial in urban contexts (Torrens and O'Sullivan, 2001). Consequently, CA have been implemented on irregular networks (Figure 23.1B), or partitions given by GIS-based coverage of land parcels or Voronoi tessellations (Figure 23.1C) (Semboloni, 2000; Shi and Pang, 2000; O'Sullivan, 2001; Benenson et al., 2002). An assortment of definitions of neighborhoods, based on connectivity, adjacency, or distance can be applied to these generalized CA, where the form of the neighborhood and the number of neighbors varies between automata.

There is no conceptual difference between irregular and regular CA; yet, an inherent weakness of the CA is the inability of automata cells to move within the lattice in which they reside. Despite repeated attempts to mimic units' mobility (Portugali et al., 1994; Schofisch and Hadeler, 1996; Wahle et al., 2001), the genuine inability to allow for automata movement in the CA framework catalyzed geographers' recent interest in MAS. This tendency is especially strong in urban geography, where the CA framework is regarded as insufficient in dealing with mobile objects such as pedestrians (Torrens, 2004a, b), migrating households, or relocating firms.

23.2.2 From General to Agent Automata

Fundamentally, agents are automata and, thus, incorporate all of the features of basic automata that have just been discussed. However, there are some important distinctions between general and agent automata, largely owing to the fact that the latter are generally interpreted for representation of autonomous *decision-makers* (Epstein, 1999; Kohler, 2000). In urban studies, the states **S** of agent automata in MAS are usually designed to represent socioeconomic characteristics; transition rules are formulated as rules of agent decision-making, and correspond to human-like *behaviors*.

In the social sciences outside geography, work in agent-based simulation is usually *non-spatial*; many of the decisions and behaviors of geographic *agents* are, however, spatial in nature. Consequently, the states **S** of *geographic* MAS should include

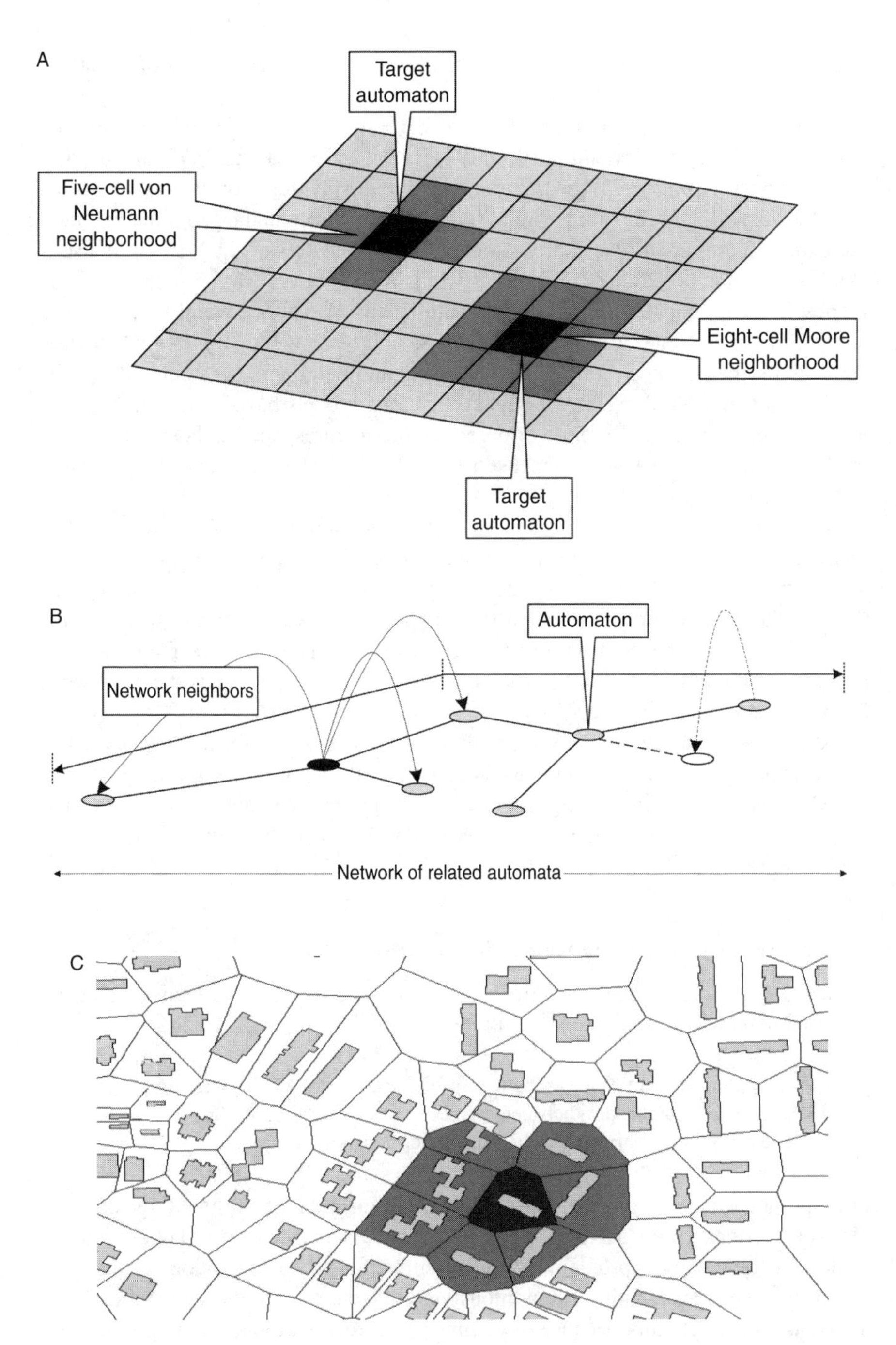

Figure 23.1 Automata defined on (A) a two-dimensional regular lattice, with von Neumann and Moore neighborhoods; (B) a two-dimensional network; (C) a Voronoi partition of two-dimensional urban space, based on property coverage.

agents' *location*, and transition rules **T** should reflect, thus, the ability of agents to relocate.

Human-based interpretations of MAS have their foundation in the work of Schelling and Sakoda (Schelling, 1969, 1971, 1974, 1978; Sakoda, 1971) and despite initial interest (Albin, 1975) the field remained relatively quiet for two decades after that. Just as with CA, the tool began to feature prominently in the geographical literature only in the mid-1990s (Portugali et al., 1997; Sanders et al., 1997; Benenson, 1999; Dijkstra et al., 2000). Until recently, the main thrust of MAS research in geography involved populating regular CA with agents of one or several kinds, which could *diffuse* between CA cells. Often, it is assumed that agents' migration behavior depends on the properties of neighboring cells and neighbors (Epstein and Axtell, 1996; Portugali, 2000). Very recently, agent-based models have been designed that locate agents in relation to real-world geographic features, such as houses or roads, the latter stored as GIS layers (Benenson et al., 2002) or landscape units — pathways and view points (Gimblett, 2002).

Despite the widely acknowledged suitability of automata tools for geographic modeling (Gimblett, 2002), there has been relatively little exploration into addressing these limitations and developing patently spatial automata tools for urban simulation. In the framework described in this chapter, spatial abilities are treated with paramount importance and we define a class of automata that is capable of supporting explicit expression of processes, as *geography* comprehends them. In what follows, we aim at formulating a *minimal framework* and, thus, formulate it on the basis of an agent as an autonomous mobile object that reacts to the environmental information by changing its state and location. In terms of MAS, these agents are called weak agents and reactive agents (Woolridge and Jennings, 1995), and we assume that they are sufficient for the vast majority of high-resolution geographic applications.

23.2.3 What Geography Needs from Automata Systems

Focusing on space, we might identify three internal geographic mechanisms that are essential to an urban automata system:

- A typology of entities regarding their use of space in which they are situated.
- The spatial relationships between entities.
- The processes governing the changes of their location in space.

Simulating spatial systems, then, involves explicit formulation of these three components and neither CA nor MAS fully provide the necessary framework. The geography of the CA framework is problematic, for example, for urban simulation because CA are incapable of representing autonomously mobile entities. At the same time, MAS are weak as a single tool because of common underestimation of the importance of space and movement behavior.

It is evident, then, that there is a need for uniting CA and MAS formalisms in such a way as to directly reflect a geographic and object-based view of urban systems. The GAS framework that we propose attempts to satisfy that demand.

23.2.4 An Idea: Urban Object ≡ Geographic Automaton

As a spatial science, geography concerns itself with the behavior and distribution of *objects in space*. In urban geography, these are urban agents — householders, pedestrians, vehicles — and urban features — land parcels, shops, roads, sidewalks, etc. In dynamic spatial systems, all these objects change their properties and/or location; the goal of a geographic model is to simulate these activities and their consequences, often at multiple scales.

In developing GAS, our aim is to infuse spatial properties into automata tools and adopt an *automata-based* view of urban systems. Objects are conceptualized as *geographic automata*, with focus on their *spatial properties and behaviors*. Under this framework, a city system can be modeled as an ensemble of geographic automata, that is, a GAS.

23.3 FORMAL DEFINITION OF GEOGRAPHIC AUTOMATA SYSTEMS

Geographic Automata Systems consist of interacting *geographic automata* of various types. In general, automata are characterized by states **S**, description of their input **I**, and transition rules **T**. In the case of geographic automata, we re-interpret and extend these components to enable the explicit consideration of *space* and *spatial behavior*.

Specification of state set **S**:

- In addition to non-spatial *states*, geographic automata are also characterized by their *locations*.
- Instead of the fixed location of CA automata, we introduce a set of *geo-referencing rules* for situating geographic automata in space.

Specification of input information **I**:

- Instead of relying on fixed neighborhood patterns that are incapable of being varied in space or time, we define *neighborhood relations* that can change in time and determine the automata providing input information for a given automaton.
- Neighborhood relations can thus change; these changes are governed by *neighborhood rules*.

Specification of transition rules **T**:

- *State transition rules* specify the changes of non-spatial states.
- The ability to change location is provided by a set of *movement rules* that allow for navigation of geographic automata in their simulated environments.

Based on these specifications, we can formulate a minimal, yet adequate, framework for geographic modeling — or geosimulation.

Let us use **K** to denote a set of *types* of automata; following this we will feature each of the previously mentioned properties and rules that determine the dynamics of those automata in space. Essentially, we are constructing a system of geographic automata from the bottom-up.

The automata basis of the system is captured by a set of *states* **S**, associated with the GAS (consisting of subsets of states $\mathbf{S}^k$ of automata of each type $k \in \mathbf{K}$), and a set of *state transition rules* $\mathbf{T_S}$, used to determine how automata states should change over time.

According to general definitions (23.1) and (23.2), state transitions and changes in location for geographic automata depend on automata themselves and on input (**I**), given by the states of *neighbors*. To specify the input information, we use **R** to denote the neighbors of the automata and $\mathbf{N_R}$ to specify the *neighborhood transition* rules that govern how automata relate to the other automata in their vicinity.

The spatial nature of the geographic automata is encapsulated by information on automata location. Let us denote this with **L**, the *geo-referencing* conventions employed to locate automata in the system, and $\mathbf{M_L}$, the *movement rules*, governing changes in their location.

Altogether, a GAS, **G**, may be thus defined as consisting of several components:

$$\mathbf{G} \sim \langle \mathbf{K}, \mathbf{S}, \mathbf{T_S}, \mathbf{R}, \mathbf{N_R}, \mathbf{L}, \mathbf{M_L} \rangle \tag{23.4}$$

Let the state of geographic automaton **G** at time t be S_t, located at L_t, and the external input, I_t, be defined by its neighbors R_t. The state transition, movement, and neighborhood rules — $\mathbf{T_S}$, $\mathbf{M_L}$, and $\mathbf{N_R}$ — define state, location, and input information of a given automaton G at time $t + 1$ as:

$$\begin{aligned}
\mathbf{T_S}: &\quad (S_t, L_t, R_t) \rightarrow S_{t+1} \\
\mathbf{N_R}: &\quad (S_t, L_t, R_t) \rightarrow R_{t+1} \\
\mathbf{M_L}: &\quad (S_t, L_t, R_t) \rightarrow L_{t+1}
\end{aligned} \tag{23.5}$$

Exploration with GAS then becomes an issue of qualitative and quantitative investigation of the spatial and temporal behavior of **G**, given all of the components defined above. In this way, GAS models offer a framework for considering the *spatially enabled* interactive behavior of elementary geographic objects in a system.

23.3.1 Tight-Coupling between GAS and Vector GIS

Many questions arise when applying the abstract framework mentioned above to modeling specific geographic systems. Geographic automata of different types can follow different topologies. How might we incorporate that into models: let householders react to their neighbors in the adjacent houses (irregular tessellation); drivers to cars ahead, behind, and on their sides (point events on the network); customers to shops, which can be far away (graphs)? How could automata represent continuous fields, if ever? How can theoretical ideas relating to spatial mobility — way-finding, spatial cognition, action-at-a-distance, etc. — be incorporated into the behavioral rules of the real-world mobile agents or immobile features and further translated into automation rules? How can the global patterns reflecting self-organization of non-elementary urban entities be recognized and validated if generated in simulations?

Answering these questions requires a tight-coupling between GIS and automata-based models.

As a first step, we could rely on vector GIS. Indeed, vector GIS as a special kind of Database Management System (DBMS). Indeed, vector GIS may provide much support for GAS models. First and foremost, vector GIS can be used to store and retrieve the location and states of spatial objects and to register spatial actions. The next step toward fusion of GIS and geosimulation exploits an *object-based* view of urban reality. Indeed, both GAS and vector GIS are object-based in their design; both deal with discrete spatial objects, which customarily represent the real world at "microscopic" scales. A geosimulation approach considers the city as a *dynamic collective* of spatially located objects. Vector GIS deals mostly with static objects and employs an *entity-relationship model* (ERM) for their representation. In our approaches detailed here, we merge GAS and vector GIS functionality by implementing GAS as a specialized object-oriented database that manages geographic automata. Automata of a certain type are interpreted as a *class*, states and location as class *properties*; and GAS state transition, movement and neighborhood transition rules of automata of different types are thus reformulated as *methods* of corresponding automata classes. Let us specify components of GAS definition (23.4) and (23.5).

23.3.2 Geographic Automata Types, K

At an abstract level, we distinguish between fixed and *non-fixed* geographic automata. Fixed geographic automata represent objects that do not change their location over time and, thus, have close analogies with CA cells. In the context of urban systems, these are objects such as road links, building footprints, parks, etc. Fixed geographic automata may be subject to state and neighborhood transition rules $\mathbf{T_S}$ and $\mathbf{N_R}$, but not of rules of motion, $\mathbf{M_L}$.

Non-fixed geographic automata symbolize entities that change their location over time, for example: pedestrians, vehicles, and households. The full range of rules for GAS can be applied to non-fixed geographic automata.

Geographic automata of fixed type evidently correspond to vector GIS features; normally the automata of a given type correspond to the features of a certain layer of vector GIS.

23.3.3 Geographic Automata States and State Transition Rules, S and T_S

State transition rules $\mathbf{T_S}$ are based on geographic automata of *all* types from **K**. In contrast to CA, the states **S** of urban fixed infrastructure objects depend on the neighboring objects of the infrastructure, but are also driven by non-fixed geographic automata — agents — that may be responsible for governing object states such as land-use, land value, etc. In this way, urban objects do not simply mutate like bacteria (O'Sullivan and Torrens, 2000); rather, state transition is governed by other objects, the latter crucial for simulating *human-driven* urban systems, in which people are affected by their environments and also change them.

Interpreted in GIS terms, automaton states are attribute values of a corresponding GIS feature, while state transition rules exploit attributes of GIS features that are in spatial and non-spatial relationships with a given automaton.

23.3.4 Geo-Referencing Conventions, L and Movement Rules M_L

Geo-referencing conventions are crucial for coupling GAS and GIS. On the one hand, they should be sufficiently flexible to enable translation of geographic perspectives on locating real world objects, both fixed and non-fixed; on the other, they should satisfy limitations of ERM, in order to be convenient for GIS management.

The GAS framework resolves these limitations by introducing two forms of geo-referencing — *direct* and *by pointing*. Direct methods of geo-referencing follow a vector GIS approach for representing reality, using coordinate lists. Such a list indicates all spatial details necessary to represent an object — automata boundaries, centroids, nodes' location, etc. Fixed geographic automata are usually located by means of direct geo-referencing. The details of the particular rules employed depend on the automata used in a modeling exercise. For typical urban objects such as buildings or street segments, two-dimensional footprint polygons or three-dimensional prisms may be used.

Non-fixed geographic automata may move; updating position coordinates might cause difficulties, which, when the automata are numerous and their shape is complex, are irresolvable by direct geo-referencing. We address these difficulties with a second method of geo-referencing — by *pointing* to other automata (Figure 23.2).

Let us consider a typical example. In the case of property dynamics, householders can be geo-referenced by address. Landlords provide a more complex case: they can be geo-referenced by pointing to the properties that they own. Indirect referencing can also be used for fixed geographic automata, for example, for apartments in a house (Torrens, 2001). Referencing by pointing is dynamic-enabled both in space and in time and compatible with the ERM database model.

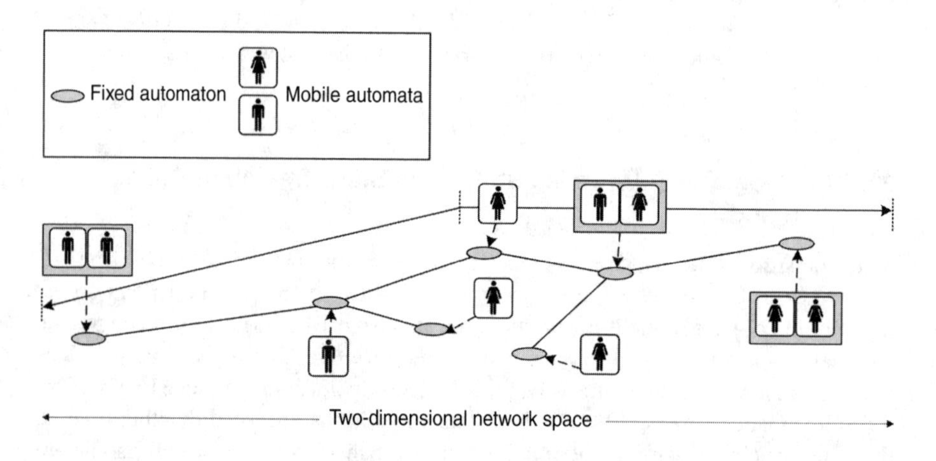

Figure 23.2 Direct and indirect geo-referencing of fixed and non-fixed GA.

GAS-based research into different formulations of $\mathbf{M_L}$ offers great potential for geographers. Realistic rules, $\mathbf{M_L}$, for encoding automata movement, based on repel–attract–synchronize interactions between close neighbors, are being developed, for example, in Animat research (Meyer et al., 2000) and in the gaming industry (Reynolds, 1999). There is much opportunity for geographers to contribute to this line of research. Traditionally, attention has focused on the generation of realistic choreographies for automata, particularly in traffic models, through the specification of rules for collision avoidance, obstacle negotiation, lane-changing, flocking, behavior at junctions, etc. (Torrens, 2004). However, there remain many relatively neglected areas of inquiry: spatial cognition, migration, way finding, navigation, etc.

23.3.5 Neighbors and Neighborhood Rules, R and N_R

The set of neighbors of automata, **R**, is necessary for determining input information **I**. In contrast to the static and symmetrical neighborhoods usually employed in CA models (Figure 23.1A), spatial relationships between geographic automata can vary in *space and time*, and, thus, rules for determining neighborhood relationships $\mathbf{N_R}$ is necessary. Neighborhood rules for fixed geographic objects are relatively easy to define (simply because the objects are static in space). There are a variety of *geographical* ways in which neighborhood rules can be expressed for them — via adjacency of units in regular or irregular tessellations, connectivity of network nodes, proximity, etc. (Figure 23.1B, C), all fit naturally in the context of GIS operations. Spatial notions related to the incorporation of human-like automata into GAS, such as accessibility, visibility, and mental maps can be formally encoded as $\mathbf{N_R}$ rules.

Non-fixed geographic automata pose more of a challenge when specifying neighborhood rules, because the objects — and hence their neighborhood relations — are dynamic in space and time. It can be done straightforwardly, via distance and nearest-neighbor relations, as used in Boids models (Reynolds, 1987), but as movement rules $\mathbf{M_L}$. This can become very heavy computationally when complex definition of, say, visibility or accessibility is involved. In this case, it is more appropriate to base neighborhood rules on indirect location, as defined in previous sections, and consider two indirectly located automata as neighbors when *the automata they point to are neighbors*.

For example, two householder agents can be established as neighbors by assessing the neighborhood relationship between the houses in which they reside. Even when these agents are physically separated in the simulation — when they go shopping or go to work, for example — they remain "neighbors" by virtue of the (fixed) relationship between their properties.

23.3.6 Temporal Dimension of GAS

A methodological base composed of fixed and non-fixed objects and direct and by-pointing locating implies specification of GASs within GIS. However, the dynamic nature of GAS also implicates *temporal* dimensions of GIS databases, and, thus entails its own limitations. Given these considerations, transition rules $\mathbf{T_S}$, $\mathbf{M_L}$, and $\mathbf{N_R}$, should be defined in a way that avoids conflicts when geographic automata are created or destroyed and their states, locations, or neighborhood relations are updated.

According to (5), a triplet of transition rules determines the states **S**, locations **L**, and neighbors **R** of automata at time $t + 1$ based on their values at time t. It is well known that different interpretations of the "hidden" — time — variable in a discrete system can critically influence resulting dynamics of the model (Liu and Andersson, 2004). There are several ways to implement time in a dynamic system. On the one hand, we consider time as governed by an *external* clock, which commands simultaneous application of rules (5) to each automaton and at each tick of the clock. On the other hand, each automaton can have its own *internal* clock and, thus, the units of time in (5) can have different meaning for different automata. Formally, these approaches are expressed as *Synchronous* or *Asynchronous* modes of updating of automata states and location. System dynamics depend strongly on the details of the mode employed (Berec, 2002; Liu and Andersson, 2004), and the spatial aspects of the problem are studied in temporal GIS (Miller, 1999; Peuquet, 2002). Tight coupling between GAS and GIS might thus demand further development of temporal GIS.

23.3.7 GAS and Raster GIS

Conceptually, GAS do not fit to modeling dynamics of continuous fields, as, say, erosion processes, or pollution transport. Nonetheless, it is formally easy to interpret raster GIS cells as geographic automata, if the resolution is established *a priori*. It is also worth noting that in the majority of situations the sources and recipients of continuously distributed characteristics *are* geographic automata. One can think about further coupling between GAS and continuous models, but such an extension is merely beyond the minimal framework we discuss.

23.4 THE SOFTWARE IMPLEMENTATION OF GAS

A software implementation of GAS for urban modeling — OBEUS — is currently in development at the Environment Simulation Laboratory of the University of Tel Aviv (Benenson et al., 2001, 2004). Recently, OBEUS was modified to include all the basic characteristics of GAS, and that was intentionally done in a minimal possible manner. Based on recent OBEUS experience, we specify — below — the main components that are necessary to implement the GAS approach as a software environment. The shareware version of OBEUS is still in development and is planned for distribution in April 2004 (see http://eslab.tau.ac.il for updates). The GAS framework is not software- or language-specific, however, and applications have been developed in different environments (Torrens, 2003).

23.4.1 Abstract Classes of OBEUS

The basic components of GAS are defined in OBEUS with respect to automata types $k \in \mathbf{K}$, its states $\mathbf{S}_k$, location **L**, and neighborhood relations **R** to other objects. These are implemented by means of the two-level structure of abstract root classes presented, partially, in Figure 23.3.

At the top level we define three abstract classes: **Population** contains information regarding the population of objects of given type k as a whole; **GeoAutomata** acts as a container for geographic automata of specific type; **GeoRelationship** facilitates specification of (spatial, but not necessarily) relationships between geographic automata. This functionality is available regardless of the degree of neighborhood and other relationships between automata, whether they are one-to-one, one-to-many, or many-to-many.

The *location* information for geographic automata essentially depends on whether the object we consider is fixed or non-fixed. This dichotomy is handled using abstract classes, **Estate** and **Agent**, which inherit **GeoAutomata**. The **Estate** class is used to represent fixed geographic automata. The **Agent** class represents non-fixed geographic automata.

Following from GeoRelationship, three abstract relationship classes can be specified: **EstateEstate**, **AgentEstate**, and **AgentAgent**. The latter is not implemented in order to avoid conflicts in relationship updating (see Section 23.4.3 below), and the only way of locating non-fixed agents modeled in OBEUS is by pointing to fixed estates; consequently, direct relationships between non-fixed objects are not allowed. This might limit possible applications of the system — we already mentioned the boids model (Reynolds, 1999) as an example where direct Agent–Agent relationships are necessary. At the same time, the majority of urban models we are aware of do not need them, and most examples in which Agent–Agent relationships are important arise in contexts beyond the intuitive limitations of the "reactive agent" idea; marriage is one example that a reviewer of this chapter suggested.

23.4.2 Management of Time in OBEUS

The OBEUS architecture utilizes both *Synchronous* and *Asynchronous* modes of update. In *Synchronous* mode, all automata are assumed to change simultaneously and conflicts can arise when agents compete over limited resources, as in the case of two householders trying to occupy the same apartment. Resolution of these conflicts depends on the model's context, a decision OBEUS leaves to the modeler. In *Asynchronous* mode, automata change in sequence, with each observing a geographic reality left by the previous automata. Conflicts between automata are thereby resolved; but the order of updating is critical as it may influence results.

OBEUS demands that the modeler sets up an order of automata update according to a template: randomly, sequence in order of some characteristic, and object-driven approaches are currently being implemented.

23.4.3 Management of Relationships in OBEUS

Relationships in GAS models can change in time and this might cause conflicts, when, in housing applications, for example, a landlord wants to sell his property, while the tenant does not want to leave the apartment. This example represents the

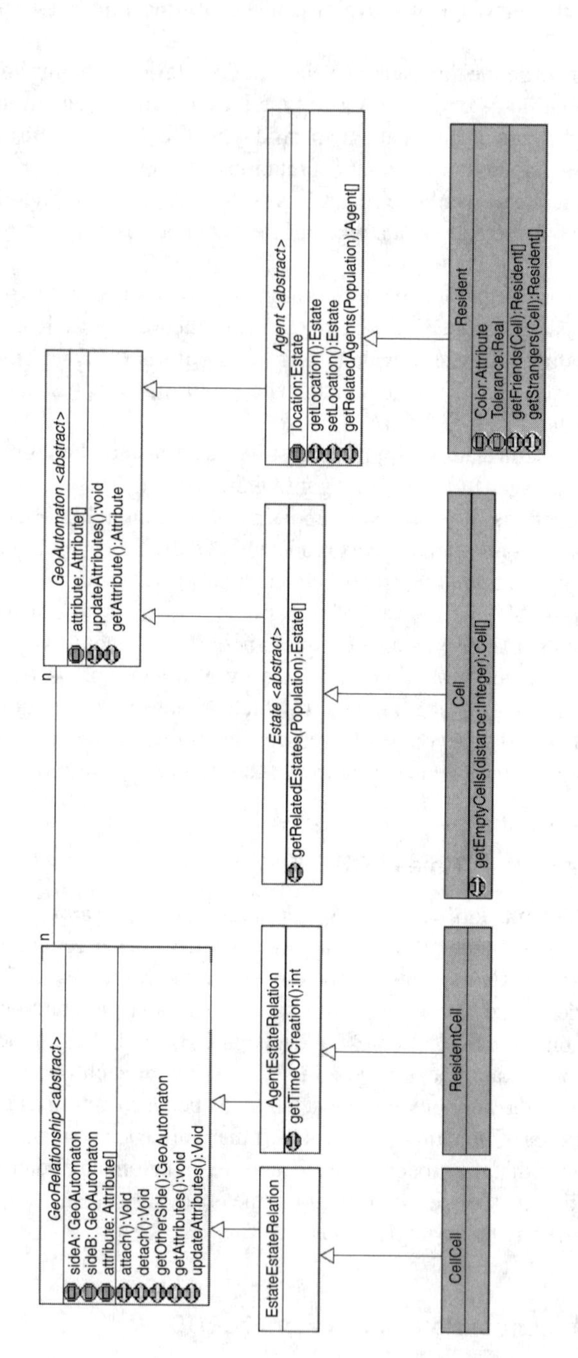

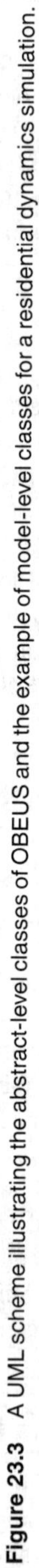

Figure 23.3 A UML scheme illustrating the abstract-level classes of OBEUS and the example of model-level classes for a residential dynamics simulation.

general *problem of consistency* in managing relationships. It has no single general solution; there are plenty of complex examples discussed in the computer science literature (Peckham et al., 1995). OBEUS aims at minimal representation of the GAS framework and thus follows the leader–follower development pattern proposed by Noble (2000). To maintain consistency in relationships, an object on one side, termed the *leader*, is responsible for managing the relationship. The other side, the *follower*, is comprised of passive objects. The leader provides an interface for managing the relationship, and invokes the followers when necessary. There is no need to establish leader or follower "roles" in a relationship between fixed objects once the relationship is established, while in relationships between a non-fixed and a fixed object, the non-fixed object is always the leader and is responsible for creating and updating the relationship. For instance, in a relationship between a landlord and her property (when ownership cannot be shared), the landlord initiates the relationship and is able to change it. We do not have evidence that the majority of real-world situations can be imitated using a *leader–follower* pattern, but we are also unaware of cases — in urban contexts — where this pattern would be insufficient.

23.4.4 Implementing System Theory Demands within OBEUS

Systems theory suggests another challenge for automata modeling in which the usefulness of the GAS–OBEUS approach appears to offer advantages. In a systems context, *many* interacting automata are often necessary for capturing the nuances of geographic reality. It is very well known that if system rules are non-linear and the system is open, then emergence and self-maintenance of entities at above-automata levels become feasible. Gentrified areas and commuting zones are examples.

The idea of self-organization is external to GAS and it is not necessary to incorporate it into the software implementation. Nonetheless, self-organization is often too important for studying urban systems to be ignored, even at the first step of GAS software implementation. To accommodate this, emerging *spatial ensembles* of geographic automata are supported in OBEUS by means of the abstract class *GeoDomain*. The simplest approach to emergence, determined by the set of *a priori* given predicates defined on geographic automata is implemented; domains are thus limited to capturing 'foreseeable' self-organization of specific types.

As with non-fixed agents, domains are always leaders in their relationships with the automata within them. These relationships can capture properties such as distance between fixed automata and the domain; several definitions of distance based on objects' and domains' centroids, boundaries, etc. can be applied.

23.4.5 Example of GAS Modeling Style — the Schelling Model in Terms of OBEUS

The well-known Schelling model of residential dynamics (Schelling, 1971; Hegselmann and Flache, 1999) reflects a basic idea of residential segregation between the members of two mutually avoiding population groups, as a self-organizing phenomenon. Thomas Schelling realized his model with black and white checkers on a chess-board and from then on, the model is formulated in terms of agents of two types, B and W, located in the cells of a regular grid, with a maximum of one agent per cell. The essence of the Schelling model is in assumption that if the fraction of agents of strange type (say, of W-type for the B-agent) within the neighborhood of a current agent's location is above the tolerance threshold of an agent, then an agent tries to relocate to a nearest unoccupied location, where the fraction of strange agents is below that threshold. The Schelling model is asynchronous and each agent observes the state of the system as left by the previous one.

As an applied example of GAS, let us formulate this model in GAS terms and then present the entities and methods necessary to implement it in OBEUS.

```
Two types of objects (K = 2)
Fixed Objects: Cells C;
  Cell states S:
     Cells have no states, i.e., S1 = Ø;
  Cell location rules L:
     Each cell C is assigned a pair of coordinates (x, y)
     for direct location,
     L1 = {Locate C(x, y) at (x, y)};
  Neighborhood rules R (Moore, as an example):
     R1 = {For a cell C(x, y), cells B(u, v), given
     max(|u − x|, |v − y|) = 1 are neighbors};
  State transition rules T:
     None, i.e., T1 = Ø;
  Movement rules M:
     None, M1 = Ø, there are no movement rules for fixed
     objects;
  Neighborhood transition rules N:
     None, N1 = Ø, neighborhood relations of fixed objects
     do not change.

Non-Fixed Objects: Residents D,
  Resident states S:
     Resident can be in one of two states, B or W,
     S2 = {B, W}
  Resident location rules L:
     Resident is located indirectly, by pointing to a cell
     it occupies, R2 = {Cell C}
  Neighborhood rules R:
     R2 = {Residents located in neighboring cells are
     neighbors}
```

```
State transition rules T:
   None, T2 = Ø;
Neighborhood transition rules N:
   None, N2 = Ø.
Movement rules (one of the versions) M:
   M2 = {If fraction of strangers fR among the neighbors
   is below threshold, i.e., fR < fTH, do nothing;
   otherwise find the closest unoccupied cell satisfying
   fR < fTH and relocate there; if there is more than one
   at the same distance choose one among them randomly;
   if no suitable location is found do nothing}
```

The OBEUS implementation of the "Schelling" GAS is as follows (Figure 23.3)

1. The abstract classes **Agent** and **Estate** are inherited by *Resident* and *Cell*, respectively.
2. The abstract relationship classes **EstateEstate** and **AgentEstate** are inherited by *CellCell* and *ResidentCell*, respectively.
3. Class *Cell* has no properties of its own.
4. Class *Cell* has one basic method of its own *getEmptyCloseCells(distance, cell)*, which employs **getRelatedEstates (estate)** and returns the list of non-occupied neighboring cells at a distance *distance*.
5. Class *Resident* has two properties of its own — Boolean *color*, with values B and W, and real *Tolerance* threshold.
6. Class *Resident* has two basic methods of its own, *getFriends(resident, cell)* and *getStrangers(resident, cell)*, which employ **getRelatedEstates(estate)**, and return lists of the neighbors of the color which is same or opposite to the color of a *resident* if located in a cell *cell*.

The goal of the Schelling GAS is to study residential segregation and, thus, its investigation demands recognition of areas populated mostly or exclusively by B- or W-agents. This can be done by specifying domain criteria based on the *CellCell* and *ResidentCell* relationship. Namely, let us consider a cell C occupied by resident R of a color L. Let us define the cell C as S_L-true (segregation-true of a color L) if the fraction of resident's neighbors of color L is above the given threshold f. For the Schelling model, if we take the value of f high enough (and higher than the tolerance threshold of residents), the sets of S_L-true cells for L = B and L = W will form continuous areas that reflect intuitive understanding of segregation. The value of f determines the density of the residents of the same color in the domain and the degree of the overlap of S_B and S_W domains. The concept of domains is considered in more details in (Benenson et al., 2004) and it is beyond the scope of this chapter. Let us note, however, that we consider the formulation of domain criteria and investigation the patterns produced by the geographic automata that satisfy the criteria as a necessary step in understanding spatial patterns that can emerge in the model.

23.5 CONCLUSIONS

We have introduced a GAS framework as a unified scheme for representing discrete geographic systems. Technically, the framework is designed to merge two popular tools used in urban simulation — CA and MAS — and specify them in a patently spatial manner. Conceptually, our assertion is that GAS forms the kernel of the system, as far as the system is spatially driven.

The minimal GAS skeleton allows for a degree of standardization between automata models and GIS. It also provides a mechanism for *transferability*. Until now, the majority of spatial simulations can be investigated only by their developers. The development of GAS software breaches this barrier, offering opportunities to turn urban modeling from art into engineering.

A few additional steps are necessary for full implementation of the GAS framework; none, we think, requires decades of research to realize. The first requirement we have identified relates to transforming the GAS framework into software. As demonstrated with reference to OBEUS, we have advanced along that line of research inquiry, in urban contexts. Development of a (preferably geography-specific) simulation language based on GAS is a second requirement that we consider. The intent, in that context, is to enable the formulation of simulation rules in terms of objects' spatial behavior. We believe that the continued development of simulation languages (Schumacher, 2001) that has gathered steam in the last decade, coupled with advances in GI Science and spatial ontology, could answer this requirement in the near future. The third requirement is development of GAS applications. We have developed GAS-based models of housing dynamics (Benenson et al., 2002) and urban growth (Torrens, 2002a, b, 2003) thus far, and the results are promising.

REFERENCES

Albin, P. S. 1975. *The Analysis of Complex Socioeconomic Systems*. Lexington, MA: Lexington Books.

Batty, M., H. Couclelis, and M. Eichen. 1997. Special issue: urban systems as cellular automata. *Environment and Planning B* 24(2).

Benenson, I. 1998. Multi-agent simulations of residential dynamics in the city. *Computers, Environment and Urban Systems* 22:25–42.

Benenson, I. 1999. Modelling population dynamics in the city: from a regional to a multi-agent approach. *Discrete Dynamics in Nature and Society* 3:149–170.

Benenson, I. and I. Omer. 2000. Multi-scale approach to measuring residential segregation and the case of Yaffo, Tel-Aviv. In *Proceedings of the Fifth Annual Conference on GeoComputation*, edited by R. Abrahart. Manchester: GeoComputation CD-ROM.

Benenson, I. and I. Omer. 2003. High-resolution Census data: a simple way to make them useful. *Data Science Journal (Spatial Data Usability Special Section)* 2:117–127.

Benenson, I. and P. M. Torrens. 2004a. *Geosimulation: Automata-Based Modeling of Urban Phenomena*. London: John Wiley & Sons.

Benenson, I. and P. M. Torrens. 2004b. Geosimulation: object-based modeling of urban phenomena. *Computers, Environment and Urban Systems* 28:1–8.

Benenson, I., S. Aronovich, and S. Noam. 2001. OBEUS: Object-Based Environment for Urban Simulations. In *Proceedings of the Sixth International Conference on GeoComputation*, edited by D.V. Pullar. Brisbane: University of Queensland, GeoComputation CD-ROM.

Benenson, I., I. Omer, and E. Hatna. 2002. Entity-based modeling of urban residential dynamics: the case of Yaffo, Tel Aviv. *Environment and Planning B: Planning and Design* 29:491–512.

Benenson, I., S. Aronovich, and S. Noam. 2004. Let's talk objects. *Computers, Environment and Urban Systems* (forthcoming).

Berec, L. 2002. Techniques of spatially explicit individual-based models: construction, simulation, and mean-field analysis. *Ecological Modelling* 150:55–81.

Box, P. 2001. Spatial units as agents: making the landscape an equal player in agent-based simulations. In *Integrating Geographic Information Systems and Agent-Based Modeling Techniques for Simulating Social and Ecological Processes*, edited by H.R. Gimblett. Oxford: Oxford University Press.

Brown, John Seely and Paul Duguid. 2000. *The Social Life of Information*. Boston: Harvard Business School Press.

Chapin, F. S. and S. F. Weiss. 1968. A probabilistic model for residential growth. *Transportation Research* 2:375–390.

Couclelis, H. 1985. Cellular worlds: a framework for modeling micro-macro dynamics. *Environment and Planning A* 17:585–596.

Dijkstra, J., H.J.P. Timmermans, and A.J. Jessurun. 2000. A multi-agent cellular automata system for visualising simulated pedestrian activity. In *Theoretical and Practical Issues on Cellular Automata*, edited by S. Bandini and T. Worsch. London: Springer-Verlag.

Epstein, J. M. 1999. Agent-based computational models and generative social science. *Complexity* 4:41–60.

Epstein, J. M. and R. Axtell. 1996. *Growing Artificial Societies from the Bottom Up*. Washington, D.C.: Brookings Institution.

Gimblett, H. Randy, Ed. 2002. *Integrating Geographic Information Systems and Agent-Based Modeling Techniques for Simulating Social and Ecological Processes, Santa Fe Institute Studies in the Sciences of Complexity*. Oxford: Oxford University Press.

Hegselmann, R. and A. Flache. 1999. Understanding complex social dynamics: a plea for cellular automata based modelling. *Journal of Artificial Societies and Social Simulation* 1(3): <http://www.soc.surrey.ac.uk/JASSS/1/3/1.html>.

Kohler, T. A. 2000. Putting social sciences together again: an introduction to the volume. In *Dynamics in Human and Primate Societies*, edited by T.A. Kohler and G. Gumerman. New York: Oxford University Press.

Liu, X.-H. and C. Andersson. 2004. Assessing the impact of temporal dynamics on land-use change modeling. *Computers, Environment and Urban Systems* 28: 107–124.

Meyer, J.-A., A. Berthoz, D. Floreano, H. L. Roitblat, and S. W. Wilson, Eds. 2000. *From Animals to Animats 6: Proceedings of the Sixth International Conference on Simulation of Adaptive Behavior*. Cambridge, MA: MIT Press.

Miller, H. J. 1999. Measuring space-time accessibility benefits within transportation networks: basic theory and computation procedures. *Geographical Analysis* 31: 187–213.

Nakajima, T. 1977. Application de la théorie de l'automate à la simulation de l'évolution de l'espace urbain. In *Congrès Sur La Méthodologie De L'Aménagement Et Du Dévelopment*. Montreal: Association Canadienne-Française Pour L'Avancement Des

Sciences et Comité De Coordination Des Centres De Recherches En Aménagement, Développement Et Planification (CRADEP).

Noble, J. 2000. Basic relationship patterns. In *Pattern Languages of Program Design 4*, edited by N. Harrison, B. Foote and H. Rohnert. New York: Addison-Wesley.

O'Sullivan, D. 2001. Exploring spatial process dynamics using irregular cellular automaton models. *Geographical Analysis* 33:1–18.

O'Sullivan, D. and P. M. Torrens. 2000. Cellular models of urban systems. In *Theoretical and Practical Issues on Cellular Automata*, edited by S. Bandini and T. Worsch. London: Springer-Verlag.

Peckham, J., B. MacKellar, and M. Doherty. 1995. Data models for extensible support of explicit relationships in design databases. *VLDB Journal* 4:157–191.

Peuquet, D. J. 2002. *Representations of Space and Time*. New York: Guilford.

Phipps, M. 1989. Dynamic behavior of cellular automata under the constraint of neighborhood coherence. *Geographical Analysis* 21:197–215.

Polhill, J.G., N.M. Gotts, and A.N.R. Law. 2001. Imitative and non-imitative strategies in a land use simulation. *Cybernetics and Systems* 32:285–307.

Portugali, J. 2000. *Self-Organization and the City*. Berlin: Springer-Verlag.

Portugali, J., I. Benenson, and I. Omer. 1997. Spatial cognitive dissonance and sociospatial emergence in a self-organizing city. *Environment and Planning B* 24:263–285.

Portugali, J., I. Benenson, and I. Omer. 1994. Socio-spatial residential dynamics: stability and instability within a self-organized city. *Geographical Analysis* 26: 321–340.

Reynolds, C. 1987. Flocks, herds, and schools: a distributed behavioral model. *Computer Graphics* 21:25–34.

Reynolds, C. 1999. Steering behaviors for autonomous characters. Paper read at Game Developers Conference, at San Jose, CA.

Sakoda, J.M. 1971. The checkerboard model of social interaction. *Journal of Mathematical Sociology* 1:119–132.

Sanders, L., D. Pumain, H. Mathian, F. Guérin-Pace, and S. Bura. 1997. SIMPOP: a multiagent system for the study of urbanism. *Environment and Planning B* 24:287–305.

Schelling, T. C. 1969. Models of segregation. *American Economic Review* 59: 488–493.

Schelling, T. C. 1971. Dynamic models of segregation. *Journal of Mathematical Sociology* 1:143–186.

Schelling, T. C. 1974. On the ecology of micro-motives. In *The Corporate Society*, edited by R. Marris. London: Macmillan.

Schelling, T. C. 1978. *Micromotives and Macrobehavior*. New York: WW Norton and Company.

Schofisch, B. and K.P. Hadeler. 1996. Dimer automata and cellular automata. *Physica D* 94:188–204.

Schumacher, M. 2001. *Objective Coordination in Multi-Agent System Engineering*. Berlin: Springer.

Semboloni, F. 2000. The growth of an urban cluster into a dynamic self-modifying spatial pattern. *Environment and Planning B: Planning and Design* 27:549–564.

Shi, W. and M. Yick Cheung Pang. 2000. Development of Voronoi-based cellular automata — an integrated dynamic model for Geographical Information Systems. *International Journal of Geographical Information Science* 14:455–474.

Tobler, W. 1970. A computer movie simulating urban growth in the Detroit region. *Economic Geography* 46:234–240.

Tobler, W. 1979. Cellular geography. In *Philosophy in Geography*, edited by S. Gale and G. Ollson. Dordrecht: Kluwer.

Torrens, P. M. 2001. New tools for simulating housing choices. Berkeley, CA: University of California Institute of Business and Economic Research and Fisher Center for Real Estate and Urban Economics.

Torrens, P. M. 2002a. Cellular automata and multi-agent systems as planning support tools. In *Planning Support Systems in Practice*, edited by S. Geertman and J. Stillwell. London: Springer-Verlag.

Torrens, P. M. 2002b. SprawlSim: modeling sprawling urban growth using automata-based models. In *Agent-Based Models of Land-Use/Land-Cover Change*, edited by D.C. Parker, T. Berger, S.M. Manson, and W.J. McConnell. Louvain-la-Neuve, Belgium: LUCC International Project Office.

Torrens, P. M. 2003. Automata-based models of urban systems. In *Advanced Spatial Analysis*, edited by P.A. Longley and M. Batty. Redlands, CA: ESRI Press.

Torrens, P. M. 2004a. Geosimulation approaches to traffic modeling. In *Transport Geography and Spatial Systems*, edited by P. Stopher, K. Button, K. Haynes, and D. Hensher. London: Pergamon.

Torrens, P. M. 2004b. Looking forward: remote sensing as dataware for human settlement simulation. In *Remote Sensing of Human Settlements*, edited by M. Ridd. New York: John Wiley and Sons.

Torrens, P. M. and David O'Sullivan. 2000. Cities, cells, and cellular automata: Developing a research agenda for urban geocomputation. In *Proceedings of the Fifth Annual Conference on GeoComputation*, edited by R. Abrahart. Manchester: GeoComputation CD-ROM.

Torrens, P. M. and David O'Sullivan. 2001. Cellular automata and urban simulation: where do we go from here? *Environment and Planning B* 28:163–168.

Wahle, J., L. Neubert, J. Esser, and M. Schreckenberg. 2001. A cellular automaton traffic flow model for online simulation of traffic. *Parallel Computing* 27:719–735.

White, R., G. Engelen, and I. Uljee. 1997. The use of constrained cellular automata for high-resolution modelling of urban land use dynamics. *Environment and Planning B* 24:323–343.

Woolridge, M.J. and N.R. Jennings. 1995. Intelligent agents: theory and practice. *Knowledge Engineering Review* 10:115–152.

CHAPTER 24

A Process-Based Ontological Approach to Simulating Behavioural Dynamics in Agent-Based Models

Pragya Agarwal

CONTENTS

24.1 INTRODUCTION

It is widely accepted in the geographical community that agent-based simulation is a useful technique to model dynamics and change encapsulated in real-world processes (Batty and Jiang, 1999). Agent-based systems have been found useful for modelling geographical phenomenon primarily because they provide the possibility to model group interactions and behaviours that emerge from individual properties (Dibble, 1996). This *emergent* property in agent-based models (ABM), since first fully exploited and demonstrated in the model *sugarscape* (Epstein and Axtell, 1996) has been seen as a revolutionary development for social science

0-8493-2837-3/05/$0.00+$1.50

(Bankes, 2002). However, behavioural dynamics include not only individual and group behaviours but also interplay of space and time to determine the spatial behaviour and decision-making in a geographical environment. Behavioural dynamics is complex because of the inherent complexity in the individual cognitive capabilities combined with the qualitative nature of space–time conceptualisations that result from these individual encodings. The capabilities of agent-based methods for modelling behavioural dynamics have not been carefully investigated yet. Although previous work has examined the role that these models play in the theoretical emergence of social constructs (O'Sullivan and Haklay, 2000), the focus of interest within this chapter is on the assessment and reconstruction of existing modelling tools within an accepted theoretical framework. To ensure that the theoretical structure emerging from a model is understood and reviewed within the framework of constraints and assumptions that an agent-based technique offers, it is crucial to be aware of the methodological and analytical foundations and underlying assumptions of an ABM.

Agent-based models have been previously stated as providing a "natural ontology" for most social problems in a National Academy of Science (NAS) document (Bankes, 2002). Ontology is most commonly defined as a "specification of a conceptualisation" (Gruber, 2001). It is being applied in geographical applications to identify the "truth contained in a domain" (Guarino, 1995) and as a way to "link models to external world and to each other" (Winter, 2001, p. 588). Although ontology was first proposed in philosophy as a means of identifying the natural categories in the real world, it has been applied as a means to define semantic constructs in design of information system that allow for inter-operability across a number of users and domains. Within this chapter, the term *ontology* is being employed in a philosophical sense as an abstract top-level view of the world in form of natural categories, concepts and entities. The plural *ontologies* is being employed at the operational level across different levels of granularity as action-oriented conceptualisation of functions, operations, actions, and events as linked to entities and concepts involved in a specific domain. The use of *ontology* or *ontologies* in all cases is to allow inter-operability of concepts across different users within the same domain or across different domains in case of an over-riding unifying upper level ontology. Process-based ontologies are a contextual research concern in GIScience, although this discussion is yet to be extended into an operational domain. ABMs have previously been applied to formalise domain-specific ontologies developed in behavioural domain. However, the basic ontological tenets available within the agent-based framework have to be investigated for their applicability to space–time dynamics in behavioural processes, and it is proposed that the development of a flexible, top-level ontology will allow for inter-operability of behavioural models and also in developing process-oriented geographical models. In this chapter, the ontological foundations of ABMs are examined, with specific reference to behavioural dynamics and process-oriented geographical applications; to identify level of natural ontology that the current framework provides and to direct the development of ontology within agent-based methods that can provide a more dynamic modelling framework.

The ontological approach, proposed here, consists of two major steps:

1. Extracting the classes, categories, relations within the domain of operation, and the scale and frame of reference that the various actions and operations are being performed within, and
2. Design a top-level ontological framework within the modelling environment to allow for incorporation of these domain categories from cognitive processes.

It is outside the domain of this chapter to discuss or influence the ontological structuring of domains for modelling purposes, although this is a wider initiative and slowly becoming part of research ethics in GIScience. However, this chapter provides directions for a higher-level conceptual model for agent-based model that is available to incorporate the domain-level ontologies and can also provide a unifying higher-level platform for the different domain ontologies to be extracted within.

24.2 AGENT-BASED MODELS, GEOGRAPHIC PROCESSES AND BEHAVIOURAL DYNAMICS

Simulation models have been found to be effective in social science because of their ability to: enhance understanding of the real-world phenomenon, for testing the applicability of theories observed and deduced at micro-scale at an aggregated scale, and for decision-making support through visualisation of transposition and alteration of rules in a model. The idea of a "computational laboratory" (Dibble, 1996) as a means to discover ways and develop tools to model people and societies in GIS, and the use of CA-based simulation within GIS to simulate space–time dynamics has been discussed by several research initiatives in GIScience (Wu, 1999). In the context of geographical and urban models, agent-based models have become popular as the limitations of cell-based structures in cellular automata was realised in the imposition of its rigid grid-based structure and the deterministic nature of other models and their limitations in modelling the heterogeneity from a real-world application. Benenson and Torrens (Chapter 23, this volume) also discuss multi-agent systems (MAS) in relation to cellular automata. With its capacity to model individual actions and behaviours, agent-based simulation has been discussed as being powerful tool for increasing dynamic modelling capabilities in GIS and to give an insight into rules of human interactions and decision-making in real-world settings (Batty and Jiang, 1999).

Swarm (SDG, 2003) is the most exhaustive object-oriented modelling environment developed and one that is most commonly applied to social and urban simulations. Other development environments such as *Repast* (University of Chicago, 2003) and Ascape (Brookings, 2003) are slowly becoming more popular in the geographical domain because of inclusion of more classes, behaviours and operations within their JAVA-based object-oriented development environment. Some of these models also have capabilities for integration with other geographical modelling platforms such as IDRISI. A detailed discussion of multi-agent based simulation techniques is not possible within the realm of this chapter; Zeigler (1976), Drogoul and Ferber (1994), Gilbert and Troitzsch (1999) can be referred to for some introductory reading in agent-based simulation in social contexts. Rodrigues et al. (1996) provides a good overview of examples of agent-based application in spatial decision support systems.

TRANSIMS (Beckman, 1997) and the STREETS model (Schelhorn et al., 1999) are examples of applications of distributed agent-based technique within a *Swarm* environment to model navigational behaviour at urban and regional level. The Amadeus framework (Timmermans et al., 2002) and its behavioural component, *A Learning-Based Transportation Oriented Simulation System*, ALBATROSS (Arentze and Timmermans, 2000), represent the most advanced agent-oriented model in the travel-demand analysis arena. This framework is moving rapidly toward the development of a large-scale agent-based activity simulation model, making significant advances in the induction of choice heuristics and in the modelling of adaptive scheduling (Joh et al., 2002). Behavioural elements for agents such as visual acuity and fixation have been included to allow for more adaptive agent properties. To date, however, this initiative has not produced a truly dynamic model of interacting agents, instead focusing more on the aggregation of individual agent–environment interactions. Portugali and Benenson (1997) and then Benenson (1998) have considered the cognitive and behavioural aspect of the individuals as determinants of the spatio-temporal dynamics in a city. This aspect was included as a uni-dimensional quantitative vector by Portugali et al. (1997) within the ABM framework. The qualitative and multi-dimensional nature of the behavioural factor becomes more defined in Benenson's model (1998) who incorporates it into his agent-based urban model as a high-dimensional binary "cultural code" vector that influences the individuals interactions. The cultural code vector is an advancement on other ABMs that are purely quantitative. However, the dynamic nature is limited as the space–time framework that the agents are acting in is a static, rigid entity common to all agents. The cognitive properties of the agents have not been extended to include the varying spatio-temporal definitions created by the individuals interacting with the urban environment. Some of these issues will become clearer in the discussion that follows in section three.

Although agent-based simulation techniques first became popular in social sciences because of their possible capacity to model human actions and dynamics within the various domains; it has been found that time is still considered as much in discrete steps within the model and the concept of space and time is very much the absolute space–time Newtonian framework. The absolute nature of space and time has been debated and discussed in geographical models (Couclelis, 1998a, b) and although ABMs are very much being used for modelling geographical processes, the underlying assumption in these models for the absolute nature of space and time and crisp discretisation of spatial and temporal entities has not been questioned or debated. Human processes rely very much on a relative spatial and temporal scale. Logic-based and qualitative reasoning based on natural language along with relational spatio-temporal models (Bennett and Galton, 2000; Bennett, 2001; Cohn and Hazarika, 2001) have been shown to be more applicable for modelling human dynamics than absolute geometric concepts. Inclusion of relative matrices rather than the absolute space–time matrices currently employed in ABMs will make them more effective for modelling human reasoning. In process algebra, differential equations have been found to be effective for dynamics modelling. This along with a Dynamic Systems Theory (DST) has formed the basis for most models of low-level cognitive processes. However, for higher-order cognitive processes, especially ones employing language processing and qualitative reasoning within the environment, the quantitative approach within a DST has been found to be limited (Jonker and Treur, 2002). In spatio-temporal reasoning, natural language is often employed as a referent and propositional structure of cognitive processes has been shown by experiments. In such cases, a version of DST that uses qualitative notions has been proposed as the way forward. The problem within such an approach lies in the application of a reasoning structure that is not based within the process itself but outside it and hence imposes an additional framework of concepts and terminology in the model. The ideal solution will be to make the framework of the agent-based models stronger to handle such tasks and reasoning approaches. A bottom-up approach will also help resolve the problem of semantic ambiguity that might arise from imposing an external framework of DST on the existing ontology within an agent-based model.

Although dynamic modelling has been positioned and discussed in biological processes and in organisational processes and a few developments have been made in that area of application, agent-based models are still struggling with the handling of the nature of space and time that are included within cognitive processes. Couclelis (2001), using the example of land-use models, maintain that agent-based models are lacking in temporal capacity to model actions and processes as the often parallel trajectories of human decision-making are equated with a one-dimensional trajectory that is present in institutional decision-making. The fact that the processes and events occurring from cognitive interactions are different from physical processes and need to be treated differently makes the need for identification of a definitive ontology even more significant to identify the different kinds of processes and actions in a particular domain. Frank (2002) brings the integration of spatial and temporal process scales within an ontological framework of operation-based semantics to integrate the actions to objects within a particular context, and hence making the meanings of actions and objects to be grounded in the context of the operation. A process-oriented approach in agent-based models can benefit from such a framework.

Most geographical agent-based models are based in an object–field duality despite the fact that the world is not distinctly divided into objects and fields. Also, when modelling spatial and temporal dynamics, an object–field structure means that the entities have to be either spatial or temporal objects or spatial and temporal fields. Although, realistically speaking, pure temporal object is yet to be designed, spatial and temporal objects imply a specific location and a concentration of a crisp value at a particular location, while fields have a range of values varying according to a set of rules. Topological and mereological relations that are specified on an object–field model are yet to be truly resolved on temporal relations, and also on spatial relations, especially in cognitive processes. An objects–field representation is thus based on a vast set of assumptions. A truly integrated spatio-temporal ontology cannot emerge from an object–field duality (Galton, 2003). Although there has been a long-standing debate on the topic, most geographic representation still exists within an object–field paradigm. Raper and Livingstone (2001) attempt to define objects and field in terms of their temporal and spatial properties in classification such as perdurantist–endurantist and substantivilist–relationist positions. This is not, however, definitive classification as the roles are inter-changeable depending on context and their roles in the process. The duality of the representation of geographic phenomenon has been seen as problematic in conceptualisation of all real-world entities, both physical (~natural) and those that emerge from social acts (~human reasoning). A need to fuse space–time within this overall ontology of geographic kinds, one that is mostly emergent from the cartographic tradition and then by digital representations and spatial databases in GIS, has been recognised. However, no definitive representation has been able to capture the four-dimensional space–time integrated geographic process structure that has been so widely talked about. Even agent-based models that appear to provide a shift away from the conventional GIS representations in allowing modelling of processes and dynamics are trapped in the same paradigm. There is enough evidence for the limitations of an object–field paradigm for modelling of dynamics and hence a need for a conceptualisation that is emergent from the process itself, both casual and contingent.

24.3 ONTOLOGICAL ASSESSMENT OF ABM FOR DYNAMIC MODELLING

The problems associated with dynamic modelling begin at the level of conceptualisation within an agent-based model. In a panel chapter emerging from the First U.K. workshop on Multi-Agent Systems (MAS) in 1997 (Fisjer, 1997), it was accepted and proposed that a bridge between theory and practice is essential in design of an ABM. Although this proposal focussed on an engineering approach to design and implementation of ABM, it re-addressed the concerns from a social science perspective and re-defined the approaches from a computer science perspective, where development of high-structure languages was evidently not linked to the theories and the kinds of behaviours that would need to be modelled within an ABM. This approach has meant that the domain alone is not defining the model constraints. Due to the fact that, to a certain degree, the operations from the real world are being forced into the operational elements defined by the language structures available for agent-based models, there is a compromise involved in modelling high-level individual and group operations from real world. As a consequence, the theories emerging from the model and the nature of results depend to a great extent on the modelling language and its ontological structure and the operational elements that the development environment has allowed. Since even the AI approach has found it difficult to mimic certain high-level knowledge structures from the real world, a bottom-up approach allows for identifying the kinds of behaviour and knowledge structures that "need" to be modelled before realising "what" is possible within the constraints of a practical development scenario.

24.3.1 Natural Ontology in Agent-Based Models

Most models in the geographical context fall into two categories: *deductive* and *inductive* (Gilbert and Doran, 1994), that can be equated to a top–down or bottom-up approach. The basic framework that such simulations are built within is that of {agent, environment, relations}. There have been few variations on this phenomenology. For example, Jiang and Gimblett (2002) provide a discussion on the four-feature nature {agents, objects, environment, communications} of MAS and their suitability to modelling space–time dynamics. Here, the inclusion of objects as a category was made primarily to distinguish between action-oriented and static agents, with agents being mobile and objects static.

Modelling environments have mostly considered agents as an object, with a set of actions and communications associated with it. The agent is a continuant; a static, non-changing entity (Nowak, 2001). There have been recent concerns in the social science community regarding the standard geometric configuration imposed on the model and the invariance in the real-world simulation that could result from it (Revilla, 2002). The agents' actions and communications are mostly prescribed, with very few exceptions, as a set of specific empirical relations. While most of these relations are set on the agent–environment communications, multi-agent models include communications between the agents in a distributed messaging environment. The agents perform a set of actions, make choices in some models, supposedly imitate human-like behaviour in models from the real world and are assigned a set of attributes that determine the actions it is allowed to perform within the model. Many models employing knowledge-engineering approach have argued for more intelligent agents. However, in such models there is a conflict between rationality and emotiveness of the agents and the two behaviours are seen as distinct. Besides that, the intelligence of the agent is not a cognitive process but in all cases is determined by an external layer comprised of rules, relations, and set of instructions.

Agents operate within a system consisting of a set of associated properties for the agent and also for the environment and rules for communication with other agents, objects, and environments. The more complex and higher-level an agent becomes, the level of complexity in the framework within which it is operating becomes increasingly difficult to design and set rules for. To simulate human actions and operations in the real world, efforts have been made to incorporate higher-level intelligence in autonomous agent systems through developing software that will allow more real-life and effective simulation of the inherent complexities in the real world. Franklin and Graesser (1997) have provided a taxonomy for proposed agents from low-level reactive agents to higher-level deliberative, intelligent agents with more adaptive behaviour through learning and communication with other agents, objects and environments, responses through internal emotional and personality traits, and remembrance for decision-making. Ferber (1999) in his "kenetic software" theory has identified two types of agents: *reactive* and *cognitive*. The cognitive agents have been designed with an inherent capacity to adapt and be influenced by the environment around and have goals and intentions that are largely self-driven. However, simplistically speaking, this distinction is based on the premise that *reactive agents* simply react to cues within the environment and the *cognitive agents* act according to their own plans. Experimental work in human cognition has shown that reactions to spatial and environmental cues are a result of the way that these cues and signs are internally processed within the cognitive domain forms a notion of spatial and temporal configuration of the reality for the individual. So this distinction needs to be revisited as a reactive agent is not devoid, epistemologically speaking, of a cognitive process and a cognitive agent will have reactive properties to be able to externalise the internal cognitive process in form of external actions. For modelling cognitive processes, the current designs and distinctions in agent-design are not adequate.

The *environment* in an agent-based model is variously termed as "*medium*" or "*surface*" (Epstein and Axtell, 1996). From an ontological perspective, this can be considered as a top-level ontology. However, the problem lies in the fact that any further categorisation or hierarchical ordering does not follow from this top-level ontology. The environment in an agent-based model has been criticised for its "checkerboard" limitations (Chattoe, 2001). This, along with setting the shape constraints in the design of the agent, which is mostly set as a square (Revilla, 2002), has implications on the topological and mereological relations and boundary conditions that can be specified within the model. Although most agent-based model in the geographical context are designed as object-oriented view of the world with the agents designed and characterised as objects, the environment is designed mostly as a rasterised surface with a cell as a basic unit and having a set of attributes attached to it. The basic cell-like real world has a specific scale set within it and follows on from a field-based view of the world. The field-based surface in an agent-based model is at times enhanced with the object-view, mostly when agent-based models are linked to a GIS database. STREETS model (Schelhorn et al., 1999) is a good example of this.

The nature of agents is such that actions and processes are linked to the transition rules outside the agent itself and not as part of the agent design itself. There has not been much consideration of the philosophical and conceptual problems that such a system proposes in modelling real-world dynamics. Because of temporal relations being defined outside the agent and not part of the agent's cognitive process, urban models that rely on experiential time, response time, and recognition time for setting behaviours and actions are not feasible within an agent-based environment. The notion of temporal reasoning based on a process philosophy (Nowak, 2001) and structured around qualitative reasoning of temporal intervals around events and occurrences (Bennett and Galton, 2000) is not feasible within such an arrangement.

In real world, the nature of processes and interactions vary at different scales and even in discretisation of space and time, the scales are very different. All agent-based models are generalisations of reality and hence the scale at which this generalisation takes place is crucial for effective simulation. Scale is a crucial issue to consider also when deciding the meanings of the actions and operations that are applicable at that particular scale and the nature of dynamics that are emerging at that particular scale. Such "multi-scale interactions" (Frank, 2001) are complex for a model, and agent-based models have dealt with these by setting the discretisation of spatial and temporal scales at a single scale throughout the whole model. Besides the scale of the overall process, a process may have objects at different scales acting within the same process with a set of operations and actions attached to them, as defined within the semantics in process-oriented ontologies as proposed by Frank (2002). In a model for a navigation system in a city, individuals moving within an airport or pedestrians on a road operate on different temporal scales, and the movement of the aeroplane and the bus is on different temporal scales as well. The action of making choices about shortest route will be taken at a larger scale with a consideration of the overall route system and goals and destination while the action of turning around a corner while driving is at a different spatial scale. Scale is, therefore, a significant factor of consideration because of the implications of understanding the process and the interactions and actions at different levels of granularity within an event, the effect of the finer-scale multiple processes on the overall larger process at the higher-level of granularity and the identification of the scale at which a model is most representative of the dynamics within the process. In geographical models, there has been an awareness of the limitations of spatially explicit models and an adoption of spatially distributed modelling approaches as representative of the different scales at which the actions and interactions are influencing the overall process. However, the issue of a *process-lag* or the sum of finer, lower level interactions not equating to the larger, higher-level process operations is an issue that is of concern in modelling geographical phenomenon. In agent-based models, the movement from one scale to another for modelling a process is limited in most cases. Either a top–down approach where larger-level interactions are deductive of lower-level dynamics or bottom-up approaches where the lower-level interactions and actions are seen as predictive of the higher-level process are adopted in agent-based models. Both these approaches have a set of underlying assumptions that do not hold true in case of modelling dynamics within geographical processes. In case of behavioural processes, such approaches are shown to be even more simplistic when the complexity in the range of spatial and temporal scales and the process-lag become more significant and the claim of MAS being able to imitate complexity in reality is contested. Specifically in the context of human process modelling, these issues of scale and meanings attached at different scales are even further complicated by the vagueness in definition of meanings attributed to the events and the fuzziness that is inherently a part of human actions and decision-making.

Specifying rules and relations as external to the agent and the environment means that the most models employ geometric specification for specifying relations between geographic kinds. This is true specially in agent-based models of geographic phenomenon where integration with a GIS model means that the geometric rigidity in standard GIS platforms determine the relational specifications. Topological and mereological relationships follow a set of co-ordinate and Euclidean geometry rules. However, it has been shown by research in spatial cognition that cognitive estimates of relations such as proximity does not necessarily follow rules of symmetry, transitivity, and reflexivity (Duckham and Worboys, 2001; Worboys, 2001). Although many more real-world experiments are needed to identify the extent to which the geometric concepts hold true in cognitive processes both in the spatial and temporal contexts, it is also obvious that models of cognitive processes cannot rely on specific geometric conceptualisations.

It follows from the previous discussion that the basic ontology within an agent-based model needs to be restructured for modelling behavioural dynamics to incorporate:

- Flexible frames of reference for resolving spatial and temporal relations.
- Non-geometric conceptualisations.
- Non-rigid parameterisation in object–field duality.
- Non-linear discretisation of spatial and temporal intervals and relations.
- Consideration of the process, event, operations and actions hierarchy within an action-oriented approach.

- Allowance for vague concepts and fuzzy membership functions.
- A top-level ontology that can set directions for domain level ontologies by providing higher-order reference concepts, and also a flexible enough framework at the same time to allow for incorporation of different domain ontologies.
- A top-level ontology that allows shift between process scales without the need to change the higher-order concepts.

24.3.2 Previous Ontological Approaches

In the last few years there have been many attempts to define an ontological models and some of the languages and models developed are DAML + OIL (McGuiness et al., 2002), Ontolingua, OntoWeb and OWL. Although these are able to provide a good overall preliminary ontology using the concepts and relations available within them, they have been found lacking in developing detailed geographic ontology especially when human behaviour and conceptualisation is to be incorporated. The five-tiered ontology proposed by Frank (Frank, 2001) and the SNAP–SPAN ontology (Smith, 2002; Grenon and Smith, 2004) have so far been unable to provide a process-oriented view of the world. SNAP–SPAN considers distinctions in real-world entities as enduring or perdurant that will have to be tested in various domains to assess its applicability in modelling entity transformation and change in processes, both physical and cognitive. The top-level ontology designed by Sowa (1995) considers distinction between *abstract* and *physical* concepts with the former being where the spatio-temporal location cannot be specified while the latter where it can be. The entities are also distinguished on the basis of their dependence relations with other entities in the model and as *continuous* or *occurant*, with these definitions being similar to discussed before. Although Sowa's ontology is more detailed and much more capable of handling process-modelling because of the spatio-temporal inclusion in the categories, these are still vague categorisations. It does not solve the problem of pre-assigning states to entities in a model. Existing top-level ontology are thereby not adequate to be directly applied to the agent-based modelling framework.

In previous agent-based models, the use of ontology has been at the operational level rather than at an overall abstract conceptualisation. Within agent-based models, work of Raubal (2001) and Raubal and Kuhn (2002) are of significance because of its specific application in consideration of spatial cognition and behavioural dimensions within the decision-making environment. An extension of the goal-oriented agent (Russell and Norvig, 1995) has been redefined as a *cognizing* agent in this context because of its capabilities to employ common-sense reasoning. Although this work does provide an ontological and epistemological standpoint to agent-based models, it does not inherently question the ontological basis and limitation of agent-based models itself for modelling behavioural and cognitive processes. There is also no discussion regarding the different temporal discretisation within cognitive processes in the model and of varying operational frames for agent and environment. The agent-based model here is seen as a vehicle for modelling the ontologies designed for the domain and adapted with its complete set of assumptions to the particular ontology on hand. In knowledge engineering approaches, there have been a few initiatives that have attempted to model the abstract all-encompassing top-level ontology as the

CYC project (Gutiérrez-Casorrán et al., 2001) for more inter-operable concepts but faces problems of hierarchical definition of processes and tangible things. This ontology is derived from Sowa's top-level ontology and captures the sequence of actions and operations as part of a process, and the environment as related to these actions and the overall process. Nevertheless, it still follows the object–field paradigm. The ontological application in this model was aimed at providing a common language and framework for all agents within the model. Although it has included concepts of social and organisational laws and some semblance of space–time integration, it is operating within an agent–environment paradigm without examining the assumptions within this framework, and the incompatibility of theoretical grounding that might be caused while imposing an externally designed ontology on it. These developments are also evidence of the lack of a strong ontological basis or "ontological thinness" of an agent-based model and evidence of a need for a better categorisation of the objects participating or being influenced in the process of a particular natural or cognitive phenomenon.

24.4 PROPOSING AN ONTOLOGICAL GROUNDING IN *IDENTITY*

An assessment of agent-based models has demonstrated their "ontological thinness" for modelling dynamics in most cases, especially for cognitive processes. It is clear from the previous discussion that most models such as agent-based models have struggled to maintain an action-oriented view or develop a process-oriented ontology. Nevertheless, where the applications have considered ontological foundations have been able to cope better with the natural and cognitive categories existing in the real world. An ontological approach to agent-based models can result in more effective theory formulation by sound conceptualisation, categorisation, and rule setting, and also in resolving representational issues in the model. An ontological framework will help resolve, to a certain extent:

1. The issue of the identification of different scales at which a process is being operationalised.
2. The kind of events that sum up to the whole overall process.
3. The nature and extent of actions and interactions happening at different scales and the objects participating in these actions and the transformations within these objects.
4. The meanings being attached to the different actions and operations within the process in that particular context. This becomes important when individuals are integral part of the process dynamics as actors or receptors.

An ontological grounding or conceptualisation is needed that could serve as a top-level abstract ontology in an agent-based model design and allow for varied domain level operational ontology to be incorporated within its structure. This ontology will also need to allow a move from rigid pre-set distinctions and labels and also a shift from object–field duality. Here, a grounding of this overall ontology in the concept of identity is being proposed.

This ontological framework is set within a process-hierarchy that is formalised as follows. In models of geographic phenomenon, commonly employed terms are *process*, *events*, *actions*, *operations*, and *dynamics*. There is some confusion as to what these terms imply and how these are applied (Raubal, 2001). From the EASG dictionary definitions (Gibson, 1999, 2000), a *process* can be interpreted as a set of *actions*, *operations*, and *events*. A *process* is a sequence of *events*; an *event* is a derivative of the overall genus *process* but is of a specified existence spatially and temporally. Spatio-temporally, an *event* can be derived as something with a more clearly defined extent than a *process* that is a set of events and is a derivative of the overall genus *sequence*. Thereby, a *process* is a sequence of events and is also defined by the set of *actions* and *operations* within it. *Actions* and *operations* are performed within the process and events and are not necessarily part of the process-event hierarchy. In the conceptual ontology proposed here, *process* is a top-level category used for an overall sequence of *events*, *actions*, and *operations* that are attached to specified objects and *event* as an occurrence or happening with a specified existence within a *process*. The *dynamics* of the *process* emerge from this overall structure and are defined by the nature of temporal and spatial scales engaging within the overall process as shown in Equations (24.1) to (24.3).

$$P \leq E_n \leq o_n + a_n \tag{24.1}$$

where P is the process, E_n is the sum total of events, a_n is the actions, o_n is the operations.

$$E_n = \sum_{t=0}^{t=n} (e_1, e_2, e_3, \ldots, e_n) \tag{24.2}$$

where $e_1, \ldots, e_n$ are the set of events within a specified time frame of $t = 0$ to $t = n$.

$$e_1 = \sum_{x_1}^{x_n} (a_n, o_n) \tag{24.3}$$

Each event is structured from the set of all actions, a_n, and operations, o_n, performed as part of each event, on all entities x_1 to x_n taking part in these. It has to be noted that the reflexivity and transitivity rule does not hold good for this hierarchical structure of process, event, operations, and actions.

It is important here to distinguish this concept of "identity" in modelling processes from the concept of "identity" as discussed previously. Raper and Livingstone (2001) have used the concept of "identity" of entities from a representation perspective and have proposed a unified formulation to state that "*the identity of phenomenon emerges through the interaction of socially driven cognitive acts with the heterogeneous structure of the world*" (p. 238). Here, although they approach the subject from an ontological perspective, the resolution is at a top-level and not at the entity level. Also, no possible directions for representation of such "identities" of these phenomena are discussed. In their discussion on representation of spatio-temporal relations between casual and contingent phenomenon, Raper and Livingstone (2001) fail to make definitive distinctions between identity of entities and that of the phenomenon itself. Hornsby and Egenhofer (2000) employed the concept of "identity" to represent change. Based on a wide body of knowledge associated with spatio-temporal representation in GIS data models, semantics primitives of change are used in applying the identity-based knowledge representation. It is, however, based in standard GIS models and, hence in an object-oriented development environment. The transition in identity is modelled as an external primitive and hence the concept of memory and change in state is not linked to the "objects" being modelled.

In the conceptual ontology being proposed here, the concept of "identity" emerged from grounding the ontology in a process-oriented view of the world, which is based loosely in the concept of formation of *sense of place* by individuals within an environment and the emergence of the environment as a spatial and temporal footprint linked to the individual's perceptions of the place. This has been discussed in further detail in Agarwal (2002), and Agarwal and Abrahart (2003). The concept of *sense of place* although borrowed from its phenomenological discussions in human geography (Tuan, 1974; Buttimer and Seamon, 1980) does not include cognitive and spatio-temporal reasoning concepts in its original meaning, and hence has been re-defined to include not only cognitive actions but also physical operations in the process. This implies that the meaning of the concept or entity (however it might be termed) lies within its current state in the process itself and the spatial and temporal trajectories can be truly inter-linked. Instead of making dualist distinctions in real-world entities, be it object–field, endurantist–perdurantist, or relationist–substantivilist, the entity is allowed to "move" between categories and labels; consequently linking each process to the identity of entities that are allowed to between categories and labels.

In this top-level ontology for agent-based models, *identity* is linked to the concept of "moderate realism" proposed by Smith (1995) where the "natural kinds" or entities are not being termed object or field but are allowed to state their current state depending on their status and meaning in the process. In a "moderate realist" *identity*, a resolution and representation of entities at several levels of granularity and resolution is possible while these are being transformed and created either through physical processes or "as shadows cast onto the spatial plane (*and temporal plane*) by human language and reasoning" (Smith and Mark, 1998, p. 308). *Identity* is defined here as the "current state" and hence comprises the notion of change and transformation within it; properties at different scales and granularity, relations set on properties, frames of reference and scales, and the meanings employed for different concepts within the process at different contexts and scales. This will allow for a move away from an object–field distinct entity and also allow for setting of properties as part of the process. This implies that even when an entity merges, splits, or transforms and forms new entities, the history of object and region association in both previous and current states is maintained. For each process, P (as defined in Equation (24.1)), the Identity, I_n, defined here in Equation (24.4), forms the basis of the modelling framework.

$$I_n \cong (C_s, M_h) \tag{24.4}$$

I_n is the Identity of all entities in the model constituted by current states and the memory of previous states for all entities.

$$C_s = [S(x_1), S(x_2), S(x_3), \ldots, S(x_n)] \tag{24.5}$$

C_s is the current state of the model as sum of current states of all individual entities in the operation as shown in Equation (24.5). M_h is the memory of previous current states of the entity as history, in Equation (24.6), where, $h_1, \ldots, h_n$ are the set of all previous histories for each current state of the model. The history of the entity up to its current state along with associated information will be maintained as the memory of the entity within the phenomenon.

$$M_h = \sum_{C_s=1}^{C_s=n} (h_1, h_2, h_3, \ldots, h_n) \tag{24.6}$$

$S(x)$, in Equation (24.5), is the current state of each entity as sum of all current states defined as relations with other entities in the operation as seen in Equations (24.7) and (24.8).

$$S(x) = \sum_{x=1}^{x=n} (x_1, x_2, x_3, \ldots, x_n) \tag{24.7}$$

$$x_1 \subseteq (l_s, l_t, r_s, r_t, R)_{\text{fr}} \tag{24.8}$$

where, R is the set of factors and relations as domain-dependent indices, l_s is the spatial location of the entity, l_t is the temporal location of the entity, r_s is the spatial relations of the entity with all other entities in the operation, r_t is the temporal relations of the entity with all other entities in the operation.

These spatial and temporal locations are determined within the frame of reference of the entity itself, fr. It is essential that the *identity* of the agent is prescribed by the frames of references that it acts within and this will need to be part of the agents' identity as well as part of its relation or association with the environment around it. In behavioural processes, often a combination of egocentric and allo-centric frames of references is employed. Within a multi-scale process, different frames of reference may be employed at different scales and also within the same process; each event or action being operationalised within a different frame of reference. In this approach, the sequence of events, actions, and operations are first identified along with the scales of operations in space and in time, and then broken down to a finer level defined by the associated properties of the actors, and the associated transformations at different scales and at different levels of the process. The semantics of the operations are linked to the properties, the meaning is derived in a particular context and it is the actors and the participators that allow the meaning to emerge through their actions and interactions and their cognitive process of engagement and assigning of meanings.

This abstract top-level ontology can be extended to include properties within the current state of the entities that are determined by the domain. Linking to *identity* will help conceptualise entities as their role in the process. Following on from Galton's (2003) desider that for spatio-temporal ontology, this will mean that instead of fixing the representation of entities within an agent–environment paradigm and giving them definitive roles as objects or environment, the dynamics within the process at each level and stage in the sequence will allow entities to transform from spatio-temporal field to object if necessary; and also allow for the same entity to partake in the process as a spatial or temporal object (or field), if necessary, depending on the actions being performed and the role in its current state. The *identity* is a relational set instead of being an absolute set. The *identity* can be extended to include the environment as well with the changes and transformations within the agents and the environment easy to resolve as these can be easily linked to the *identity*. Until the designs of agent-based models are enhanced to incorporate the multi-scale dynamics within such processes, this ontological approach through structuring and identification of the categories, rules, and relations within the process might provide us with indicators of the scale that is most representative of the overall process and at which a single-scale modelling approach will be most appropriate.

24.5 CONCLUSIONS

Previous discussion has demonstrated that the basic ontology for agent-based models is limited, especially for modelling processes and dynamics. Although some models have attempted to extend the ontology by defining their own categorisation for their own domains, these have not been widely accepted resulting in "ontological thinness" of the models and theories that emerge from the use of the {agent, environment, rules} ontology. An overall ontology in an agent-based model needs to allow for flexibility and adaptability to requirements of cognitive processes. Since behavioural research, both theoretical and empirical, has shown that cognitive spatio-temporal models influence human spatial behaviour and reasoning in urban environments, simulated models can be made more effective by making the framework more suitable for incorporating cognitive rules and behaviours for geometry, topology, and relations in space and time. However, this is limited by the "ontological thinness" of the agent-based models and the lack of process-oriented foundations in this technique. In this chapter, limitations of existing framework are identified and an ontological approach proposed to resolve it. The ontological approach allows the categories of the process and the rules to be incorporated within the modelling framework constituted of an overall ontology of *identity*. Grounding the ontology of an agent-based model in an overall framework of *identity* also permits a shift from rigid constraints imposed by an object–field structure and the rigid geometric constraints imposed by such designs. It allows entities to progress, transform, merge, aggregate, and split in the process and allows previous states to be stored in the memory. The concept of *identity* also draws parallels from the creation of an identity of entities in an environment through the process of cognition and through their relation with other entities in the environment. The relations and communications here can be part of this overall ontology and hence are allowed to influence the current state of the entity. This ontological approach and the conceptualisation of the model into a top-level ontology of identity also makes it easier for qualitative reasoning languages and knowledge representation methods to be incorporated in the model and hence more effective models of cognitive processes in an geographical environment.

Although the proposed top-level conceptualisation based in the concept of *identity* is abstract and needs further formalisation before application, it suggests significant ways forward and away from the object–field duality and provides directions for making agent-based models more theoretically sound. Formalisation of this approach and then testing through application in different behavioural processes is needed to assess the applicability to different categorisation and dynamics. Besides operationalising behavioural and cognitive matrices within modelling environments and consequently extending the design framework of GIS platforms, an ontological approach to agent-based models also provides a stable framework for making agent-based techniques much more applicable to modelling complexities of human behaviour.

ACKNOWLEDGEMENTS

I am thankful to Dan Montello (University of California, Santa Barbara) and Paul Aplin (University of Nottingham) for their comments on the initial drafts of the chapter. I also wish to thank the editors of this volume for all their help and suggestions and the two anonymous reviewers for their comments. The ideas presented in this chapter were developed during my PhD funded by a University Research Scholarship at The University of Nottingham, U.K.

REFERENCES

Agarwal, P., 2002, Perceptual places. In D. Mark and M. Egenhofer (Eds.), *GIScience 2002 Proceedings*, GIScience 2002, September 24–26, Boulder, USA.

Agarwal, P. and Abrahart, R.J., 2003, Agent-based simulation of cognitive neighbourhoods in urban environments. In *Proceedings of International Conference on GeoComputation*, September 8–10, 2003, Southampton, U.K.

Arentze, T. and Timmermans, H., 2000, ALBATROSS: a learning based transportation oriented simulation system, European Institute of Retailing and Service Studies, Eindhoven.

Bankes, S.C., 2002, Agent-based modelling: a revolution? *Proceedings of the National Academy of Science*, 99(Suppl. 3), 7199–7200.

Batty, M. and Jiang, B., 1999, Multi-agent simulation: new approaches to exploring space-time dynamics within GIS, Paper 10, Working Papers, CASA, UCL, U.K., http://www.casa.ucl.ac.uk/multi_agent.pdf

Beckman, R., 1997, TRANSIMS-Release 1.0-The Dallas-Fort Worth case study, Los Alamos Unclassified Report (LA-UR), 97-4502, http://transims.tsasa.lanl.gov

Benenson, I., 1998, Multi-agent simulations of residential dynamics in the city. *Computer, Environment and Urban Systems*, 22, 25–42.

Bennett, B., 2001, Application of supervaluation semantics to vaguely defined spatial concepts. In D.R. Montello (Ed.), *Spatial Information Theory: Foundations of Geographic Information Science*, In *Proceedings of COSIT'01*, LNCS, 2205, pp. 76–91.

Bennett, B. and Galton, A., 2000, A Unifying Semantics for Time and Events QSR Group Technical Report, School of Computing, University of Leeds, U.K.

Box, P., 2002, Spatial units as agents: making the landscape an equal player in agent-based simulations. In H.R. Gimblett (Ed.), *Integrating Geographic Information Systems and Agent-Based Modeling Techniques for Simulating Social and Ecological Processes*, Oxford University Press, New York.

Brookings, 2003, Ascape documentation, The Brookings Institute, http://www.brook.edu/dybdocroot/es/dynamics/models/ascape/

Buttimer, A. and Seamon, D., 1980, *The Human Experience of Space and Place*, Croom Helm, London.

Chattoe, E., 2001, The role of agent-based systems in demographic explanation, http://users.ac.uk/~econec/rostock-0.pdf

Cohn, A. and Hazarika, S., 2001, Qualitative spatial representation and reasoning: an overview, *Fundamenta Informaticae*, 46(1–2), 1–29.

Couclelis, H., 1998a, Geocomputation and space. *Environment and Planning B: Planning and Design*, 25(Anniversary Issue), 41–47.

Couclelis, H., 1998b, Aristotelian spatial dynamics in the age of GIS. In M. Egenhofer and R.G. Golledge (Eds.), *Spatial and Temporal Reasoning in Geographic Information Systems*, Oxford University Press, New York, pp. 109–118.

Couclelis, H., 2001, Why I no longer work with agents: a challenge for ABMS of human–environment interactions. In Dawn C. Parker, Thomas Berger, and Steven M. Manson (Eds.), *Agent-Based Models of Land-Use and Land-Cover Change, Proceedings of an International Workshop*, October 4–7, 2001, Irvine, California, U.S.A.

Dibble, C., 1996, Representing individuals and societies in GIS. In *Proceedings of the Third International Conference on Integrating GIS and Environmental Modelling*, January 21–25, 1996, National Centre for Geographic Information and Analysis, Santa Barbara, CA.

Donnelly, M. and Smith, B., 2003, Layers: a new approach to locating objects in space. In W. Kuhn, M. Worboys, and S. Timpf (Eds.), *Spatial Information Theory: Foundations of Geographic Information Science, Proceedings of International Conference on Spatial Information Theory*, COSIT 2003, Springer, pp. 46–61.

Drogoul, A. and Ferber. J., 1994, Multi-agent simulation as a tool for studying emergent processes in societies. In N. Gilbert and J. Doran (Eds.), *Simulating Societies: The Computer Simulation of Social Phenomenon*, UCL Press, London, pp. 127–142.

Duckham, M. and Worboys, M., 2001, Computational structures in three-valued nearness relations. In D.R. Montello (Ed.), *Spatial Information Theory: Foundations of Geographic Information Science; Proceedings of COSIT'01*, LNCS, 2205, Springer, Morro Bay, pp. 108–123.

Epstein, J. and Axtell, R., 1996, *Growing Artificial Societies: Social Science from the Bottom-Up*. Princeton University Press, Princeton.

Fabrikant, S.I., Ruocco, M., Middleton, R., Montello, D.R., and Jörgensen, C., 2002, *The First Law of Cognitive Geography: Distance and Similarity in Semantic Space*, GIScience 2002, September 25–28, 2002, Boulder, CO, pp. 31–33.

Ferber, J., 1999, *Multi-Agent Systems: An Introduction to Distributed Artificial Intelligence.* Addison-Wesley, Rading, MA.

Fisher, P.F. and Orf, T.M., 1991, An investigation of the meaning of near and far on a University Campus. *Computers, Environment and Urban Systems*, 15, 23–35.

Fisjer, M., 1997, Representing and Executing Multi-Agent systems, report on Panel discussion, First UK Workshop on Foundations of Multi-Agent Systems (MAS'97).

Frank, A.U., 2001, Tiers of ontology and consistency constraints in geographic information systems. *International Journal of Geographical Information Science*, 15, 667–678.

Frank, A.U., 2002, Operation based semantics, Position Paper, Workshop on Action-Oriented Approaches in Geographic Information Science, ACTOR 2002, November 2–4, Maine, U.S.A.

Frank, A.U., Bittner, S., and RAUBAL, M., 2001, Spatial and cognitive simulation with multi-agent systems. In D.R. Montello (Ed.), *Spatial Information Theory — Foundations of Geographic Information Science, International Conference COSIT 2001*, Morro Bay, USA, September Lecture Notes in Computer Science, Vol. 2205, Springer-Verlag, Berlin Heidelberg, pp. 124–139.

Franklin, S. and Graesser, A., 1997, Is it an agent, or just a program?: A taxonomy for autonomous agents. In J.P. Muller, M.J. Wooldridge, and N.R. Jennings (Eds.), *Intelligent Agents III: Agent Theories, Architectures and Languages*, Springer, Berlin, pp. 21–35.

Galton, A., 2003, Desiderata for a Spatio-temporal geo-ontology. In W. Kuhn, M. Worboys, and S. Timpf (Eds.), *Spatial Information Theory: Foundations of Geographic Information Science, Proceedings of International Conference on Spatial Information Theory*, COSIT 2003, Springer, pp. 1–13.

Gibson, F.C., 1999, 2000, The Organon-EASG Conceptual Dictionary, Conceptual Hierarchy Program using CLOS.

Gilbert, N. and Doran, J., (Eds.), 1994, *Simulating Societies: The Computer Simulation of Social Phenomena*. UCL Press, London.

Gilbert, N. and Troitzsch, K.G., 1999, *Simulation for the Social Scientist*, Open University Press. Buckingham, U.K.

Gimblett, H.R., 2002, Integrating geographic information systems and agent-based technologies for modelling and simulating social and ecological phenomenon. In H.R. Gimblett (Ed.), *Integrating Geographic Information Systems and Agent-Based Technologies for Modelling and Simulating Social and Ecological Phenomenon*, Oxford University Press, New York, pp. 1–20.

Grenon, P. and Smith, B., 2004, SNAP and SPAN: Towards dynamic spatial ontology, *Spatial Cognition and Computation*, 14, 69–103.

Gruber, T., 2001, What is an ontology? http://www-ksl.stanford.edu/kst/what-is-an-ontology.html.

Guarino, N., 1995, Formal ontology, conceptual analysis and knowledge representation. *International Journal of Human and Computer Studies*, 43, 625–640.

Gutiérrez-Casorrán, J., Fernandez-Breis, J.T., and Matrinez-Bejar, R., 2001, Ontological modelling of natural categories-based agents: an ant colony, Workshop on Ontologies in Agent Systems, *5th International Conference on Autonomous Agents*, Montreal, Canada, 2001. http://cis.otago.ac.nz/OAS2001/. Accessed 06 September 2003.

Hirtle, S., 2001, Dividing up space: creating cognitive structures from unstructured space. Meeting on Fundamental Questions in GIScience, Manchester, U.K.

Hornsby, K. and Egenhofer, M., 2000, Identity-based change: a foundation for spatio-temporal knowledge representation. *International Journal of Geographical Information Science*, 14, 207–224.

Jiang, B. and Gimblett, H.R., 2002, An agent-based approach to environmental and urban systems within geographic information systems. In H.R. Gimblett (Ed.), *Integrating Geographic Information Systems and Agent-Based Technologies for Modelling and Simulating Social and Ecological Phenomenon*, Oxford University Press, New York, pp. 171–190.

Joh, C-H., Arentze, T., and Timmermans, H., 2002, Modeling individual's activity-travel rescheduling heuristics: theory and numerical experiments. In *Proceedings of the 81st annual meeting of the Transportation Research Board*, Washington, D.C.

Jonker, C.M. and Treur, J., 2002, A dynamic perspective on an agent's mental states and interaction with its environment. In C. Castelfranchi and W.L. Johnson (Eds.), *Proceedings of the First International Joint Conference on Autonomous Agents and Multi-Agent Systems, AAMAS'02*, ACM Press, pp. 865–872.

Kray, C., 2001, The benefits of multi-agent system in spatial reasoning (abstract). In *Proceedings of FLAIRS* 2001.

Mark D., Egenhofer, M., and Hornsby, K., 1997, Formal Models of Commonsense Geographic Worlds, National Center for Geographic Information and Analysis, NCGIA Technical Report 97-2.

McGuiness, D., Fikes, R., Stein, A.L., and Hendler, J., 2002, DAML-ONT: An ontology language for the semantic web. In Dieter Fensel, James Hendler, Henri Lieberman, and Wolfgang Walhster (Eds.), *The Semantic Web: Why, What, and How*? The MIT Press.

Nowak, C., 2001, A note on process ontology for agents,Workshop on Ontologies in Agent Systems, *5th International Conference on Autonomous Agents*, Montreal, Canada, 2001. http://cis.otago.ac.nz/OAS2001/. Accessed 10 September 2003.

O' Sullivan, D., 2000, Exploring the structure of space: towards geo-computational theory, Working Chapter Series, Centre for Advanced Spatial Analysis, London.

O' Sullivan, D. and Haklay, M., 2000, Agent-based models and individualism: is the world agent-based? *Environment and Planning A*, 32, 1409–1425.

Parkes, D. and Thrift, N., 1978, Putting time in its place. In T. Calstein, D. Parkes, and N. Thrift (Eds.), *Making Sense of Time*, Vol. 1, London, Edward Arnold, pp. 119–129.

Portugali, J. and Benenson, I., 1997, Human agents between local and global forces in a self-organising city, In F. Schweitzer (Ed.), *Self-Organisation of Complex Structures: From Individual to Collective Dynamics*, Gordon and Breach, London, pp. 537–546.

Portugali, J., Benenson, I., and Omer, I., 1997, Spatial cognitive dissonance and socio-spatial emergence in a self-organising city. *Environment and Planning B*, 24, 263–285.

Raper, J. and Livingstone, D., 2001, Let's get real: spatio-temporal identity and geographic entities. *Transaction of the Institute of British Geographers*, 26, 237–243.

Raubal, M., 2001, Ontology and epistemology for agent-based wayfinding simulation. *International Journal of Geographical Information Science*, 15, 653–665.

Raubal, M. and Kuhn, W., 2002, Enriching Formal Ontologies with Affordances, Position Paper, Workshop on Action-Oriented Approaches in Geographic Information Science, ACTOR 2002, November 2–4, Maine, U.S.A.

Revilla, C., 2002, Invariance and universality in social agent-based simulations. *Proceedings of National Academy of Science*, USA. 2002 May 14; 99(Suppl. 3): 7314–7316.

Rodrigues, A., Grueau, C., Raper, J., and Neves, N., 1996, Environmental planning using spatial agents. In S. Carver (Ed.), *Innovations in GIS 5*, Taylor and Francis, London, pp. 108–118.

Russell, S. and Norvig, P., 1995, *Artificial Intelligence: A Modern Approach*. Prentice Hall, Inc.

Sadalla, E. K. and Burroughs W.J., 1980, Reference points in spatial cognition. *Journal of Environmental Psychology: Human Learning and Memory* 6, 516–528.

Schelhorn, T., O'Sullivan, D., Haklay, M., and Thurstain-Goodwin, M., 1999, STREETS: an agent-based pedestrian model, CASA Working Chapter 9, University College London, Centre for Advanced Spatial Analysis, London.

SDG, 2003, Swarm Development Group, www.swarm.org

Smith, B., 1995, Formal ontology, common-sense and cognitive science. *International Journal of Human-Computer Studies*, 43, 641–647.

Smith, B., 2002, SNAP and SPAN, Presentation, Workshop on Action-Oriented Approaches in Geographic Information Science, ACTOR 2002, November 2–4, Maine, U.S.A.

Smith, B., and Mark, D.M., 1998, Ontology and geographical kinds, *Proceedings, 8th International Symposium on Spatial Data Handling*. Vancouver, Canada, 12–15 July 1998, pp. 308–320.

Sowa, J.F., 1995, Top-level ontological categories. *International Journal of Human–Computer Studies*, 43, 669–685.

Takeyama, M. and Couclelis, H., 1997, Map dynamics: integrating cellular automata and GIS through geo-algebra. *International Journal of Geographic Information Science*, 11, 73–91.

Timmermans, H., Arentze, T., Dijst, M., Dugundji, E., Joh, C-H., Kapoen, L., Krijgsman, S., Maat, K., and Veldhuisen, J., 2002, Amadeus: a framework for developing a dynamic multi-agent, multi-period activity-based micro-simulation model of travel demand, In *Proceedings of the 81st annual meeting of the Transportation Research Board*, Washington, D.C.

Tuan, Yi-Fu, 1974, *Topophilia: A study of Environmental Perception, Attitudes and Values.* Prentice Hall Inc, Englewood Cliffs, NJ.

UNIVERSITY OF CHICAGO, 2003, *Repast: REcursive Porous Agent Simulation Toolkit*, University of Chicago Social Science Research Computing. http://repast.sourceforge.net/index.php

Winter, S., 2001, Ontology: buzzword or paradigm shift in GIScience? *International Journal of Geographic Information Science*, 15, 587–591.

Worboys, M.F., 2001, Nearness relations in environmental space. *International Journal of Geographical Information Systems*, 15, 633–651.

Wu, F., 1999, GIS-based simulation as an exploratory tool for space-time processes. *Journal of Geographical Systems*, 1, 199–218.

Zeigler, B.P., 1976, *Theory of Modelling and Simulation*, John Wiley and Sons, New York.

CHAPTER 25

GeoDynamics: Shaping the Future

Peter M. Atkinson, Giles M. Foody, Stephen E. Darby, and Fulong Wu

CONTENTS

25.1 BACKGROUND

While this final chapter concludes this book, the subject of this chapter (GeoDynamics) is in its infancy. Thus, this chapter provides a general review of some of the most important research strands emerging within GeoDynamics (as evident from the chapters of this book) and a look forward to the future. The objective of the chapter is to encourage researchers to explore and model Earth surface processes and predict their outcomes using the tools of GeoDynamics.

25.2 WHY GEODYNAMICS?

As suggested in the preface, GeoComputation research arose from the application of computer power to solve geographical problems. The uppercase C in GeoComputation was retained to emphasise that each part (geography, computation) was equally important (Gahegan, 1999; Openshaw and Abrahart, 2000). Much was written at

0-8493-2837-3/05/$0.00+$1.50

the end of the 1990s about the definition of GeoComputation (see Longley, 1998; MacMillan, 1998) and several books have emerged on the topic (Longley et al. 1998; Atkinson and Martin, 2000; Openshaw and Abrahart, 2000). Importantly, whereas early proponents emphasised the efficacy of computational solutions, almost as an antidote to the strictures of model-based statistics (Openshaw and Abrahart, 2000), Longley (1998) was at pains to retain dynamics (and, thereby, process) as a component of GeoComputation. In many ways, this book represents that component.

As GeoComputation arose from the merger of two strands, so GeoDynamics has evolved from the application of spatially explicit measurement techniques (e.g., remote sensing) and spatially distributed dynamic models to geographical (Earth surface) processes. GeoDynamics is set to take centre-stage amongst a broad range of numerate researchers working in quite disparate disciplines. For example, it already provides a key focus for computer-literate researchers from subjects such as climatology, oceanography, geology, hydrology, hydraulic engineering, geomorphology, ecology, epidemiology and urban dynamics, not to mention computer science. As a consequence, a minority of researchers in each of these fields applying GeoDynamics models have found that, perhaps for the first time, they use the same formal language as their counterparts across the research sector. Biologists are talking to geographers and oceanographers. Computer scientists are collaborating with applied scientists across the board. GeoDynamics research is truly inter-disciplinary, and therein lies its strength: (i) often the best ideas are found at the boundaries between disciplines and (ii) the potential opportunities uncovered by communication between researchers occupying quite different positions in intellectual space can be profound.

Why has GeoDynamics research proliferated into so many subjects? Or, more generally, why GeoDynamics? There are, of course, many possible answers to these questions, but the following arguments are important among them. First, much research in sciences related to Earth surface phenomena (particularly, spatially distributed phenomena) involves analysis of static data, for example, using spatial statistics. Such research focuses on form and pattern. While important, such research fails to take full account of variation in these forms over time and, more importantly, the underlying processes that lead to this variation. Temporal remote sensing allows the researcher to ask questions relating to change (e.g., change detection and monitoring). However, more than this, spatially distributed dynamic modelling (SDDM) actually provides simple models of the underlying processes. Modelling process (cf. form) is important because it allows:

(i) prediction into the future given current data
(ii) prediction under alternative input (boundary) conditions (e.g., possible catchment "what-if" scenarios)
(iii) prediction under possible future conditions (e.g., possible climate change "what-if" scenarios) and
(iv) some development of understanding (answering questions of "why").

Of these, (i) is useful for real-time management (e.g., early warning, rapid response), (ii) can be used to plan or manage landscape change and (iii) facilitates prediction made using scientific outputs, and (iv) is the fundamental objective of science.

The first benefit depends to some extent on the difficulty of the prediction task (e.g., how far into the future), the accuracy required and, interestingly, the complexity of the SDDM. Complex SDDMs in which a multitude of physical (or socio-economic) processes are represented do not necessarily lead to more accurate or reliable prediction than simpler SDDMs in which a few basic (but important) processes are represented.

The fourth benefit is also rather difficult to argue. The level of abstraction present in many landscape-scale SDDMs is so great that often the "understanding" that they encapsulate was accepted by the science community decades and, often, centuries ago. True, great strides have been made in the last few decades in dynamic modelling, but this has arguably been limited primarily to developing the SDDM as a predictive tool, and the fundamental understanding that the model encapsulates about the real world (real world mechanics) has not advanced as a function of the modelling.

Counter to the above argument is the view that SDDM becomes the environment in which "experiments" can be conducted to test hypotheses made about the real world. SDDMs have been used fruitfully in this regard, even though the basic point-to-point mechanics on which they are founded are usually derived from well-established theory. The key is the scale of the phenomenon or process being investigated. While, at least in the natural sciences, physics provides laws at the micro-scale, interactions between aggregations of points and through time are so complex that our "understanding" of landscape-scale phenomena is fragmentary at best. GeoDynamics has the potential to deliver such understanding, whatever the Earth surface phenomenon studied.

25.3 REMOTE SENSING

A plethora of environmental properties may be predicted from remotely sensed imagery at a range of spatial scales. Such properties include land cover, but also continua such as biophysical (e.g., biomass) and biochemical (e.g., nitrogen, lignen) properties of vegetation (Curran, 1998) and a range of climatological, oceanographical, geological, hydrological, geomorphological, and anthropogenic properties. Scales of measurement (specifically length on a side of a pixel, that is, measurement cell) range from centimetres through to tens of kilometres. In addition, remote sensing provides complete coverage, synoptically (in the form of an image) and without the smoothing inherent in interpolation. As a result, remote sensing is a primary source of spatial data for a wide range of applications.

Remote sensing is invaluable for monitoring and change detection. In fact, this was the objective of the original NASA Landsat programme in the 1970s. Despite this, most remote sensing analysis to date has focused on single-date imagery. This is partly because the problems inherent in single-date analysis were found to require considerable research effort, but also because two basic requirements for multi-temporal analysis were difficult to meet. First, it is preferable that each image in a temporal sequence is calibrated to absolute reflectance (atmospherically corrected).

This is hampered by a requirement for either *in situ* ground-based reflectance measurements or data on atmospheric conditions at the time of the sensor overpass. Second, it is necessary to transform each image in the temporal sequence to the same co-ordinate system. Perfect atmospheric correction and geometric registration is impossible, meaning that any change analysis will be affected by uncertainties in the basic pre-processing steps. Despite these problems, progress is being made (see Brown, Chapter 8, this volume). Perhaps one of the most encouraging changes in recent years is the increasing number of researchers who have acknowledged the long-term benefits of atmospheric correction (e.g., archiving images corrected to reflectance facilitates usage by others in years to come) (Woodcock, 2002).

SDDMs are always fitted to data. No process model is independent of data. The question is, "what data are appropriate and how should such data be provided?." In the simplest case, a spatial property may be measured at a series of points with a given spatial sampling strategy. For example, if elevation is required as a boundary condition for flood inundation modelling then contour data provide a possible source of information. For use in a SDDM such data must be interpolated to the required grid or cell arrangement, and a variety of procedures are available for this (Wilson and Atkinson, Chapter 14, this volume). However, interpolation imparts several undesirable properties (uncertainty, smoothing) on the predicted values, such that caution should be exercised when using them in SDDMs. Increasingly, boundary conditions for use in SDDMs are provided by remote sensing. Given the benefits of remote sensing as described above it is no wonder that remote sensing and SDDMs have been converging over the last decade (Goodchild et al., 1996; Raper and Livingstone, 1996; Bates et al., 1998). It is within this context that this book is placed.

25.4 SDDMS

Spatially distributed dynamic models may be distinguished in terms of the fundamental data model on which they are built. In geographical information systems (GIS), two data models are common: the raster (grid or image) and vector (points, lines and areas) data models. Somewhere between these lies a third model; the triangular irregular network (TIN), an irregular arrangement of points connected to form triangular patches completely covering the region of interest. The raster data model tends to be used in conjunction with random field (RF) models (spatially continuous field, e.g., suitable for modelling water flow). The vector data model tends to be used in conjunction with object-based modelling (e.g., suitable for modelling movements of individuals).

Some of the most useful SDDMs make use of the raster data model (Burrough, 1998; Burrough and McDonnell, 1998). In ecology and urban modelling the cellular automata (CA) has gained popularity (Batty, 1997). The CA model, as promoted by Wolfram (2002), usually comprises very simple rules that allow transitions to occur between neighbouring cells. Now CA models have been developed to support decision making in city planning and environmental conservation. The basic principle of raster-based manipulation together with GIS visualization capacities are attractive features of these spatially explicit rule-based models.

Early hydraulic and geomorphological modelling was often one-dimensional (e.g., Kirkby et al., 1987; Chow et al., 1988). However, related to the CA approach, raster-based routing models have emerged in hydraulic modelling (Aitkenhead et al., 1999; Bates and De Roo, 2000). While computationally simple, benefits include (i) computational efficiency (e.g., matrix manipulation), (ii) ease of handling and (iii) straightforward coupling with remotely sensed image inputs.

In hydraulic modelling, the computational fluid dynamics (CFD) approach (essentially the finite element or finite difference model) is based on the principles of the TIN (Lane et al., 1999; Hardy et al., 2000). Advantages of this approach include (i) the ability to characterize the boundary (e.g., elevation) accurately locally, (ii) the use of partial differential equations based on Newtonian mechanics and (iii) the straightforward application of the model to three dimensions (e.g., modelling suspended sediment dynamics in an estuary, chlorophyll dynamics in lakes, within-channel flow in rivers) (Horritt and Bates, 2001, 2002).

Both the raster and TIN data models are used to characterize landscape variables (i.e., those that vary across a continuous space). An alternative type of model is required to model individuals (e.g., animals, including humans) and their dynamics within a continuous landscape. Here, it is the object-based models that provide a solution. In particular, agent-based models provide an exciting new method for modelling the behaviour of individuals and utilizing, in a bottom-up fashion, such behaviour to dictate emergent landscape-scale dynamics. Such a bottom-up approach is clearly predicated on the availability of character profile data; and therein lies a challenge for the future. While currently we are limited to questionnaires and similar traditional methods of survey, location-based services (LBS) and mobile telecommunications technology are set to provide unprecedented opportunities for acquiring personal behaviour data in the next five years or so, though the ethical issues involved in doing so will also need to be confronted.

25.5 RESEARCH ISSUES FOR THE FUTURE

Whatever the type of data-model implemented (raster, TIN, vector), two issues remain top of the agenda of many GeoDynamics researchers: scale and uncertainty.

Scale is defined here as the "size" of something. Strictly, scale is a ratio (e.g., cartographic scale), but in everyday language (and in physics and ecology) scale is synonymous with size. Thus, a large-scale phenomenon is simply a large phenomenon. Two kinds of scale are of interest: scale(s) of measurement and scale(s) of variation (and process). Scales of measurement (e.g., spatial resolution, spatial extent) are important because all observations of the real world are a function of both the underlying property measured and the sampling framework (Atkinson and Tate, 2000). Thus, changing the spatial resolution results in a change in the data analysed. While this is not surprising, many examples can be found where the effect is ignored (e.g., how often is the standard deviation reported without the support — the size, geometry and orientation of the space on which an observation is defined — even though it is a function of it?). The important point is that the results of any analysis relate to the data and not reality.

In terms of remote sensing, the effects of spatial resolution are paramount leading to many investigations on the subject (Curran and Atkinson, 1998). For SDDMs, spatial resolution can have important effects on model prediction (Coulthard, 1998; Hardy et al., 1999). This has led to the notion of grid independence: model predictions are comparable for a threshold spatial resolution and finer spatial resolutions, but divergent at coarser ones. Models should always be checked for grid independence. Issues of spatial measurement scale are compounded when different input variables or boundary conditions are represented at different spatial or temporal resolutions in the same dynamic model.

Before treating the subject of uncertainty briefly, it is worth noting that here uncertainty is defined as a general concept and is not measurable (Atkinson and Foody, 2002). Accuracy, on the other hand, which is comprised of unbiasedness and precision, can be represented by measurable quantities such as the root mean squared error (inaccuracy) and mean error (bias).

As described above, uncertainty is important in remote sensing, particularly for change analysis. For SDDMs uncertainty is important in many different ways. Here, we consider calibration and validation, sensitivity analysis, error propagation and model-based estimates of posterior distributions.

Fundamental to SDDMs are calibration and validation. Classically, calibration involves comparing model predictions to known spatial distributions and using the error to adjust (estimate) selected model parameters. Once a model is calibrated it should then be tested (validated) on unseen data by, again, comparing prediction to known distributions. The problems and issues involved in these two processes are complex and have concerned researchers in GeoDynamics for decades (e.g., Beven, 1995). Some of the most important are:

(i) it may be important to calibrate using a sample from the complete space–time dimension of the prediction,
(ii) it is often preferable to calibrate using certain variables (e.g., continua) rather than others (e.g., binary response),
(iii) it may be beneficial to re-calibrate frequently (e.g., assimilation techniques such as Kalman filtering provide a neat means of combining model and data in real-time) and
(iv) accuracy statistics should not favour the model (e.g., inflated accuracy because it is easy to predict certain locations where the response is null. An example is flooding where non-flooded cells predicted accurately inflate the overall accuracy).

In socio-economic applications of the CA model, the issue of calibration has now become critical for the development of decision-support systems (Wu, 2002). Two different approaches have been examined. The first is to develop a more behaviourally oriented model, tied up with statistical calibration (e.g., discrete choice modelling). The second is to use various data mining and neural network approaches without developing explicit rules. Both are still under development.

Calibration and validation as described above suggests that a single model is to be determined and once fitted, produces a single prediction (for given input conditions). It is also useful to estimate the characteristics of the posterior distributions (in the first instance of the predictions, but also of the parameters themselves). One way to estimate the posterior predictive distribution is to model various sources of uncertainty in an error propagation model (e.g., Veregin, 1996). First, measurement errors are expected to be present in input data. Compounded with such errors is the generalization (smoothing, meaning a loss of information) imposed through the spatial and temporal resolutions. If the data were provided by interpolation (or similar inference) then the uncertainties associated with spatial prediction (including further smoothing) can be added. Models exist to propagate such errors through the SDDM to predict output error (Heuvelink, 1998). Of the various approaches, the most widely used is the Monte Carlo approach in which the input uncertainty is modelled using a cumulative distribution function (cdf) and alternative draws or realizations are propagated through the model to build a cdf of possible outputs. Recently, researchers have realized the potential of geostatistical models of input uncertainty (Journel, 1996; Goovaerts, 1997; Deutsch and Journel, 1998) for this purpose (see Wilson and Atkinson, Chapter 14, this volume).

Related to the above approach for error propagation is sensitivity analysis in which the interest is in estimating the sensitivity of model predictions to choice of model parameters. It is important to know, for example, whether a parameter has a small or large effect on the output. If the effect is large, more effort can be put into estimating it, for example.

It is important to realise that a statistical model (e.g., the Bayesian technique of Markov chain Monte Carlo, MCMC) can allow predictions of the posterior distributions of both predictions and parameters as part of the fitting of the model. This means, importantly, that it is not necessary to build an explicit error propagation model: all the desired information on uncertainty will be estimated by the model. Both simple and (to a lesser extent) complex models can be built within a statistical framework. Arguably, until this is done, it is not possible to evaluate the model fully: both the ability to predict and the ability to estimate uncertainty should be validated. The benefits of estimating uncertainty should be obvious both in relation to prediction and the running of what-if scenarios (e.g., under uncertain future climate change).

Problems relating to scale and uncertainty are so many and complex that these two issues continue to dominate many research agendas.

25.6 CONCLUSION

Research into Earth surface phenomena (both physical and socio-economic) is increasingly focused on change, dynamics and process where previously it was focused on static form. This shift in emphasis has occurred in large part because the former is increasingly facilitated by computational resource. Further, the role of the latter static analysis has not diminished. Such analysis is still required in a range of circumstances, not least the requirement to feed data hungry process models.

In remote sensing, the problems associated with change detection analysis and monitoring are being addressed allowing the original monitoring objectives of early remote sensing missions to be fulfilled. In SDDModelling, increases in computational power and ease of access to such power have facilitated increases in both the sophistication of modelling (e.g., CFD models) and the spatial resolution of modelling (through brute force number crunching, for example, as applied to CA and raster routing models). However, the real power of GeoDynamics is realised when spatial data such as provided by remote sensing are combined with spatially distributed process models. Realisation of the utility and power of such spatially distributed processing machines has lead to an explosion of research and operational interest in GeoDynamics in recent years. There is much excitement in this inter-disciplinary field presently as the rate of new discoveries is high and researchers across (essentially arbitrary) old-discipline boundaries share knowledge and commonalities. As understanding increases (particularly of the limitations of models, analyses and predictions) it will be the responsibility of researchers to convey that understanding to practitioners who are eager for application.

REFERENCES

Aitkenhead, M.J., Foster, A.R., Fitzpatrick, E.A., and Townend, J., 1999, Modelling water release and absorption in soils using cellular automata. *Journal of Hydrology*, **220**, 104–112.

Atkinson, P.M. and Foody, G.M., 2002, Uncertainty in remote sensing and GIS: fundamentals, in Foody, G.M. and Atkinson, P.M. (Eds.), *Uncertainty in Remote Sensing and GIS* (Chichester: Wiley), pp. 1–18.

Atkinson, P.M. and Martin, D., 2000, Introduction, In Atkinson, P.M. and Martin, D. (Eds.), *Innovations in GIS VII: GIS and Geocomputation* (London: Taylor and Francis), pp. 1–7.

Atkinson, P.M. and Tate, N.J., 2000, Spatial scale problems and geostatistical solutions: a review. *Professional Geographer*, 52, 607–623.

Bates, P.D. and De Roo, A.P.J., 2000, A simple raster-based model for flood inundation simulation. *Journal of Hydrology*, **236**, 54–77.

Bates, P.D., Horritt, M.S., Smith, C.N., and Mason, D., 1998, Integrating remote sensing observations of flood hydrology and hydraulic modelling. *Hydrological Processes*, **11**, 1777–1795.

Batty, M., 1997, Cellular automata and urban form: a primer. *Journal of the American Planning Association*, **63**, 266–274.

Beven, K., 1995, Linking parameters across scales — subgrid parameterizations and scale-dependent hydrological models. *Hydrological Processes*, **9**, 507–525.

Brown, K., Chapter 8, this volume.

Burrough, P.A., 1998, Dynamic modelling and geocomputation, In Longley, P.A., Brooks, S.M., McDonnell, R. and MacMillan, W. (Eds.), *Geocomputation: a Primer* (Chichester: Wiley), pp. 165–191.

Burrough, P. and McDonnell, R., 1998, *Principles of Geographical Information Systems. Spatial Information Systems and Geostatistics* (Oxford: Oxford University Press).

Chiles, J.P. and Delfiner, P., 1999, *Geostatistics. Modelling Spatial Uncertainty* (Chichester: Wiley).

Chow, V.T., Maidment, D.R., and Mays, L.W., 1988, *Applied Hydrology* (New York: McGraw-Hill Inc).

Couclelis, H., 1998, Geocomputation in context, In Longley, P.A., Brooks, S.M., McDonnell, R., and MacMillan, W. (Eds.), *Geocomputation: A Primer* (Chichester: Wiley), pp. 17–29.

Coulthard, T.J., Kirkby, M.J., and Macklin, M.G., 1998, Non-linearity and spatial resolution in a cellular automaton model of a small upland basin. *Hydrology and Earth System Sciences*, **2**, 257–264.

Craig, P.S., Goldstein, M., Rougier, J.C., and Seheult, A.H., 2001, Bayesian forecasting for complex systems using computer simulators. *Journal of the American Statistical Association*, **96**, 717–729.

Curran, P.J. and Atkinson, P.M., 1998, Remote sensing and geostatistics. *Progress in Physical Geography*, 22, 61–78.

Deutsch, C.V. and Journel, A.G., 1998, *GSLIB Geostatistical Software Library and User's Guide*, 2nd ed. (New York: Oxford University Press).

Gahegan, M., 1999, What is geocomputation?. *Transactions in GIS*, **3**, 203–206.

Goodchild, M., Steyaert, L., Parks, B.O., Johnston, C.O., Maidment, D.R., Crane, M.P., and Glendinning, S., 1996, *GIS and Environmental Modelling: Progress and Research Issues* (Cambridge: GeoInformation International).

Goovaerts, P., 1997, *Geostatistics for Natural Resources Evaluation* (New York: Oxford University Press).

Hardy, R.J., Bates, P.D., and Anderson, M.G., 1999, The importance of spatial resolution in hydraulic models for floodplain environments. *Journal of Hydrology*, **216**, 124–136.

Hardy, R.J., Bates, P.D., and Anderson, M.G., 2000, Modelling suspended sediment deposition on a fluvial floodplain using a two-dimensional dynamic finite element model. *Journal of Hydrology*, **229**, 202–218.

Heuvelink, G.B.M., 1998, *Error Propagation in Environmental Modelling with GIS* (London: Taylor and Francis).

Horritt, M.S. and Bates, P.D., 2001, Predicting floodplain inundation: raster-based modelling versus the finite element approach. *Hydrological Processes*, **15**, 825–842.

Horritt, M.S. and Bates, P.D., 2002, Evaluation of 1D and 2D numerical models for predicting river flood inundation. *Journal of Hydrology*, **268**, 87–99.

Journel, A.G., 1996, Modelling uncertainty and spatial dependence: stochastic imaging. *International Journal of Geographical Information Systems*, **10**, 517–522.

Kirkby, M.J., Naden, P.S., Burt, T.P., and Butcher, D.P., 1987, *Computer Simulation in Physical Geography* (Chichester: Wiley).

Lane, S.N., Bradbrook, K.F., Richards, K.S., Biron, P.A., and Roy, A.G., 1999, The application of computational fluid dynamics to natural river channels: three-dimensional versus two-dimensional approaches. *Geomorphology*, **29**, 1–20.

Longley, P.A., 1998, Foundations, In Longley, P.A., Brooks, S.M., McDonnell, R., and MacMillan, W. (Eds.), *Geocomputation: A Primer* (Chichester: Wiley), pp. 3–15.

Longley, P.A., Brooks, S.M., McDonnell, R., and MacMillan, W. (Eds.), 1998, *Geocomputation: a Primer* (Chichester: Wiley).

MacMillan, W., 1998, Epilogue, In Longley, P.A., Brooks, S.M., McDonnell, R., and MacMillan, W. (Eds.), *Geocomputation: A Primer* (Chichester: Wiley), pp. 257–264.

Openshaw, S. and Abrahart, R. (Eds.), 2000, *Geocomputation* (London: Taylor and Francis).

Raper, J. and Livingstone, D., 1996, High level coupling of GIS and environmental process modelling, In Goodchild, M.F., Steyaert, L.T., Parks, B.O., Johnston, C., Maidment, D., Crane, M., and Glendinning, S. (Eds.), *GIS and Environmental Modelling: Progress and Research Issues* (Fort Collins, CO: GIS-World Books), pp. 387–390.

Romanowicz, R. and Beven, K., 2003, Estimation of flood inundation probabilities as conditioned on event inundation maps. *Water Resources Research*, **39**, 1073.

Veregin, H., 1996, Error propagation through the buffer operation for probability surfaces. *Photogrammetric Engineering and Remote Sensing*, **12**, 419–428.

Wolfram, S., 2002, *Cellular Automata and Complexity* (New York: Perseus Publishing).

Woodcock, C.E., 2002, Uncertainty in remote sensing, in Foody, G.M. and Atkinson, P.M. (Eds.), *Uncertainty in Remote Sensing and GIS* (Chichester: Wiley), pp. 19–24.

Wu, F., 2002, Calibration of stochastic cellular automata: the application to rural–urban land conversions. *International Journal of Geographical Information Science*, **16**, 795–818.

Index